PHYSICS OF QUANTUM WELL DEVICES

SOLID-STATE SCIENCE AND TECHNOLOGY LIBRARY

VOLUME 7

Aims and Scope of the Series

The aim of this series is to present monographs on semiconductor materials processing and device technology, discussing theory formation and experimental characterization of solid-state devices in relation to their application in electronic systems, their manufacturing, their reliability, and their - limitations (fundamental or technology dependent). This area is highly interdisciplinary and embraces the cross-section of physics of condensed matter, materials science and electrical enginee-ring.

Undisputedly during the second half of this century world society is rapidly changing owing to the revolutionary impact of new solid-state based concepts. Underlying this spectacular product development is a steady progress in solid-state electronics, an area of applied physics exploiting basic physical concepts established during the first half of this century. Since their invention, transistors of various types and their corresponding integrated circuits (ICs) have been widely exploited covering progress in such areas as microminiaturization, megabit complexity, gigabit speed, accurate data conversion and/or high power applications. In addition, a growing number of devices are being developed exploiting the interaction between electrons and radiation, heat, pressure, etc., preferably by merging with ICs.

Possible themes are (sub)micron structures and nanostructures (applying thin layers, multi'layers and multi-dimensional configurations); micro-optic and micro-(electro)mechanical devices; high-temperature superconducting devices; high-speed and high-frequency electronic devices; sensors and actuators; and integrated opto-electronic devices (glass-fibre communications, optical recording and storage, flat-panel displays).

The texts will be of a level suitable for graduate students, researchers in the above fields, practitioners, engineers, consultants, etc., with an emphasis on readability, clarity, relevance and applicability.

The titles published in this series are listed at the end of this volume.

Physics of Quantum Well Devices

by

B.R. Nag
INSA Senior Scientist,
Institute of Radio Physics and Electronics,
Calcutta University, Calcutta, India

KLUWER ACADEMIC PUBLISHERS
DORDRECHT / BOSTON / LONDON

A C.I.P. Catalogue record for this book is available from the Library of Congress.

ISBN 1-4020-0360-9
Transferred to Digital Print 2001

Published by Kluwer Academic Publishers,
P.O. Box 17, 3300 AA Dordrecht, The Netherlands.

Sold and distributed in North, Central and South America
by Kluwer Academic Publishers,
101 Philip Drive, Norwell, MA 02061, U.S.A.

In all other countries, sold and distributed
by Kluwer Academic Publishers,
P.O. Box 322, 3300 AH Dordrecht, The Netherlands.

Printed on acid-free paper

Printed in the Netherlands

CONTENTS

CONTENTS

PREFACE

Quantum well devices have been the objects of intensive research during the last two decades. Some of the devices have matured into commercially useful products and form part of modern electronic circuits. Some others require further development, but have the promise of being useful commercially in the near future. Study of the devices is, therefore, gradually becoming compulsory for electronics specialists. The functioning of the devices, however, involve aspects of physics which are not dealt with in the available text books on the physics of semiconductor devices. There is, therefore, a need for a book to cover all these aspects at an introductory level. The present book has been written with the aim of meeting this need. In fact, the book grew out of introductory lectures given by the author to graduate students and researchers interested in this rapidly developing area of electron devices.

The book covers the subjects of heterostructure growth techniques, band-offset theory and experiments, electron states, electron-photon interaction and related phenomena, electron transport and the operation of electronic, opto-electronic and photonic quantum well devices. The theory as well as the practical aspects of the devices are discussed at length.

The aim of the book is to provide a comprehensive treatment of the physics underlying the various devices. A reader after going through the book should find himself equipped to deal with all kinds of quantum well devices.

The book may serve as a text-book for advanced level graduate courses. New entrants into researches or developments in the area should also find the book useful.

Indian National Science Academy helped the author to complete the book by awarding him the position of INSA Senior Scientist for three years (1998-2000).

B. R. Nag
Calcutta
June, 2000

ACKNOWLEDGMENTS

The author wishes to record his indebtedness to C. T. Foxon, M. J. Ludowise, H. Asahi, H. Kinoshita, M. V. Pessa, M. Notomi, C. G. Van de Walle, J. S. Roberts, M. L. Cohen, R. L. Greene, M. Sturge, D. A. B. Miller, A. C. Gossard, M. Jaros, Der-San Chuu, A. Forchell, L. F. Eastman, H. Ruda, W. Ted Masselink, C. T. Bulle Liewema, H. Morkoc, P. K. Bhattacharya, Ming Hu, P. S. Zory, N. K. Dutta, B. F. Levine, J. Y. Andersson, J. S. Harris, J. Bowers and their coauthors; to the American Institute of Physics, the American Physical Society, the Institution of Electrical and Electronic Engineers, IOP Publishing Ltd.,The Institute of Pure and Applied Physics, Japan,Elsevier Science ; and to the Editors of the journals, Physica Status Solidi, Japanese Journal of Applied Physics and Physics Letters, for their permission to use the Figures 2.1, 2.2, 2.3, 2.4, 2.5, 2.6, 3.3, 3.5(a), 3.5(b), 4.2, 4.4, 4.8, 5.4, 5.5, 5.6, 5.7, 5.9, 5.10, 5.11, 5.12, 5.14, 5.15, 6.1, 6.2, 6.3, 6.4, 6.5, 6.6, 7.4, 7.5, 7.6, 9.4, 9.5, 10.2, 10.3, 10.4, 10.5, 10.6(a), 10.6(b), 10.8 and 10.10. Sources of the figures are listed in the Appendix at the end of the book.

The author is indebted to (Ms) Charon Duermeijer and (Ms) Margaret Deignan, Dr Sabine Freisem and (Ms) Vaska Krabbe of Kluwer Academic Publishers, who gave guidance in the preparation of the manuscript and looked after its production.

He is indebted to (Ms) Dyuti Bhattacharya ,who typed the first draft of the book, to Dr S. R. Bhattacharya, who helped him by providing the Latex software and the Guide book, to Professor B. P. Sinha for help in the use of the Latex, to Professor S. Sen for help in the preparation of the manuscript and to his students for fruitful discussions.

The book is dedicated to the memory of the author's deceased Parents, to his teachers and students and to the students of the future.

The author will appreciate receiving comments and criticisms from the readers, which may help him to improve the book if it ever goes into a revised edition.

CHAPTER 1

INTRODUCTION

1.1. Quantum Well Devices

The seed of quantum well devices was planted when Esaki and Tsu [1.1,2] suggested in 1969 that a heterostructure consisting of alternating ultrathin layers of two semiconductors with different band gaps should exhibit some novel useful properties. The band-edge potential varies from layer to layer as a result of the difference in the band gaps and a periodically varying potential is produced in the structure with a period equal to the sum of the widths of two consecutive layers. For layer thicknesses of the order of ten nanometer, the wavelength as well as the mean free path of the electrons extend over several layers and the periodic potential transforms the energy bands of the host lattice into minibands. Phenomena like Bloch oscillation and low-field negative differential resistance may be produced by the electrons in such minibands. Attempts to fabricate the proposed structure and to demonstrate the predicted phenomena were only partially successful[1.3,4]. Interest was, however, generated in fabricating heterostructures with transition regions extending over a few atomic layers. Structures were initially grown by using the technique of molecular beam epitaxy [1.5],(MBE) but soon several other techniques were developed. Heterostructures may now be grown of any composition with crystalline perfection at the interfaces. Such structures form the basis of quantum well devices.

Experiments were done on the proposed heterostructures in 1974 with gallium arsenide (GaAs) and aluminum gallium arsenide ($Al_xGa_{1-x}As$). The mismatch in the lattice constant of GaAs and AlAs being only 0.12% it is easy to grow layers of GaAs and mixed compounds of GaAs and AlAs on each other with crystalline perfection. Experiments were done with structures consisting of GaAs layers sandwiched between $Al_xGa_{1-x}As$ layers, which form potential-barrier layers as the band gap of $Al_xGa_{1-x}As$ is larger than that of GaAs. The potential profile in these structures are as envisaged in textbook problems of one-dimensional potential wells and potential barriers. The component of the electron wave vector is quantized in such structures in the direction of potential variation.

Two experiments were reported in which the effects of quantization were explored in order to establish that quantization really occurs. In one experiment[1.6] optical absorption was measured in a multi-quantum well (MQW) structure. The absorption was found to increase in steps for certain wavelengths. This result indicated that the component of the wave vector was quantized in the direction perpendicular to the interfaces and the electron and the hole gases were two-dimensional. The stair-case-like density of states of the two-dimensional gas explained the experimental absorption characteristic. In the second experiment[1.7], the barrier layers on the two sides of a sandwiched GaAs layer were provided with ohmic contacts by growing on them degenerate GaAs layers. Current between the contacts was measured for different voltages. The current-voltage characteristic exhibited peaks, which could be explained by resonant tunneling from the contact layer to the quantized levels in the GaAs well layer. These two experiments confirmed the conjecture that the energy levels are indeed quantized in the potential wells formed due to the band gap difference.

The next important experiment demonstrated[1.8] in 1979 the formation of a two-dimensional electron gas (2DEG) on the GaAs surface of a $Al_xGa_{1-x}As/GaAs$ heterojunction. The GaAs layer was undoped, while the $Al_xGa_{1-x}As$ layer was doped n-type with a donor concentration of 10^{18} cm^{-3}. Electrons migrated to the lower lying conduction band of GaAs to form a space charge layer at the interface. The conduction -band edge of GaAs was so bent by this space charge that a narrow near-triangular well was formed. The energy levels were quantized as a result. Quantization was confirmed by the observed anisotropy of the Shubnikov-de Haas oscillations. More importantly, mobility of the confined electrons was found to be high even though the areal electron concentration was about 1.1×10^{12}cm^{-2}. The high mobility was mainly due to the segregation of the impurity atoms which were confined to the $Al_xGa_{1-x}As$ barrier layers, whereas the electrons were in the GaAs surface layer.

A transistor was realized[1.9] by using a $Al_{0.7}Ga_{0.3}As/GaAs$ heterostructure in 1980. The $Al_{0.7}Ga_{0.3}As$ layer was doped n-type ($N_D = 6.6 \times 10^{17}$cm^{-3}) and the GaAs layer was undoped. The electron concentration in GaAs and the in-plane current could be controlled by using a conventional metal-oxide-semiconductor field-effect-transistor (MOSFET)-like structure with source, drain and gate contacts. The high mobility of the electrons[1.10,11] in the structure gave at 77 K a transconductance, three times higher than that of a conventional Schottky-gate GaAs field-effect transistor (GaAs FET). The operation of the AlGaAs/GaAs FET depends on the capability to change the space charge by the gate voltage and quantization of the energy levels is not a necessary condition. However, the carrier concentration in the devices is such that the potential well formed by the space-charge layer is very narrow and quantization results. Electrons can have, therefore, only in-plane motion, the perpendicular motion being restricted due to quantization. The quantization, although not necessary, affects significantly the device characteristics. That energy levels may be quantized in surface space-charge layers was predicted by Schreiffer[1.12] as early as 1957 and quantization in SiO_2/Si system has been studied extensively[1.13]. The conditions in silicon MOSFET's are, however, such that the effects of any quantization are not significant. On the other hand, in the transistors realized by using AlGaAs/GaAs structures, quantization cannot be ignored and the device should be treated as a quantum well device. It has been given different names TEGFET, MODFET, SDFET, HFET, but it is mostly referred as HEMT (High Electron Mobility Transistor). The performance characteristics have been improved through intensive research and the devices today[1.14] may have cut-off frequencies of 455 GHz, noise figure of 1.7 dB with 7.7 dB gain at 44 GHz. HEMT's are extensively used in low-noise receivers, digital IC's and microwave circuits.

In parallel with the development of a transistor, research was also being done for realizing a laser by using a quantum well structure, as some advantages were envisaged for such a laser. The density of states being stair-case-like, the concentration of carriers is large at the band edges in quantum wells, whereas in bulk material the concentration is zero at the band edges. The density of carriers is also larger in quantum wells for the same injection current as the carriers are confined in a narrower region. These two conditions were expected to cause a lowering of the threshold current density in quantum well laser(QWL)'s. An additional expected advantage was the possibility of tuning the laser wavelength by controlling the well width. The quantized energy levels being related to the inverse square of the well width, the energy difference between the lowest energy level in the conduction band and the highest energy level in the valence band is determined by the well width. A desired lasing wavelength can, therefore, be realized by a suitable design of the well structure.

An optically pumped QWL was reported in 1975,[1.15] but the required pumping power was 36 kW/cm². Large pumping power was required as the quality of the available MBE material was poor at that time. The interfaces were uniform and sharp, but the radiation efficiency was low due to interface defects. Performance of the QWL's was significantly improved with materials grown by the technique of metal-organic chemical vapor deposition (MOCVD). A threshold current density of 3 kA/cm² was reported[1.16] for a MOCVD-grown QWL in 1978. Further improvement was reported[1.17,18] in 1979 and the threshold current density was reduced to 1.66 kA/cm². The quality of the MBE material was also improved in the mean time. The non-radiative defects in AlGaAs interfaces were removed by raising the growth temperature. An MBE-grown multi-quantum well laser (MQWL) was reported[1.19] in 1979, which had a room temperature threshold current density of 2 kA/cm².

The successful developments initiated extensive research and development in QWL's. Various heterostructure systems using lattice-matched as well as lattice-mismatched (the so-called strained layer) materials have been evolved. The compositions of the structures have also been varied[1.20,21] to cover the wavelength range 1.55 μm to 417 nm. Different configurations have also been developed to obtain emission from the edges or from the surfaces. The threshold current has also been lowered[1.22] to 0.35 mA and the characteristic temperature has been raised[1.23] to 240 K.

Quantum well laser may be considered to be the most successful quantum well device. It is now extensively employed in optical communication systems, bar-code scanners, laser arrays and erasable optical discs. Short wavelength QWL's are also being developed currently, which have promise of increasing the storage capacity of data storage systems by an order of magnitude.

HEMT and quantum well laser are the two well-developed quantum well devices which are now extensively used. However, various other QW devices are also being studied in different laboratories. Among these devices resonant tunneling devices have received much attention.

The resonant tunneling experiment[1.5]of 1974 demonstrated the quantization of energy levels in the wells formed by semiconductor heterostructures. But, the magnitude of the peak in the voltage-current characteristic and the negative differential resistance were very small, although much higher peak-to-valley current ratio was expected from theory. The discrepancy was found to be due to the poor quality of the MBE material used in the experiment. Interest in the devices was, however, renewed in 1983 when it was shown[1.24] that the negative resistance of the diode could be used to construct oscillators for terahertz frequencies. As the quality of MBE material was also improved, it became possible to realize this prediction. Intensive work has been carried out on resonant tunneling diodes (RTD)'s in the last decade. The figures for the peak-to-valley current ratio and valley current have been much improved[1.25]. In addition to the AlGaAs/GaAs system $In_{0.52}Al_{0.48}As/In_{0.53}Ga_{0.47}As$ systems[1.26] have been used to realize a large peak-to-valley current ratio. A few other heterostructure systems[1.27] have also been used in which interband resonant tunneling occurs. Significant improvement in the characteristics of RTD's have resulted from these studies.

A transistor using a resonant tunneling structure was also suggested [1.28] in 1963 and such a device was realized [1.29] at 77 K in 1985. Room temperature operation of a resonant tunneling transistor with a double barrier in the base was demonstrated[1.30] in 1986. Various other devices using resonant tunneling structures have also been suggested[1.31], but practical applications of these devices are not yet much reported.

Another mature and promising quantum well device is the quantum well infrared photodetector (QWIP). Quantum wells may be so designed that the energy separation between the ground state and the continuum or a second quantized level near the top of the well is equal to the photon energy of infrared radiation. Infrared radiation may be absorbed in such doped wells to cause excitation of electrons from the ground state to the upper level or the continuum and to produce current. The suitability of such a device for infrared detection was first suggested[1.32] in 1977 and experiments were conducted[1.33] in 1983. Strong intersubband absorption was also demonstrated[1.34] experimentally in 1985. These initial studies prompted work on the development of a practical detector. A detector was reported[1.35] in 1987 for the wavelength of 10.8 μm with a detectivity of 0.52 A/W. Further development occurred in course of time and today 128×128 high sensitivity staring arrays are available,[1.36] which use QWIP's. Although major part of the work was done by using AlGaAs/GaAs system other structures have also been studied. For example, a QWIP using $GaAs/In_{0.2}Ga_{0.8}As$ has been reported[1.37] for the wavelength of 16.7 μm with a detectivity of 1.8×10^{10} cm(Hz)$^{1/2}$/W at 40 K. The QWIP has advantages compared to the HgCdTe detectors which are currently used for infrared detection. QWIP's are highly reproducible and can be made uniform so that large-area low-cost staring arrays may be easily realized. The composition of the detectors is also more stable and the detectors may be integrated with other GaAs devices. At the same time, the spectral response may be controlled by choosing the structural dimensions. It is predicted that because of these advantages QWIP's would soon be a competitor to HgCdTe detectors.

Infrared detectors are also being developed[1.38] by using GaInSb or AlGaSb /InAs superlattices and Si/Si_xGe_{1-x} quantum wells. The Si/Si_xGe_{1-x} detectors are based on the same principles as AlGaAs/GaAs detectors but holes are used instead of electrons. On the other hand, the GaInSb/InAs or AlGaSb/InAs superlattice detectors are based on different principles. In these superlattices valence- band levels are higher in GaInSb or in AlGaSb and conduction- band levels are lower in InAs. Incident radiation is detected by exciting electrons from the valence band of GaInSb or AlGaSb to the conduction band of InAs. Significant progress has been made with these two kinds of detectors, but the materials technology requires further development.

Devices are also being developed by using quantum wells for the modulation of light. Absorption of light in a quantum well may be significantly modified by applying a voltage across the well; the effect is known as quantum- confined Stark effect (QCSE). The effect was demonstrated[1.39] in 1985 by using AlGaAs/GaAs quantum wells. Extensive development work was done in the last decade on modulators and switches, based on QCSE. The modulator may be of the transmission type[1.40]. But, better performance characteristics have been obtained in the reflection type modulators using Fabry-Perot(FP) resonators[1.41]. A multi-quantum well structure is provided with two reflectors on the two sides to form a FP resonator. The device acts as a $p - i - n$ diode and the modulating voltage is applied by reverse biasing the diode. The percentage modulation per unit driving voltage is about 20 %/V and the contrast ratio, the ratio of the highest magnitude and the lowest magnitude of controlled light ranges[1.42] between 15:1 to 100:1. The modulation frequency may also be as high[1.43] as 37 GHz. The same kind of structure has also been used to realize optic switches and voltage-tunable optic detector. Development of these devices is progressing at a very fast pace and the performance characteristics are outdated in a few months.

Quantum well structures have also large nonlinearity. Optical bistable devices[1.44], all-optic directional couplers[1.45] and degenerate four-wave mixers[1.46] have been realized by using the nonlinearity. Progress in this area has been, however, rather slow and only a few reports are available in the literature.

The successful quantum well devices have so far used size-limited quantization in one direction only or the so-called quasi two-dimensional electron gas. Attempts are being made to realize quantum well devices with size-quantization in two or even in all the three dimensions. It is predicted that the resulting one-dimensional and zero-dimensional electron gas will yield more efficient devices.

The quantum well devices use the same basic principles as the corresponding bulk-material devices. The physics of the devices may, therefore, be considered to have been discussed extensively. However, formation of the potential wells, quantization of the electron momentum and the transport and optical interaction properties of electrons with reduced dimensionality involve aspects of physics, which are not as yet much discussed at an introductory level. The present text is an attempt to fill this gap. The scope of the book is detailed below.

1.2. Scope of the Book

Quantum wells are realized by heterostructures which consist of different kinds of semiconductor layers, one grown on top of another. Techniques used for the growth of the structures are described in Chapter 2 with suitable diagrams. Although the book is mainly concerned with devices using 2DEG, techniques for growing structures to realize one-dimensional electron gas (1DEG) or zero-dimensional electron gas (0DEG) are also briefly discussed, as the subjects are of great current interest.

In Chapter 3 are discussed the theoretical models of band offset and also the experimental methods for its determination. Values of band offsets for common heterostructures are also discussed in this chapter.

Behavior of electrons in quantum wells of semiconductors is very nearly the same as elaborated in text book problems of quantum mechanics. However, the basic characteristics of semiconductors add some new features which affect the performance of the devices. Energy levels and energy-wave-vector dispersion relation in quantum wells are discussed in Chapter 4, by considering the special aspects of wells in semiconductors.

The optoelectronic quantum well devices use the special optical properties of the quantum well heterostructures. The basic theory of electron-photon interaction and its applications to the various interaction phenomena are discussed in Chapter 5.

Performance of transistor-like quantum well devices depend on the electron transport properties. The reduced dimensionality of the electron gas makes these properties very different from those of the bulk materials. Chapter 6 deals with electron transport in quantum wells giving emphasis on both the low-field and the high-field properties.

In Chapter 7 through 10 are discussed the principles of operation and the performance characteristics of the various electronic and optoelectronic devices, viz., HEMT, RTD, QWL, QWIP, OBD and optical modulators and switches. The subject has already grown in vast proportions and for economy of space, discussion has been limited to the essential features and current realizations. Detailed analysis of all the devices could not be given, but the analysis related to the basic features are presented.

CHAPTER 2

HETEROSTRUCTURE GROWTH

Composite semiconductor structures consisting of two or more layers of different materials, one grown on another, are commonly referred as heterostructures. Heterostructures have been used effectively in semiconductor devices since about 1969[2.1-5]. It was by using heterostructures that the critical temperature for diode lasers could be increased beyond room temperature. Heterostructures have been used also in transistors[2.6] to realize better performance characteristics. The structures were mostly grown by the techniques of liquid phase epitaxy[2.7,8] or chemical vapor deposition techniques[2.9]. The junctions yielded by these techniques are not very sharp and the transition region from one layer to another may have extents of 10 nm or more. Since the device dimensions were much larger and the wavelength of light of interest was a few nanometer, the transition did not affect the performance of the devices in any significant way. Heterostructures for quantum well devices are, however, required to have much sharper junctions, since for quantization the dimensions are required to be comparable to the electron wavelength, which is of the order of a few nanometers.

Several epitaxy techniques have been developed to grow heterostructures with transition region as thin as a monolayer. Work was initially started by using the technique of molecular beam epitaxy (MBE)[2.10,11]. This technique was perfected with time and in addition, were developed two other techniques: metalorganic chemical vapor deposition (MOCVD)[2.12-14] or metalorganic vapor phase deposition (MOVPD)[2.15] and chemical beam epitaxy (CBE)[2.16,17] or metalorganic molecular beam epitaxy (MOMBE). These techniques of growth are described in this chapter.

2.1. Molecular Beam Epitaxy

Molecular beam epitaxy is a kind of ultrahigh vacuum evaporation in which the atoms or molecules containing the desired atoms are directed from effusion cells to a heated substrate. The atoms on arriving at the substrate combine on the lattice sites. The flux of the incident beams are determined by the temperature of the effusion cells. The composition of the grown layer may be controlled by opening and closing the shutters in front of the effusion cells, which are kept at predetermined temperatures.

A schematic diagram of an MBE system is given in Fig. 2.1. The effusion cells, made of boron nitride or pure graphite are individually provided with heat shields, alumina- coated tungsten heater and a thermocouple to monitor and control their temperature. All the effusion cells are kept inside a shroud, cooled with liquid nitrogen. The gases effusing from the cells are collimated by holes in the shroud. The shutters in front of the cells may be opened or shut to start or stop a beam within 0.1-0.3 s. The substrate is heated by resistive heaters or by radiation, to a temperature of 400 -700 ^{0}C. The chamber, a bell jar, is of stainless steel and is evacuated by an ultrahigh vacuum system consisting of sorption pump, sublimation (titanium) pump and a ion or a closed-cycle helium pump. The pressure in the chamber is initially in the 10^{-9}-10^{-11} torr range, which is achieved by long-time baking (8 hrs or more at a temperature around 180 ^{0}C). This high vacuum is required for ensuring low pressure of ambient impurities. The vapor pressure of evaporating atomsis in the 10^{-9}-10^{-8} torr range. The substrate is scrupolously cleaned before inserting in the growth chamber by etching and sometimes by ion bombardment. It is introduced via two or three-chamber system connected to the sorption pump, cryo-pump and the ultrahigh vacuum (UHV) system to reduce the pressure in steps from 10^{-4} to 10^{-6} and then to 10^{-11} torr.

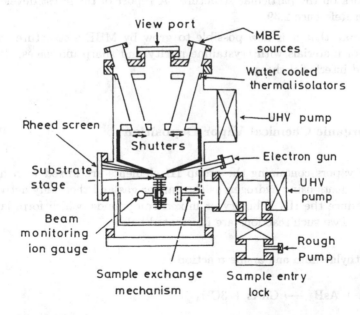

Figure 2.1. Schematic diagram of a molecular beam epitaxy system. Components of the system are labelled in the diagram. (After C. T. Foxon and B. A. Joyce in *Growth and Characterization of Semiconductors*, R. A. Stradling and P. C. Klipstein, eds., Adam and Hilger, New York, 1990, p.36; Copyright: IOP Publishing Ltd).

The system is also provided with quadrupole mass spectrometer (QUAD), a reflection high-energy electron diffraction (RHEED) system and a beam monitoring ion gauge. These are used to monitor respectively the ambient gases, growth rate and intensity of arriving beams.

The system allows growth rate of a monolayer per second and it is possible to grow layers with composition changing within a few monolayers. The heterojunction interfaces have the sharpness of monolayers but laterally there may be variation in the thickness of one or two monolayers.

Heterostructures have been grown by MBE of various compositions, as detailed below:
$GaAs/Ga_x Al_{1-x}As$[2.18], $GaAs/AlAs$[2.19], $In_x Ga_{1-x}As/InP$[2.20], $In_{0.53}Ga_{0.47}$ As/ $In_{0.52}Al_{0.48}As$[2.21], $In_x Ga_{1-x}As/GaAs$[2.22], $In_{0.53}Ga_{0.47}As/AlAs$[2.23], InAs/ GaSb[2.24], InAs/AlSb[2.25], $InAs/In_{0.52}Al_{0.48}As$[2.26], InAs/GaAsSb[2.27], InGa AlAs/InP[2.28], $GaAs/Ga_x Al_{1-x}Sb$[2.29], InGaAsSb/AlGaAsSb[2.30], InGaP/InGa AlP[2.31], $In_x Al_{1-x}P/GaAs$[2.32], $In_{0.47}Ga_{0.5}Al_{1-x}P_x/GaAs$[2.32], $In_x Al_{1-x}P/ GaAs$[2.33], $Ge_x Si_{1-x}/Si$[2.34], ZnSe/ZnMnSe[2.35], CdTe/ZnTe[2.36], HgTe/ CdTe[2.37], $Ga_{1-x}In_x Sb/GaSb$[2.38].

Literature on the subject is vast. The references, given above, are only illustrative of the work on the particular structure. A report of the latest developments is available in Reference 2.39.

We may note that it is now possible to grow by MBE a structure with any combination of materials with crystalline purity and sharp interfaces, devoid of roughness and interfacial charge.

2.2. Metalorganic Chemical Vapor Deposition

Metalorganic vapors containing the group III elements are mixed with hydrides of group V elements using hydrogen as the carrier gas and thermally activated to react and produce the III- V binary or mixed compounds, which form layers on the substrate. Two such reactions are illustrated below.

(a) Trimethylgallium and arsine reaction :

$$(CH_3)_3Ga + AsH_3 \longrightarrow GaAs + 3CH_4$$

(b) Triethylgallium and arsine reaction

$$(C_2H_5)_3Ga + ASH_3 \longrightarrow GaAs + 3C_2H_6$$

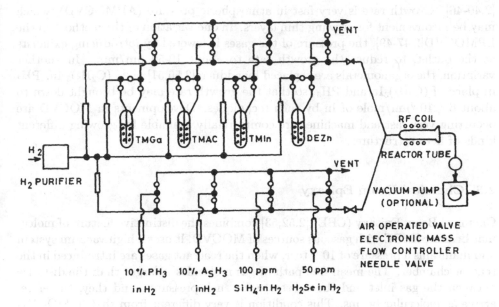

Figure 2.2. Schematic diagram of a metalorganic chemical vapor deposition system. Group III elements are supplied by methyl compounds stored in the tubes and the Group V elements are supplied by the gases. The composition is controlled by controlling the flow rate of the gases and the hydrogen gas by mass flow controllers symbolized by rectangles. The mixed gas reacts in the reactor tube. [After M. J. Ludowise, *J. Appl. Phys.* **58**, R31 (1985); Copyright: American Institute of Physics].

The apparatus used for this purpose is illustrated in Fig. 2.2. Sources of the different vapors are connected to the reactor furnace through mass flow controllers and vents. The substrate is so mounted in the reactor that the arriving gases are well mixed and form an atmosphere near its surface. A continuous flow at near atmospheric pressure is maintained with the right composition. The desired layer grows with a rate, mostly controlled by the arrival rate of Group III element and weakly determined by temperature. Group V elements have practically little effect on the growth rate. Arrival of the group III precursor is determined by a diffusion process through a boundary layer which forms on the substrate and is controlled by its pressure. MOCVD technique gives very sharp interfaces like the MBE technique. The impurity concentration was, however, difficult to control initially as it is determined by the purity of the precursor gas which may be of the order of 1 ppm. Currently, however, layers with a concentration of 10^{15}-10^{16} cm^{-3} are possible to grow by this technique and materials containing phosphorus are produced better in quality than produced by MBE. The technique has been used to grow combinations of mostly III-V binaries, ternaries and quaternaries

[2.40-46]. Growth rate is very fast in atmospheric pressure (APMOCVD), which may be inconvenient for growing thin layers. In one variation of the method, in the LPMOCVD[2.47-49], the pressure of the gases is lowered by introducing exhausts at the outlet, to reduce the growth rate to about 10-50 nm/min. In another variation, the organometals are replaced by adducts[2.50,51], e.g., $(C_2H_5)_3$ In: PH_3 in place of $(C_2H_5)_3$In and PH_3, so that the growth rate may be brought down to about $6\times 10^{-3}\mu$m/mole of In by such techniques. Developments in MOCVD are occurring very fast and machines are commercially available for growing different kinds of heterostructures.

2.3. Chemical Beam Epitaxy

Chemical Beam Epitaxy (CBE) [2.52,53] combines the distinctive feature of molecular beams of MBE with gaseous sources of MOCVD. It uses a high vacuum system and maintains a pressure of 10^{-4} torr, when the reactant gases are introduced in the reactor chamber. The mean free path of gas molecules is larger than the distance between the gas inlet and the substrate for such pressures, and they, therefore, arrive as molecular beams. This condition is very different from that in MOCVD reactor, where the reactant gases diffuse through the overlying gas mixture to the substrate.

A schematic diagram of a CBE system is given in Fig. 2.3. The group V elements are obtained by thermally cracking their hydrides at about 920 °C, while the group III elements are obtained by the pyrolysis of the organometallic compounds at the heated substrate, as in MOCVD. The flow of incoming gases, the hydrides and the organometallic gases, with the carrier hydrogen gas, are controlled as in MOCVD by precision electronic mass flow controllers. The group III atoms and the organometallic gases arrive as molecular beams at the substrate heated to about 550-580 °C and combine to produce the layers. Layers may be grown at the rate of 1.5-3.65 μm/hr. Dopant is introduced as atomic beams or as organometallic gas molecules.

The CBE technique was developed around 1986, while work on the other two techniques was initiated from about 1968. It was, therefore, a late-comer in the field and combined the individual advantages of the two techniques. The advantages in comparison to MBE are: semi-infinite source supply with instant flux response, beams of well-mixed elements, ensuring uniformity of composition, no oval defects, high growth rates. On the other hand, in comparison to MOCVD it offers the advantages of very sharp interfaces and ultrathin layers, no stagnation or flow pattern of gases, clean growth environment and possibility of in-situ monitoring of the growth as in MBE. The technique has been used extensively by Tsang and his collaborators[2.53] to produce InP/InGaAs quantum wells. Wells

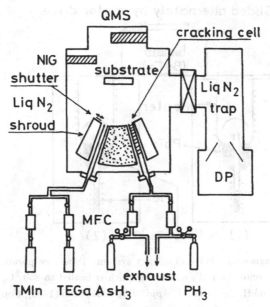

Figure 2.3. Schematic diagram of a chemical beam epitaxy system. Group III elements are supplied by the methyl and ethyl compounds, while the Group V elements are supplied by the gases stored outside the deposition chamber. The composition is controlled by controlling the flow rate with mass flow controllers [After Y. Kawaguchi and H. Asahi, *Appl. Phys. Lett.* **50**, 1243 (1987); Copyright: American Institute of Physics].

as narrow as 0.6nm have been produced. The technique is reported to give wells of uniform width and of very pure material.

2.4. Other Techniques

A few other techniques have also been reported for the growth of heterostructures. Mention may be made of hot-wall epitaxy (HWE)[2.54,55], gas-source MBE (GS-MBE)[2.56], atomic layer epitaxy[2.57] and RF or ECR MBE[58].

2.4.1 HOT-WALL EPITAXY (HWE)

In HWE, the compounds for the layers are heated in quartz crucibles with long necks, walls of which are also heated and the heated substrate is pushed mechanically from the front of one crucible to another to obtain the layers with

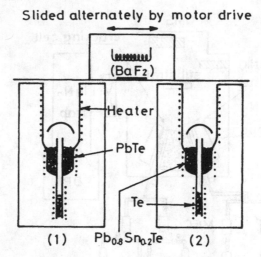

Figure 2.4. Schematic diagram of a hot-wall epitaxy system. Tube 1 evaporates PbTe and Tube 2 evaporates $Pb_0.8Sn_0.2Te$ onto the BaF_2 substrate which is heated to 250 ^{0}C. The walls of the tubes [After H. Kinoshita and H. Fujiyashu, *J. Appl. Phys.* **51**, 5845 (1988); Copyright: American Institute of Physics].

different composition The deposition takes place in a pressure of about 10^{-6} torr. A schematic diagram is given in Fig. 2.4. Multiple layers with thicknesses of 100-200 Å have been grown of PbTe-PbSnTe, by this method. The growth rate is is about 2-2.4 μm/hr.

2.4.2 GS-MBE

The technique of GS-MBE is in between MBE and CBE. The group V elements are supplied in this system from outside the deposition chamber, instead of by heating solids in effusion cells in the chamber. Group III elements are, however, supplied by effusion cells. This arrangement has the advantage of an infinite supply of As, which otherwise requires opening the vacuum at frequent intervals.

2.4.3 LASER-ASSISTED MBE

In some MBE systems, deposition has been done by using laser heating of either the source material or of the beam impinging on the substrate. This technique is often referred as laser- assisted MBE and is reported to improve the quality of deposit as the decomposition temperature is lowered[2.59].

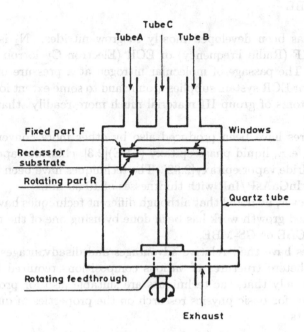

Tube C
Tube A   Tube B

Fixed part F   Windows
Recess for
substrate
Rotating part R
                                Quartz tube

Rotating feedthrough

Exhaust

Figure 2.5. Schematic diagram of an atomic layer epitaxy system showing the growth chamber and the susceptor. The susceptor has a fixed part F, a rotating part R and a recess to hold the substrate. Tubes A and B are the inlets for reactant gases and C is for the input of H_2 to prevent mixing of the reactant gases [After C. L. Goodman and M. V. Pessa, *J. Appl. Phys.* **60**, R65 (1986); Copyright: American Institute of Physics].

2.4.4 ATOMIC LAYER EPITAXY

In atomic layer epitaxy, layers are grown by alternately depositing the anions and cations, so that layers of one kind of atom first deposits on the substrate, and then the atoms of the other kind deposit on it and form bonds to yield a monolayer of the compound. Alternate deposition of the two kinds of atoms may be arranged by introducing the beams, usually hydrides and organometallics, through different inlets, and starting and stopping them alternately. In a variation of the method, the substrate is placed on a rotating mount (see Fig. 2.5), which is inside a susceptor with holes aligned to the inlet tube. For growing a layer, the substrate is rotated and exposed to the constituent gases alternately. The layer grows at the rate of one monolayer per cycle. GaAs/InAs and few other systems have been grown by this technique.

2.4.5 RF-ECR MBE

This technique has been developed mostly to grow nitrides. N_2 is supplied in this systrem as RF (Radio Frequency) or ECR (Electron Cyclotron Resonance) assisted plasma. The passage of molecular nitrogen at a pressure of about 10^{-4} Torr through RF or ECR system supplies atomic and to some extent ionic N_2 which reacts with the atoms of group III material much more readilyt than neutral N_2 molecules.

Heterostructures have been produced also by other more conventional techniques of epitaxy, e.g., liquid phase epitaxy (LPE)[2.3], chloride vapor phase epitaxy[2.60,61], hydride vapor epitaxy[2.62]. The techniques have been used to grow quantum wells of InGaAsP/InP with thicknesses of 10-30 nm.

It should, however, be noted that although different techniques have been tried, bulk of the reported growth work has been done by using one of the techniques of MBE, MOCVD, CBE or GS-MBE.

The techniques have their relative advantages and disadvantages but may be used to produce heterostructures of various compositions required for quantum well devices. Not only that, the techniques are suitable also for producing well-trimmed structures for basic physics research on the properties of quantum wells and semiconductors.

2.5. 1D Structures

Several reports have appeared, in which the fabrication of the so-called quantum wires (1DEG structures) and quantum boxes (0DEG structures) have been described[2.63-79]. Growth of quantum wires has been attempted by using MBE and MOCVD for the growth of the layer materials, but different techniques have been used for reducing the lateral dimensions. These techniques are: etching and regrowth, growth on vicinal substrates, interdiffusion of group III element and growth on patterned non-planar substrates. These techniques are briefly described in the following subsections.

2.5.1. ETCHING AND REGROWTH

There have been several reports of the realization of quantum wires by this technique. For example[2.63], InGaAs quantum wires were realized by using a GS-MBE grown quantum well structure, consisting of an InP substrate, InGaAs well and an InP layer covered by another InGaAs layer. The desired wire pattern was oproduced on the top InGaAs layer by using a resist and electron beam lithography.

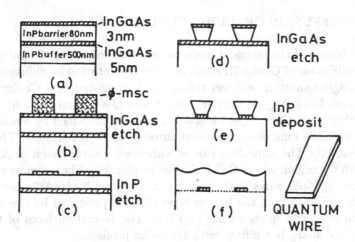

Figure 2.6: Schematic diagram showing the steps in the fabrication of quantum wire by the technique of etching and regrowth [After M. Notomi, M. Naganuma, T. Nishida, T. Tamamura, H. Iwamura, S. Nojima and M. Okamoto, *Appl. Phys. Lett.* **58**, 720 (1991); Copyright : American Institute of Physics].

A selective etch ($H_2SO_4/H_2O_2/H_2O$) was then used to etch the unmasked part of the InGaAs layer. The resist was removed and the InP layer now masked by the remaining InGaAs layer (as shown in Fig. 2.6) was etched by HCl/H_2O. The selective etch was used again and the exposed part of the InGaAs well layer as well as the InGaAs layer masking the upper InP layer were removed. InP was grown again by MOCVD to cover the exposed parts. The unetched part of the InGaAs well layer formed quantum wires with nearly rectangular geometry. The wires were reported to have exhibited the expected energy level shifts due to quantization.

$In_{0.53}Ga_{0.47}As$ quantum wires have been constructed [2.64] also by deep wet chemical etching using high voltage electron beam lithography and dry plasma etching with CH_4, H_2 or Ar.

2.5.2. GROWTH ON VICINAL SUBSTRATES

A 2^0-4^0 misoriented (100) substrate is taken in this technique as the starting material. The misorientation produces in subsequent growths a sequence of alternating steps and terraces. Alternate submonolayers of GaAs and GaAlAs are then grown on this surface. GaAs wires surrounded by GaAlAs are obtained in step advancement. This happens because GaAs grows faster in the lateral direction at the steps while GaAlAs grows uniformly in all directions. Successful growth of stacks of quantum wires has been reported by this technique with a monolayer thickness and lateral dimension of a few nm[2.66-68].

2.5.3. INTERDIFFUSION OF GROUP-III ELEMENTS

Effective shortening of the lateral dimension was realized in this technique by enhancing interdiffusion of Group III elements via implantation or diffusion through a mask. A single quantum well was taken which consisted of a Cr-doped GaAs substrate, 50 nm $Ga_{0.65}Al_{0.35}As$ barrier, 5 nm GaAs QW, 50nm $Ga_{0.65}Al_{0.35}As$ barrier and another 5 nm GaAs layer. Metal masks were produced by electron beam lithography and Ga^+ ions were implanted through the open regions. The sample was then annealed. The annealing caused enhanced interdiffusion of Al and Ga atoms through the region where Ga^+ ions were implanted. The ultimate result was the production of GaAs wires in the well layer, surround by GaAlAs produced by enhanced interdiffusion from the barrier layers. Wires produced by this technique gave evidence of lateral quantization[2.69,70]. The lateral surfaces of the wires are, however, not sharp but diffuse with Gaussian profiles.

2.5.4. GROWTH ON PATTERNED NON-PLANAR SUBSTRATE

This technique relies on the difference in the growth rate of GaAs in different crystallographic directions. It grows faster in the <100> direction in comparison to other directions. Hence, when GaAs is grown on a <100> oriented grooved substrate, the grown layers are thicker at the centre in comparison to the sides and behave like quantum wires. Different techniques have been used to realize non-planar substrates.

In one technique[2.72-74], V-shaped grooves are etched on the <100> oriented GaAs substrate along the <0$\bar{1}$1> direction. An AlGaAs layer is grown on the substrate, which has a sharp corner between two (111)A planes. GaAs, grown on this GaAlAs layer, has a triangular thick region at the bottom and thin regions on the sides as illustrated in Fig. 2.7(a).

In a second technique, wires are grown on patterned substrates[2.75-77]. A SiO_2 layer is first formed by plasma chemical-vapour deposition on the (100) substrate. The SiO_2 layer is then etched to form gratings oriented along the <0$\bar{1}$1> directions by using resist and wet chemical etching. GaAs grown on this masked substrate has triangular surfaces with smooth (111)A faces, due to the directional dependence of the growth of GaAs. The growth of GaAs is continued in one report to join over the SiO_2 masks. GaAs forms a corrugated surface, as shown in Fig. 2.7(b). Quantum wires are realized as in the earlier case by growing first AlGaAs and

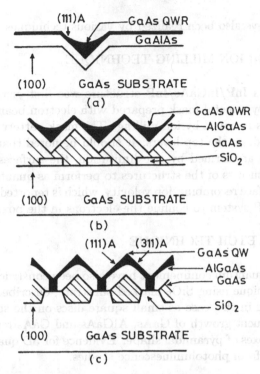

Figure 2.7. Schematic diagram showing quantum wires grown by the technique of patterned substrate. (a) Growth on grooved substrate. (b) Triangular quantum wire grown by SiO₂ masked substrate. (c) Arrowhead quantum wire grown by SiO₂ masked substrate.

then GaAs on this corrugated surface. In a second report[2.78], the SiO₂ masks were left uncovered by GaAs. AlGaAs grown on the resulting surfaces, had side walls parallel to the (111)A planes, but the top surface consisted of (311)A planes. Subsequent growth of GaAs resulted in the formation of GaAs layers with thick arrow-head tops and thin sides as shown in Fig. 2.7(c). The thick region behaved as quantum wires.

The technique of growth on patterned substrates has been used also to grow SiGe/Si wires[2.79].

The three techniques of construction of quantum wires, described above, have yielded wires with fairly clean and sharp interfaces, and have been tried also in devices.

2.6 0D structures

Quantum boxes have also been realized by various techniques.

2.6.1 ETCHING OR ION MILLING TECHNIQUE

In one case[2.80], a InP/InGaAs/InP structure was first grown. The top layer was then masked by a metal mask prepared with electron beam lithography. The unmasked part was removed by ion milling. The mask pattern produced wires or dots, as was desired, on the top layers. It should be noted that in this technique, the wires or boxes are formed by size limitation, the surfaces being exposed to the ambient. The success of the structures to perform as quantum well, therefore, depends on the surface recombination velocity, which is reported to be good enough for the InGaAs/InP system to confine the electrons in the box.

2.6.2 SELECTIVE ETCH TECHNIQUE

Structures for 3D quantum confinement have also been constructed[2.81,82] by the selective etch technique using SiO_2 masked substrate, described above. The SiO_2 masks are prepared in this case as small square discs on the surface of the GaAs substrates. Subsequent growth of GaAs, AlGaAs and GaAs results in the formation of quantum boxes of pyramidal shape. Evidence for 3D quantum confinement has been obtained from photoluminescence studies.

2.6.3 SELF-ORGANIZED GROWTH

More recently, quantum boxes have been realized[2.83-88] by controlled growth of the epitaxial layer of lattice-mismatched InGaAs on GaAs or AlGaAs substrate by using the so-called Stranski-Krastanow growth conditions. A few monolayers of $In_xGa_{1-x}As$ are deposited on a GaAs or a AlGaAs substrate. These layers grow as islands with dimensions of a few nanometers to relieve the accumulated strain arising from the lattice mismatch. The islands are then embedded in GaAs or GaAlAs by overgrowing a few layers of these materials. The embedded islands of InGaAs behave as quantum dots. Quantum dots of SiGe[2.89] on Si and InP on $In_{0.085}Ga_{0.515}As$ have also been grown by using the same technique[2.90].The technique is being used currently to realize quantum dots based on various material systems [2.91].

Arrays of quantum dots having approximately 20-30 nm diameter with a density of 10^{11} cm^{-2} may be grown [2.92] by monitoring the transition from the two-dimensional (layered) to the three-dimensional (islanded) growth by Reflection High Energy Electron Diffraction (RHEED). The variation in the size of such arrsays is within $\pm 7\%$. The luminous efficiency of such dots have been shown to be very high.

2.7. Conclusion

In this chapter has been described the various techniques, currently employed to grow heterostructures with interfaces good enough for quantum well devices. Emphasis in the early years was on lattice-matched heterostructures. Currently, however, a major part of the work is concerned with the growth of non-lattice-matched strained-layer heterostructures of various compositions[2.93], which give better performance characteristics for some quantum well devices. As a further improvement, compensated heterostructures[2.94] are also being grown. In these structures, a compressively strained layer may be sandwiched between two tensile-strained barrier layers, which are again enclosed by two lattice-matched layers.

Effort is also being made to grow quantum wires and quantum dots by using different techniques.

Centres for the growth of heterostructures proliferated as commercial MBE and MOCVD machines became available. Heterostructures with various compositions and complicated structures are being grown and studied extensively in these centres.It appears at the present stage that techniques of growth do not introduce any limitation in trying any idea that may be conceived. Various kinds of devices with very novel ideas are being proposed and tried. It is not yet clear how many of them will be ultimately accepted commercially. Only those, which show promise currently, will be of interest for further discussion in this book.

CHAPTER 3

BAND OFFSET

Heterostructures, grown by the techniques described in Chapter 2, are structurally pure, although of varying composition and consequently of varying band gap. The structural purity is achieved by selecting constituents having the same lattice constant and crystal structure. In Fig. 3.1 are plotted the direct band gap and lattice constant of the various III-V and II- VI compounds[3.1,2].

There are a few binaries, which have nearly equal lattice constant but different band gap, e.g., GaP and AlP; Ge, GaAs, ZnSe and AlAs; InAs, ZnTe, GaSb and AlSb; αSn, InSb and CdTe. Heterostructures with good crystalline interfaces have been grown using some of these binaries. Examples are : GaAs/AlAs, InAs/GaSb, InAs/AlSb, InSb/CdTe.

Ternaries having different band gap and lattice constant may be obtained by combining two binaries. The lines joining the binary compounds in Fig. 3.1 give the values for the ternaries which may be obtained by combining the end-point binaries. The ternary will have the same lattice constant when the component binaries have nearly equal lattice constant. The band gap, however, may be adjusted in between the two end values by choosing the proportion of the binaries. For example, $Al_xGa_{1-x}As$ has nearly the same lattice constant as GaAs and AlAs; the direct band gap may, however, be varied between 2.79 eV and 1.51 eV by changing x. Binaries with different lattice constants may also be combined to obtain a ternary with a lattice constant which is equal to that of a third binary. $Ga_{0.47}In_{0.53}As$ and $Al_{0.48}In_{0.52}As$ are examples of ternaries which are lattice-matched to InP. The band gap of the ternary cannot be chosen in this case. The lattice constant and the band gap may, however, be chosen independently by combining four elements to realize quaternaries. $Ga_xIn_{1-x}As_yP_{1-y}$ is an important example of such a quaternary which has the same lattice constant as InP and a band gap which may be adjusted between 0.73 eV and 1.35 eV when x and y are chosen according to the following relation :

$$x = 0.1894\,y/(0.4184 - 0.013\,y) \qquad (3.1)$$

Heterostructures with good crystalline quality may also be grown with non-lattice-matched compounds, but only for limited dimensions. The lattice-mismatch causes strains which are accommodated up to a thickness of about 4 nm. Larger thickness results in the appearance of stacking faults due to the strain. Strained-layer heterostructures with limited thickness have also been grown to realize chosen band gaps. The interfaces in the heterostructures are mostly free of crystal defects or interfacial charge density. In early days, there used to be surface roughness, but such roughness has been mostly eliminated, although there remain variations of widths with extended areas. Electron states in the heterostructures are, therefore, altered from the bulk states only by the potential profile resulting from the variation in the band gap.

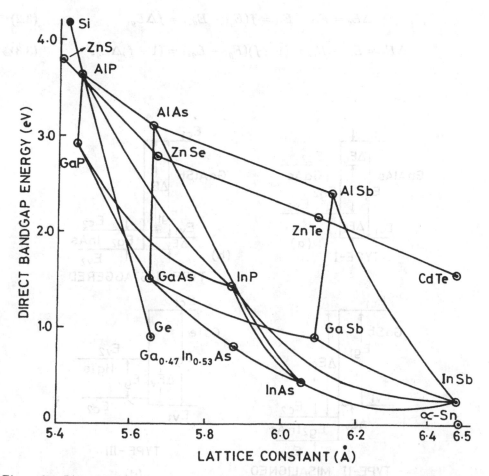

Figure 3.1. Direct band gap-lattice constant diagram for III-V and II-VI compounds. The symbol ⊙ gives the values of direct band gap and the lattice constant of the associated binary. The lines joining two binaries give the combination of values of band gap and lattice constant of ternaries that may be obtained by combining the end-point binaries as exemplified by $Ga_{0.47}In_{0.53}As$.

3.1. Types of Heterosructures

The potential profile near the heterojunction has step- discontinuities in the absence of accumulated charges as shown in Fig. 3.2. Heterostructures are named[3.3] as Type I, Type II and Type III according to the alignment of the bands producing the discontinuity.

In Type I heterostructures, illustrated in Fig. 3.2(a), the band gap of one material overlaps that of the other and the potential discontinuities for the conduction band, ΔE_c, and for the valence band, ΔE_v, may be expressed as,

$$\Delta E_c = E_{c1} - E_{c2} = f(E_{g1} - E_{g2}) = f\Delta E_g, \tag{3.2}$$

$$\Delta E_v = E_{v1} - E_{v2} = (1-f)(E_{g1} - E_{g2}) = (1-f)\Delta E_g, \tag{3.3}$$

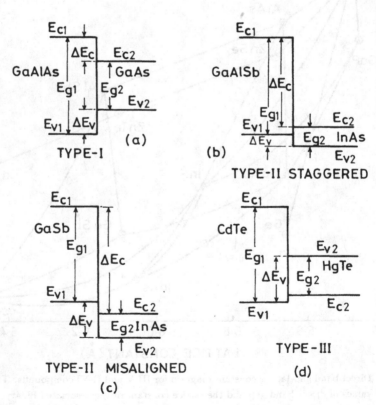

Figure 3.2. Band alignments in heterojunctions. (a) Type I heterostructure. (b) Type II-staggered heterostructure. (c) Type II-misaligned heterostructure. (d) Type III heterostructure.

$E_{g1} > E_{g2}$ and ΔE_g indicates the band gap difference. The factor f gives the ratio $\Delta E_c / \Delta E_g$; it depends on the pair of materials forming the heterostructure and its magnitude ranges between 0 and 1. Both electrons and holes are confined in the lower-band-gap material in these heterostructures. $Al_x Ga_{1-x} As/GaAs$, $InP/Ga_{0.47}In_{0.53}As$, $Al_{0.48}In_{0.52}As/Ga_{0.47}In_{0.53}As$ and $InP/Ga_x In_{1-x} As_y P_{1-y}$ heterostructures are of this type.

In Type II heterostructures, $E_{v1} > E_{v2}$ and ΔE_c may or may not be larger than E_{g1}. Also, E_{g2} is not necessarily smaller than E_{g1}. Hetrostructures of Type II, (illustrated in Fig. 3.2(b)) with $\Delta E_c < E_{g1}$ have been named Type II-staggered. Both the conduction-band edge and the valence-band edge of one material being lower than the corresponding band edges of the other material, electrons are confined in one material, while holes are confined in the other material. This type of band alignment occurs, for example, in $InAs/Al_{0.4}Ga_{0.6}Sb$ heterostructures.

Type II hetrostructures, (illustrated in Fig. 3.2(c)), with $\Delta E_c > E_{g1}$ have been named Type II-misaligned. In these heterostructures also, electrons and holes are confined separately in the two materials. But, as the valence band of the material, in which the holes are confined, overlaps the conduction band of the other material some novel phenomena result. $InAs/GaSb$ heterostructure belongs to this class.

Type III hetrostructures are formed by the combination of a semimetal with inverted bands and a semiconductor as shown in Fig. 3.2(d). $HgTe/CdTe$ is an example of this type of hetrostructure. Valence bands of the two materials strongly interact in this structure. It should be mentioned that there is some confusion in the literature about the nomenclature of Type III; Type II-misaligned has been named Type III by some authors[3.4].

Quantum well devices have been realized mostly by using the Type I heterostructures, either lattice matched or strained- layer systems. $GaAs/Ga_x Al_{1-x}As$, $Ga_{0.47}In_{0.53}As/InP$, $Ga_{0.47}In_{0.53}As/Al_{0.48}In_{0.52}As$ and $InP/Ga_x In_{1-x} As_{1-y} P_y$ are commonly used as lattice- matched systems while $Al_x Ga_{1-x} As/Ga_x In_{1-x} As$, $Ga_x In_{1-x} As/ Al_y In_{1-y} As$, $Ga_{0.47}In_{0.53}As/AlAs$, $Ga_x In_{1-x} As/InP$ and a few other combinations are used as strained layer systems.

Type II or Type III structures are not much used in quantum well devices. Only $InAs/GaInSb$ or $InAs/AlInSb$ is used in QWIP's. The $Si_x Ge_{1-x}/Ge$ system is another type II system which has been finding much use in recent years.

The potential discontinuity plays a crucial role in quantum well devices as it is this discontinuity or the so-called band offset which causes the quantum behavior of the structures. Models are first discussed for estimating the band offset from the bulk properties of the constituents. Experimental methods are then described for the determination of its values, and the best available values are also listed at the end.

3.2. Emperical Rules

Two empirical rules[3.5-8] have been used to estimate the band offset, namely, the electron affinity rule and the common anion rule. Although the relevance of these rules has been much reduced by the development of accurate theoretical methods, these are briefly discussed below considering their extensive use in the past.

3.2.1. ELECTRON AFFINITY RULE

The oldest rule is that given by Anderson[3.5,6]. According this rule, the discontinuity in the conduction band edges is the difference of the electron affinities for the constituents, i.e., the conduction band offset,

$$\Delta E_c = \chi_1 - \chi_2, \tag{3.4}$$

where χ_1 and χ_2 are the electron affinities of material 1 and 2 constituting the heterojunction.

It has been argued that this simple rule should not be necessarily applicable, since the electron affinity relates to the free surface with a large discontinuity, whereas in heterostructures, the discontinuity is much smaller and the surface conditions are much different. Particularly, it is thought that the contribution of dipoles with negative charge on the outer surface, which form on it due to quantum mechanical tunneling of the electrons, should be much different in heterojunctions than in free surfaces. Further, values of the band offset is a small fraction of the electron affinity, and experimental values of the latter are not accurate enough to yield non-anomalous values of the former.

However, in spite of these criticisms, the affinity rule has been widely used, and in many cases, the values given by the rule are found to be close to those obtained by other experimental methods.

3.2.2. COMMON ANION RULE

This rule[3.7,8] is based on the empirical results obtained from the Schottky contacts to III-V and II-VI compounds. The Schottky barrier for the valence band of these compounds to gold has been found to be the same for different compounds containing the same anion. This result has been explained by considering that the theoretical band structure calculations indicate that the valence band develops out of the p-levels of the anions, contribution of the cations being comparatively negligible. Hence, the position of the valence band minimum with reference to the vacuum, or any other common reference level, e.g., the Fermi level of a metal, should depend only on the electronegativity of the anion. Valence-band offset for compounds with the same anion should, therefore, be zero and those for different anions may be derived from the Schottky barrier valence-band offsets. Although, the band offsets for compounds with the same anion are found to be closer than those with different anions, the rule does not apply strictly. Variation of the band offset of the order of 0.2- 0.4 eV is observed between compounds with the same anion. The rule cannot, therefore, be applied to obtain accurate values of the band offset, but has its usefulness as a guide for selecting the anion for the compound. The offset may be expected to increase in the order of As, P, Te, Se and S.

3.3. Theoretical Methods

Several theoretical methopds [3.9-16] haqve been reported for the calculation of band offset. Two of these methods, the Tersoff method and the Van de Walle-Martin method are explained below as these methods give values agreeing with the experiments to within 0.1 eV.

3.3.1 TERSOFF METHOD

The method is based on the argument that there are states in the gap, produced by the interface, and an energy level, E_M, may be identified for each semiconductor swuch that it divides the gap states into two groups. Those above are conduction-band-like while those below are valence-band-like. When a heterojunction is formed, dipoles are produced if the E_M's of the constituents are not aligned.

The alignment of the bands is assumed to be dictated by the condition that the interface dipoles will be minimized, and therefore the midgap energies, E_M's shall be aligned. Knowing the position of E_M's with reference to the conduction-band edges, the band offset may be determined. The value of E_M may be theoretically calculated to obvtain the band offset. However, E_M of a semiconductor also determines the pinning of the Fermi energy of a metal when it is contacted to it. This provides a means of determining the relative position of E_M of two semiconductors from the Schottky barrier with respect to a common metal.

Band offsets determined experimentally for different binaries with respect to germanium and silicon have been compared with the theoretical values and found to agree within 0.15 eV.

3.3.2. VAN DE WALLE -MARTIN METHOD

Band offsets are obtained in this method from first- principles calculations[3.13-18]. In such calculations, an infinite crystal is considered to be made up of a periodically repeated supercell consisting of the two kinds of layers including the

interface region. The pseudo-potentials are evaluated self-consistently for the supercell geometry by using the local density-functional approximation (LDA) (applied in the momentum space formalism) and non-local norm-conserving *ab initio*
pseudopotentials,i.e., potentials generated by theoretical calculations on atoms.
The potentials , averaged over planes parallel to the interface,are determined by
using the pseudopotentials as a function of the distance from the interface. The
average in the central regions of the two sides of the interface are considered to be
bulk-like. Individual bulk-band calculations are then made to fix the valence-band
maxima with respect to these average potentials and the band offset is calculated
therefrom, as illustrated in Fig.3.3 for AlAs/GaAs system. Calculated values have
been given for the heterostructures of the III-V/III-V, III-V/IV, II-VI/III-V , II-
VI/II-VI compunds and III-nitride/ III-nitride system.

The calculated values are in most cases within the spread of the experimental results. Values obtained by using two different techniques[3.13-15] of band

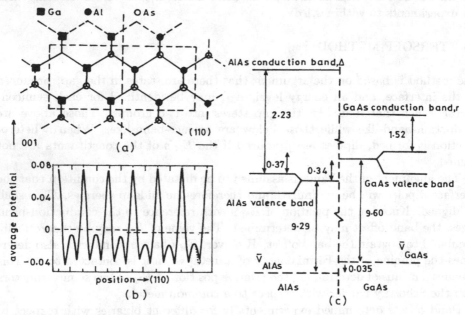

Figure 3.3. (a) A schematic representation of a GaAs/AlAs interface showing the supercell by
the dashed lines. (b) Variation of the average potential in the directiuon normal to the interface.
Dotted lines are its values on its two sides which coincide with the bulk values far from the interface.
(c) Illustrates the derivation of the band line-up for AlAs/GaAs junction. The values are given
with reference to the $l = 1$ angular momentum component potential. The valence-band offset
is found to be 0.34 eV. The value increases to 0.37 eV when the effects of spin-orbit splitting is
included.[After C. G. Van de Walle and R. M. Martin, *Phys. Rev. B* **35**, 8154 (1987); Copyright
(1987) by the American Physical Society].

structure calculations, however, differ significantly for some structures, notably for the important AlAs/GaAs structure. One method[3.13] gives a value of 0.39 eV, while the other method[3.14] gives a value of 0.53 eV for the valence-band offset.

3.4. Experimental Methods

Design of a quantum well device requires more accurate values of band offset than may be obtained from the theoretical models. Attempts have been made to obtain such values from experiment.

Many methods have been used to obtain the band offset from the analysis of data on experimental phenomena. These are based on: absorption measurements[3.19], photoluminescence studies in quantum wells[3.20-24], photoemission spectroscopy[3.25-28], current-voltage[3.29] measurements, capacitance-voltage measurements[3.30,31], charge transfer analysis[3.32] and X-ray photoemission[3.33-40] core level spectroscopy (XPS).

The first method was used to establish that energy levels are quantised in potential well structures, while the second one has been extensively used to obtain the band offset for different heterostructures. The XPS technique is believed to give more accurate values and does not involve any assumption, approximation or knowledge of other physical constants. These three methods are described below.

3.4.1. ABSORPTION MEASUREMENT

Quantum wells are realized by sandwiching a thin layer of the lower-band-gap constituent, between two layers of the larger- band-gap material. The component of momentum in the direction perpendicular to the interfaces being quantized, the absorption spectrum of such structures show peaks corresponding to transitions from the hole levels to the electron levels at near- liquid- helium temperature as illustrated in Fig. 3.4. Band offsets are determined by fitting the transition energies with the calculated values for different offsets. The transition energies do not, however, change sharply with the band offset and the method has

therefore yielded erroneous results. In fact, this method was first applied[3.19] to a 315 Å wide quantum well of the GaAs/GaAlAs system and gave a value of 0.85 for $\Delta E_c/\Delta E_g$, the ratio of the conduction-band offset and the band gap difference. This value was used for many years for the analysis of the properties of quantum wells and superlattices, but was ultimately revised to 0.6 by other experiments.

The accuracy of the method becomes particularly poor, if wide wells are used, since the energy levels in such wells do not deviate much from the values for infinite barrier height. If, however, narrow wells are used, much better results may be obtained, as the energy levels then vary sharply with the band offset.

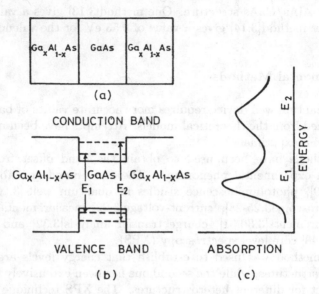

Figure 3.4. Schematic diagram showing the absorption spectrum of a quantum well. (a) The heterostructure. (b) Quantized energy levels. (c) Schematic absorption coefficient for different photon energy.

3.3.2. PHOTOLUMINESCENCE MEASUREMENT

Photoluminescence experiments give the difference in energy between the first level in the conduction band and the first level of the heavy-hole band. Strong light of an intensity of 1-5 W/cm² from a argon laser is shone on the sample to excite electrons from the valence band to the conduction band. These excited electrons finally recombine to produce the luminescence radiation, giving effectively the shift in band gap due to quantization. Such luminescence lines, shown in Fig. 3.5(a) are recorded for different well widths. The band offset is then obtained by fitting the calculated energy shifts with the measured values as illustrated in Fig. 3.5(b).

The method has been extensively applied[3.21] to study the GaInAs/InP system, growing wells as thin as 3 Å. The values of band offset obtained by different authors are, however, much different. The difference arises mostly from the errors in the determination of well widths and the incompleteness of the theory of energy

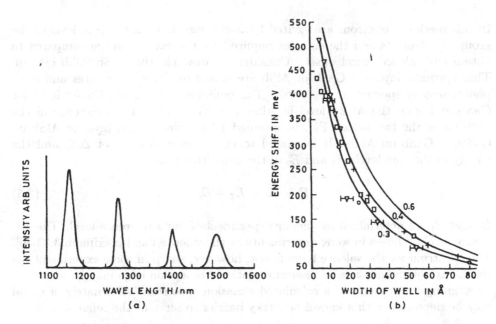

Figure 3.5. Photoluminescence spectra for quantum wells. (a) Successive spectrum from widths of 10, 20, 40 and 60 Å [After J. H. Marsh, J. S. Roberts and P. A. Claxton, *Appl. Phys. Lett.* **46**, 1161 (1985); Copyright: American Institute of Physics]. (b) Energy shift in $In_{0.53}Ga_{0.47}As/InP$ wells for different well widths [After B. R. Nag and S. Mukhopadhyay, *Appl. Phys. Lett.* **58**, 1056 (1991); Copyright: American Institute of Physics]. Solid lines give the calculated values of energy shift for different values of conduction-band offset, indicated by the numbers in eV against each curve. The data points are the experimental values.

levels. The theory is incomplete in two ways. First, the energy levels being deep within the conduction band, energy-band nonparabolicity plays an important role. Proper formulation of this aspect of the theory is still being debated in the literature. Second, it is not yet clearly established, how accurate is the effective mass formalism when the well is so thin that it has a few monolayers. The method should, therefore, be considered to be of limited accuracy like the absorption or excitation spectroscopy method. A variation of the method has also been used[3.22], in which is analyzed the splitting of the spectra found for narrow wells due to the variation in the thickness. This method also has the same problem, as the analysis of the basic spectra. Photoluminescence spectra have also been used to obtain the band offset in the currently emerging GaN/AlGaN material system[3.23,24].

3.4.3. X-RAY CORE LEVEL PHOTOEMISSION SPECTROSCOPY (XPS)

In this method , electrons are excited by X-ray signals from the core level of the group III elements and the energies required for the excitation are compared to obtain the valence- band offset. Consider for example the GaSb/AlSb system. Thin epitaxial layers of GaSb or AlSb are grown on GaSb substrates and their photoemission spectra are recorded. The emission is from the Ga $3d$ level for GaSb and from the Al $2p$ level in AlSb (see Fig. 3.6). The difference in the position of the two peaks, F_3 , is obtained by growing a thin layer of AlSb on GaSb or GaSb on AlSb. It is related to the valence-band offset ΔE_v and the energy of the core levels E_1 and E_2 of the constituents as

$$E_3 = E_2 - E_1 + E_v. \tag{3.5}$$

E_2 and E_1 are determined by similar experiments from a reference level. The reference level is chosen in some experiments as the valence band maximum[3.33,36]. The spectrum for the valence band is not, however, sharp in some experiments as the X-ray source may not be monochromatic. Absorption edge may be identified by aligning the features of a calculated emission spectrum. Alternately, a metal may be deposited with a known Schottky barrier to serve as the reference. In

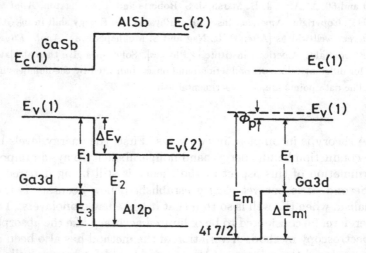

Figure 3.6. Energy bands and core energy levels in, (heterojunction of GaSb and AlSb and (b) a Au Schottky barrier on GaSb. $E_c(1)(E_c(2))$ and $E_v(1)(E_v(2))$ are respectively the conduction and valence band levels of GaSb(AlSb). E_{Fm} is the Fermi level and $\phi_p(1)$ is the Schottky barrier height of Au on GaSb. Ga $3d$,Al $2p$ and Au $4f_2$ are respectively the $3d$ level of Ga, $2p$ level of Al and $4f_2$ level of Au. The significance of E_1, E_2, E_3, E_M and ΔE_m are indicated in the figure.

one experiment[3.34], gold was deposited on the GaSb and AlSb layers and the differences between the Au $4f_{7/2}$ and the Ga $3d$ or the Al $2p$ levels were determined from the photoemission data. If now E_{mi} and E_i denote these energies, for one material (i=1 for Ga and i=2 for Al), then the measured energy difference may be expressed as,

$$\Delta E_{m1} = E_m - \phi_1 - E_1, \tag{3.6}$$

ϕ_1 being the Schottky barrier for gold on the particular p-type material. The valence-band offset is now given by

$$\Delta E_v = E_3 - (E_2 - E_1) = E_3 + (\Delta - E_{m2} - \Delta E_{m1}) + \phi_2 - \phi_1. \tag{3.7}$$

Using the predetermined values of ϕ_1 and ϕ_2 and the values of ΔE_{m1} and ΔE_{m2} obtained in this experiment, ΔE_v may be determined.

This method has been used to measure the valence-band offset in the GaSb/AlSb system[3.34]. A variation of the method was used to determine ΔE_v for the InAs/GaSb[3.35] system ,in which a common Fermi level served as the reference . On the other hand, ΔE_v for the HgTe/CdTe system[3.38] and the GaAs/AlAs system[3.39] was obtained by this method using the valence-band edge as the reference. More recently, the method has been used to determine ΔE_v for the GaN/AlN system[3.37] and InN/GaN, GaN/InN, InN/AlN, AlN/InN, GaN/AlN and the AlN/GaN system[3.40].

3.5. Values of Band Offset

It should be evident from the preceding discussion that accurate values of band offset are difficult to determine. Attempts for determining its value from band structure calculation or from the analysis of the experimental results often yielded widely varying values of band offset. For example, initially the ratio $\Delta E_c/\Delta E_v$ for the GaAs/Ga$_{0.7}$Al$_{0.3}$As system was fixed[3.19] to be 85:15, but later experiments established a value of 60:40. The value of $\Delta E_c/\Delta E_v$ for Ga$_{0.47}$In$_{0.53}$As/InP is reported to be between 36:64 and 100:0[3.41], and for the Ga$_{0.5}$In$_{0.5}$P/GaAs system between 8:92 and 90:10[3.42-49]. Apparently, the value depends very much on the interpretation of the experiment and also on the conditions of preparation of a sample. Consequently universally acceptable values are not available for all the heterojunctions. In Table 3.1 are collected the currently accepted values of valence-band offset for some common heterojunctions.

In many cases, experimental results may not be available for the combinations of materials of interest. Data available for the valence band position of the constituents with reference to a common reference valence band as obtained in photoemission experiment may be used for this purpose. Such data have been

Table 3.1 Valence-Band offset, ΔE_v, in heterojunctions

Heterojunction material	ΔE_v(eV)	Reference
GaAs/Al$_x$Ga$_{1-x}$As	0.46x	3.14
GaAs/Ga$_{0.5}$Al$_{0.5}$As	0.21	3.30
GaAs/AlAs	0.31	3.36
GaAs/InAs	0.35	3.36
AlAs/Al$_{0.37}$Ga$_{0.63}$As	0.34	3.31
AlAs/InAs	0.35	3.36
InP/Ga$_{0.47}$In$_{0.53}$As	0.32	3.15
Al$_{0.48}$In$_{0.52}$As/Ga$_{0.47}$In$_{0.47}$As	0.75	3.29
GaAS/ALSb	0.40	3.25
GaSb/InAs	0.51	3.32
GaSb/InAs$_{0.95}$Sb$_{0.05}$	0.67	3.32
CdTe/HgTe	0.34	3.36
CdTe/ZnTe	0.40	3.36
ZnTe/HgTe	0.30	3.36
GaN/AlN	0.7±0.24	3.30
AlN/GaN	0.57±0.22	3.40
InN/AlN	1.81±0.2	3.40
AlN/InN	1.32±0.14	3.40
InN/GaN	1.05±	3.40
Ga$_{0.5}$In$_{0.5}$P/GaAs	0.32	3.45

given in Table 3.2. It should be evident that the table lists values only for the binaries. However, the data may be used as a guide for ternaries or quaternaries by linear interpolation or by using a relation analogous to the Vegard's law.

Table 3.2 Valence-Band offset, ΔE_v, with Reference to Germanium and Silicon*

Material	ΔE_v(eV)	Material	ΔE_v(eV)
Si-Ge	0.17	InP-Ge	0.64
AlAs-Ge	0.95	GaAs-Si	0.05
GaAs-Ge	0.35	InAs-Si	0.15
InAs-Ge	0.33	GaSb-Si	0.05
GaSb-Ge	0.20	GaP-Si	0.80
GaP-Ge	0.80	InP-Si	0.57

* Values are quoted from Reference 3.12.

3.6. Conclusion

In this chapter has been discussed the theories of band offset and the experimental methods available for its determination. Band offset forms the quantum wells and determines how well are the electrons confined in the well. It plays a very important role in some devices. Its values are therefore required to be known with greater accuracy than is possible at present for some heterojunctions. However, till such values are available analysis will proceed with the best current values quoted here.

It should be noted also that in considering the band line-ups, attention should be given to the location of the minimum of the junction materials in the Brillouin zone. The minimum of the two materials may not have the same locations in the Brillouin zone and it becomes important to consider which combination of the minimum should be considered as forming the barrier. There are not many studies, but the indications of the few studies so far reported[3.50], are that it is the same kind of minimum which forms the barrier. For example, in the GaAs/AlAs combination, the minimum in GaAs is at the Γ point, while in AlAs, it is at the X point. It is, however, found that the offset between the Γ-point minimum of GaAs and that of the Γ-point minimum of AlAs form the barrier. In fact the X-point minimum of the two materials also form wells[3.51], but with a different characteristic. Such studies are, however, required to be extended also to other heterojunctions.

CHAPTER 4

ELECTRON STATES

Heterostructures for quantum well devices are constructed in three forms : the single-junction structures, often referred as simply heterostructures, the double - junction structures, mostly referred as quantum wells and - multi-junction structures, called superlattices. These are shown in Fig. 4.1. Electron states in the structures are evaluated by assuming that the bulk band structures remain applicable for the constituents, even though the physical dimension in one or more directions may be comparable to the lattice constant. Electron states in the structures are obtained by solving the wave equation for the potential distributions in the structure by using the bulk physical constants and by applying the so-called effective-mass approximation and suitable boundary conditions, as explained in the following section.

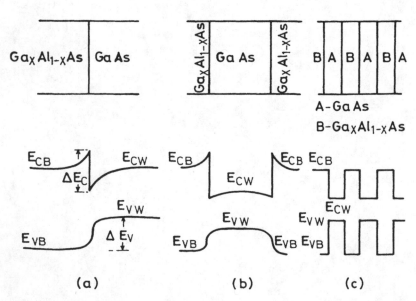

Figure 4.1. Heterostructures. (a) Single-junction heterostructure. (b) Double-junction heterostructure. (c) Superlattice.

4.1. Effective-Mass Approximation

Time-independent Schrödinger wave equation for the electrons in a solid is

$$[(-\hbar^2/2m_0)\nabla^2 + V_c(\mathbf{r})]\psi(\mathbf{r}) = E\psi(\mathbf{r}), \qquad (4.1)$$

where m_0 is the free electron mass, $-i\hbar\nabla$ is the momentum operator, $V_c(\mathbf{r})$ is the operator for the potential arising from the ions including all the bound electrons and also in the one- electron approximation, partly from free electrons except the one under consideration, $\psi(\mathbf{r})$ is the wave function and E is the eigenvalue of energy. The solution of the equation is the Bloch function, having the form,

$$\psi_{n\mathbf{k}}(\mathbf{r}) = (1/N^{1/2})\, U_{n\mathbf{k}}(\mathbf{r})\exp{(i\mathbf{k}.\mathbf{r})}, \qquad (4.2)$$

where N is the number of unit cells in the crystal. The wave vector is represented by $\mathbf{k}$. The function $U_{n\mathbf{k}}(\mathbf{r})$ has the periodicity of the lattice i.e., $U_{n\mathbf{k}}(\mathbf{r}+\mathbf{a}) = U_{n\mathbf{k}}(\mathbf{r})$, where $\mathbf{a}$ is a lattice translation vector. The function is referred as the cell periodic part of the Bloch function and is normalized over a unit cell. The subscript n indicates the band to which the particular state belongs.

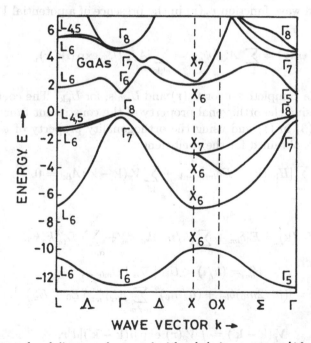

Figure 4.2. Energy band diagram of a crystal with sphalerite structure. [After J. R. Chelikowsky and M. L. Cohen, *Phys. Rev.* B **14**, 556 (1976); Copyright(1976) by the American Physical Society].

Different methods have been developed for solving Eq. (4.1) to obtain the relation between e and $\mathbf{k}$. These band structure calculations yield diagrams as illustrated in Fig. 4.2 for a representative III-V binary compound, having the sphalerite crystal structure. The conduction band and the valence band have several extrema, some of which may be degenerate. In electronic, optoelectronic or photonic applications of semiconductors, these extrema are mostly of importance, as the carriers are concentrated near them. The full details of the band structure, as revealed by the band-structure calculations are not involved. It is, therefore, convenient if the wave equation in the presence of an external potential is recast in terms of the behavior of the electrons at these extrema. This is done in the effective- mass approximation of the wave equation by using the $\mathbf{k.p}$ method[4.1-4] outlined below.

4.1.1. EFFECTIVE-MASS APPROXIMATION FOR DEGENERATE EXTREMA

We assume that the electron states are required to be evaluated near an r-fold degenerate extremum with energy E_0 and other extrema are assumed to be separated by large energies from this extremum.

The perturbed wave function $\psi_p(\mathbf{r})$ in the presence of a potential $V_p(\mathbf{r})$ may be expressed as,

$$\psi_p(\mathbf{r}) = \sum_{\mathbf{k}'} A_{n\mathbf{k}'} \psi_{n\mathbf{k}'} = \sum_{n',\mathbf{k}'} A_{n'\mathbf{k}'} U_{n'0} \exp{(i\mathbf{k}'.\mathbf{r})}, \qquad (4.3)$$

since $\psi_{n\mathbf{k}}$'s form a complete set for $\psi_p(\mathbf{r})$ and $U_{n'0}$'s, for $U_{n\mathbf{k}}$. The coefficient $A_{n'\mathbf{k}'}$ is evaluated by using the orthogonal property of the constituent functions.

Substituting (3) in (1) and using the orthogonality property of $\psi_{n\mathbf{k}}$'s the following equation is obtained for the coefficients.

$$\sum_m [H_{lm}(\mathbf{k}) - E\delta_{lm}]A_{m\mathbf{k}} + \sum_{\mathbf{k}'} \tilde{V}_p(\mathbf{k} - \mathbf{k}')A_{l\mathbf{k}'} = 0, \qquad (4.4)$$

where,

$$H_{lm}(\mathbf{k}) = E_0\delta_{lm} + \sum_{\alpha=1}^{3} (\hbar/m_0)k_\alpha p_{lm}^\alpha + \sum_{\alpha,\beta=1}^{3} D_{lm}^{\alpha\beta}k_\alpha k_\beta, \qquad (4.5)$$

$$p_{lm}^\alpha = (\hbar/i) < U_{l0} \mid \nabla_\alpha \mid U_{m0} >, \qquad (4.6)$$

$$D_{lm}^{\alpha\beta} = [(\hbar^2/2m_0)\delta_{lm} + (\hbar/m_0)^2 \sum_{n'\neq m} p_{ln'}^\alpha p_{n'm}^\beta (E_0 - E_{n'})^{-1}, \qquad (4.7)$$

$$\tilde{V}_p(\mathbf{k} - \mathbf{k}') = \int V_p(\mathbf{r}) \exp{[i(\mathbf{k} - \mathbf{k}')]}d^3r, \qquad (4.8)$$

α, β indicate the coordinates; 1,2,3 indicate respectively x, y, z.

Taking the Fourier transform of (4.4) and putting,

$$F_l(\mathbf{r}) = \sum_{\mathbf{k}} A_{l\mathbf{k}} \exp(i\mathbf{k}.\mathbf{r}),\qquad(4.9)$$

we obtain,

$$\sum_{m=1}^{r} [H_{lm}(-i\nabla) + V_p(\mathbf{r})\delta_{lm}]F_m(\mathbf{r}) = EF_l(\mathbf{r}).\qquad(4.10)$$

The wave function $\psi_p(\mathbf{r})$ may be written in terms of $F_m(\mathbf{r})$ as follows.

$$\psi_p(\mathbf{r}) = \sum_{m=1}^{4} [U_{m0}F_m(\mathbf{r}) + \sum_{n'\neq m} -(i\hbar/m_0)\nabla F_m(\mathbf{r}).\mathbf{p}_{n'm}(E_0 - E_{n'0})^{-1}U_{n'0}].\qquad(4.11)$$

The solution $F_m(\mathbf{r})$ is a slowly varying function like $V_p(\mathbf{r})$ and gives the envelope of the cell-periodic functions. The function is referred as the envelope function and Eq. (4.10) as the envelope-function equation.

In quantum well devices, we are mostly concerned with the conduction-band minimum or the valence-band maximum. The envelope-function equation for these extrema are obtained from Eq.(4.10) by using the cell-periodic functions for these extrema as explained in the following sections.

4.1.2. ENVELOPE-FUNCTION EQUATION FOR ELECTRONS

Electrons occupy the minima of the conduction band. These are nondegenerate (excluding the spin-degeneracy) and separated by the band gap energy from the valence band maxima and by larger amounts of energy from other extrema. For the purpose of the calculation of electron states near a particular conduction-band minimum, the contribution of only that minimum is considered significant in the first approximation. It is assumed that $A_{c\mathbf{k}} \gg A_{n'\mathbf{k}}$ where the subscript c indicates the conduction band minimum.

The conduction-band minimum is located either at the Γ-point i.e., at $\mathbf{k} = 0$, or at a point in the Δ or Λ i.e., in the <100> or <111> direction. We first consider the Γ-point minimum.

Using the expression for the Hamiltonian, given in Eq. (4.4) we get for the energy in the absence of the perturbing potential,

$$E_c(\mathbf{k}) = E_{c0} + \sum_{\alpha,\beta=1}^{3} k_\alpha k_\beta[(\hbar^2/2m_0)\delta_{\alpha\beta} + (\hbar/m_0)^2 \sum_{n\neq c} p_{cn}^\alpha p_{nc}^\beta (E_{c0} - E_n)^{-1}].\qquad(4.12)$$

The $E - \mathbf{k}$ relation for values of $\mathbf{k}$ close to the minimum is expressed as,

$$E_c(\mathbf{k}) = E_{c0} + (\hbar^2/2)\mathbf{k}.\mathbf{M}^{-1}\cdot\mathbf{k},\qquad(4.13)$$

where $\mathbf{M}^{-1}$ is defined as the inverse effective mass tensor, elements of which are given by

$$1/m_{\alpha\beta} = (1/\hbar^2)(\partial^2 E/\partial k_\alpha \partial k_\beta), \tag{4.14}$$

Comparing (4.14) with (4.12) we find that the sum of the matrix elements may be identified with the components of the effective mass tensor as follows,

$$1/m_{\alpha\beta} = (1/m_0)\delta_{\alpha\beta} + (2/m_0^2)\sum_{n\neq c} p_{cn}^\alpha p_{nc}^\beta (E_{c0} - E_n)^{-1}. \tag{4.15}$$

The envelope-function equation for the electrons is written by using this equivalence as,

$$[(\hbar^2/2)(i\nabla).\mathbf{M}^{-1}\cdot(i\nabla) + V_p(r)]\,F_c(\mathbf{r}) = (E - E_{c0})F_c(\mathbf{r}). \tag{4.16}$$

For cubic crystals, the effective mass m_c^* for electrons at the Γ point is a scalar and Eq. (4.16) simplifies to,

$$[(-\hbar^2/2m_c^*)\nabla^2 + V_p(\mathbf{r})]F_c(r) = (E - E_{c0})F_c(\mathbf{r}). \tag{4.17}$$

In quantum wells, $V_p(\mathbf{r})$ varies only in one direction. Choosing this direction as the z direction $F_c(\mathbf{r})$ may be expressed as,

$$F_c(\mathbf{r}) = F(z)\exp{(i\mathbf{k}_t.\rho)}, \tag{4.18}$$

where $\mathbf{k}_t$ and ρ represent respectively the in-plane wave vector and the position vector.

Substituting (4.18) in (4.17) we obtain

$$[(-\hbar^2/2m_c^*)\nabla_z{}^2 + V_p(z)]F(z) = (E - \hbar^2 k_t{}^2/2m_c^* - E_{c0})F(z). \tag{4.19}$$

Electron states for most of the quantum well devices, using n- type III-IV compounds, are evaluated by using Eq. (4.19). The value of the effective mass, m_c^* is taken from experiments such as cyclotron resonance, magneto-phonon oscillation, Faraday rotation. etc.

The wave function $\psi_p(\mathbf{r})$ is given in this case by,

$$\psi_p(\mathbf{r}) = U_{c0}F_c(\mathbf{r}) + \sum_{n\neq c} -i(\hbar/m_0)\nabla F_c(\mathbf{r}).\mathbf{p}_{cn}(E_0 - E_n)^{-1}U_{n0}. \tag{4.20}$$

The second term in Eq. (4.20) may be neglected in the analysis of most of the phenomena. Only for the case of intersubband absorption, the contribution from the second term becomes significant (see Section 5.2.2).

The wave function is simplified in the normalized form as

$$\psi_p(\mathbf{r}) = (\Omega/ < F \mid F >)^{1/2}U_{c0}F_c(\mathbf{r}), \tag{4.21}$$

where $F_c(\mathbf{r})$ is the solution of Eq. (4.17) and Ω is the volume of a unit cell. It should be noted that the cell-periodic function is normalized over the volume of a unit cell, Ω, while the wave function is normalized over the volume of the structure.

Equation (4.21) may be generalized for non-degenerate minima with energy $E_{c0}(k_0)$, located at $\mathbf{k} = \mathbf{k}_0$. The cell-periodic functions $U_{n\mathbf{k}_0}$ form a complete set like U_{n0} and $\psi_p(\mathbf{r})$ may be expressed as,

$$\psi_p(\mathbf{r}) = \sum_{n',\mathbf{k}'} A_{n',\mathbf{k}'}[U_{n'\mathbf{k}_0} \exp{(i\mathbf{k}'.\mathbf{r})}] \qquad (4.22)$$

Substituting (4.22) in the Schrödinger equation and proceeding as for $\mathbf{k}_0 = 0$, we get

$$[(\hbar^2/2)(i\nabla + k_0).\mathbf{M}^{-1} \cdot (i\nabla + k_0) + V_p(\mathbf{r})]F_c(\mathbf{r}) = [E - E_c(\mathbf{k}_0)]F_c(\mathbf{r}), \qquad (4.23)$$

The elements of the tensor $\mathbf{M}^{-1}$ are defined so as to replace the momentum matrix elements, according to the following relation,

$$1/m_{\alpha\beta} = (1/m_0)\delta_{\alpha\beta} + (2/m_0^2) \sum_{n \neq c} \mathbf{p}_{cn}^{\alpha}(\mathbf{k}_0).\mathbf{p}_{nc}^{\beta}(\mathbf{k}_0)[E_c(\mathbf{k}_0) - E_n]^{-1}, \qquad (4.24)$$

where $m_{\alpha\beta}$ may be identified as the elements of the effective mass tensor, defined as

$$1/m_{\alpha\beta} = (1/\hbar^2 k_\beta)(\partial E/\partial k_\alpha). \qquad (4.25)$$

The values of $m_{\alpha\beta}$ may be obtained from cyclotron resonance and other experiments. These are used in Eq. (4.23) to analyze quantum well devices using the indirect-gap materials, e.g., Si or $Si_x Ge_{1-x}$.

We note that in the theory, as discussed above, the minimum is taken as isolated. However, indirect-gap materials have multiple equivalent minima located either in the <100> directions, viz., in Si or in the <111> directions, viz., in Ge. The perturbing potential may affect the different minima differently, depending on the direction of variation of $V_p(\mathbf{r})$ in relation to the crystallographic directions. In such cases, each minimum may be treated as being independent. On the other hand, for some directions of variation of $V_p(\mathbf{r})$, more than one minimum may be affected identically. The minima cannot then be considered as independent. The effect of the possible coupling between such minima have not been evaluated. It is presumed that the coupling may cause only finite splitting of the minima, without materially affecting other results and the minima may be treated independently even when the energies are the same.

4.1.3. ENVELOPE-FUNCTION EQUATION FOR HOLES

The valence band of all the cubic semiconductors, except the chalcogenides, have three valence band maxima located at the Γ point. Two maxima are four-fold degenerate (including the spin degeneracy). The third maximum is lower in energy from these maxima by the spin-orbit splitting, Δ. The Hamiltonian for the holes is evaluated by assuming that only the degenerate maxima contribute significantly, while the contribution of the third maximum and all other extrema may be treated as perturbation.

The cell-periodic parts of the Bloch functions for these maxima are like the p-type atomic orbitals corresponding to the angular momentum number $J = 3/2$ and the magnetic quantum numbers $J_z = 3/2, 1/2$. This identification is done by considering that the effective mass obtained from the cyclotron resonance experiments agree with the $E - k$ relation, calculated by using this assumption.

The cell-periodic functions for these bands are therefore chosen to be

$$U_{v0}(3/2, 3/2) = U_1 = -(1/\sqrt{2})(X + iY) \uparrow, \qquad (4.26)$$

$$U_{v0}(3/2, 1/2) = U_2 = (1/\sqrt{6})[(X - iY) \downarrow +2Z \uparrow], \qquad (4.27)$$

$$U_{v0}(3/2, -1/2) = U_3 = -(1/\sqrt{6})[(X + iY) \uparrow -2Z \downarrow], \qquad (4.28)$$

$$U_{v0}(3/2, -3/2) = U_4 = (1/\sqrt{2})(X - iY) \downarrow. \qquad (4.29)$$

The numbers in the parenthesis indicate the values of J and J_z. The prefactors and phases are chosen so as to normalize the functions and to ensure time reversal. Arrows indicate the two opposite spins. X, Y and Z indicate the x, y and z type p-orbitals, x, y, z being chosen along the crystallographic directions. Using the above expressions for U_1, U_2, U_3 and U_4 and taking into account the symmetry of the functions, we get for $H(k)$ the following matrix[4.2].

$$H(k) = \begin{bmatrix} P+Q & R & -S & 0 \\ R^* & P-Q & 0 & S \\ -S^* & 0 & P-Q & R \\ 0 & S^* & R^* & P+Q \end{bmatrix} \qquad (4.30)$$

where,

$$P \pm Q = E_{v0} - (\hbar^2/2m_0)[(\gamma_1 \pm \gamma_2)(k_x^2 + k_y^2) + (\gamma_1 \mp 2\gamma_2)k_z^2], \qquad (4.31)$$

$$S = (\hbar^2/m_0)\sqrt{3}\gamma_3(k_x - ik_y)k_z, \qquad (4.32)$$

$$R = -(\hbar^2/2m_0)(\sqrt{3}/2)[\gamma_2 + \gamma_3)(k_x - ik_y)^2 - \gamma_3 - \gamma_2)(k_x + ik_y)^2]. \qquad (4.33)$$

The constants $\gamma_1, \gamma_2, \gamma_3$, known as Luttinger parameters[4.5], have been substituted for the momentum matrix elements (see Section 4.1.2) as detailed below.

$$\gamma_1 = -(2/3)(A + 2B)m_0, \gamma_2 = -(1/3)(A - B)m_0, \gamma_3 = (1/3)Cm_0,, \qquad (4.34)$$

where

$$A = (1/2m_0) + (1/m_0^2) \sum_{n'} p_{xn'}^x p_{n'x}^x (E_{v0} - E_{n'})^{-1}, \qquad (4.35)$$

$$B = (1/2m_0) + (1/m_0^2) \sum_{n'} p_{xn'}^y p_{n'x}^y (E_{v0} - E_{n'})^{-1}, \tag{4.36}$$

$$C = (1/m_0^2) \sum (p_{xn'}^z p_{n'y}^y + p_{xn'}^y p_{n'y}^z)(E_{v0} - E_{n'})^{-1}. \tag{4.37}$$

$$p_{xn'}^\alpha = (\hbar/i) < X \mid \nabla_\alpha \mid U_{n'0} >, \tag{4.38}$$

$$p_{yn'}^\alpha = (\hbar/i) < Y \mid \nabla_\alpha \mid U_{n'0} >, \tag{4.39}$$

$$p_{zn'}^\alpha = (\hbar/i) < Z \mid \nabla_\alpha \mid U_{n'0} > . \tag{4.40}$$

α represents $x, y,$ or z. The summation is carried out over all the extrema except the degenerate valence band maxima, under consideration. The equivalence between the matrix elements (e.g., $p_{xn'}^z = p_{yn'}^y = p_{zn'}^z$), which are evident from the the symmetry of the functions are not listed, but the constants A, B, C are to be interpreted accordingly. We note that $\gamma_1, \gamma_2, \gamma_3$ give the inverse of the effective mass ratio of holes and the signs in the above expressions have been chosen so that these parameters are positive.

We find that all the envelope functions F_1, F_2, F_3 and F_4 are coupled in the equations given by the Hamiltonian of Eq. (4.30). The equations may be decoupled into two sets of two coupled equations using the unitary transformation given below[4.6,7].

$$\bar{U} = \begin{bmatrix} \alpha^* & 0 & 0 & -\alpha \\ 0 & \beta^* & -\beta & 0 \\ 0 & \beta^* & \beta & 0 \\ \alpha^* & 0 & 0 & \alpha \end{bmatrix} \tag{4.41}$$

where

$$\alpha = (1/\sqrt{2}) \exp\left[i(3\pi/4 - 3\phi/2)\right], \tag{4.42}$$

$$\beta = (1/\sqrt{2}) \exp\left[i(-\pi/4 + \phi/2)\right], \tag{4.43}$$

$$\phi = \arctan\left(k_y/k_x\right) \tag{4.44}$$

The transformed envelope functions, F_1', F_2', F_3' and F_4' are given by,

$$\mathbf{F} = \bar{U} \cdot \mathbf{F}'. \tag{4.45}$$

The equation for the transformed functions is therefore

$$(\bar{U} \cdot H \cdot \bar{U}^\dagger) \cdot \mathbf{F}' = E\,\mathbf{F}', \tag{4.46}$$

where $\bar{U}^\dagger$ is the hermitian conjugate of $\bar{U}$ and H is the matrix operator. The transformed operator is

$$\bar{U} \cdot H \cdot \bar{U}^\dagger = \begin{bmatrix} P+Q & \tilde{R} & 0 & 0 \\ \tilde{R}^* & P-Q & 0 & 0 \\ 0 & 0 & R-Q & \tilde{R}^* \\ 0 & 0 & \tilde{R} & P+Q \end{bmatrix}, \tag{4.47}$$

$$\tilde{R} = |R| - i|S|.\tag{4.48}$$

Evidently, equations for F_1', F_2' are decoupled from those for F_3', F_4'.

The two sets of coupled equations are identical and may be written in the expanded form as,

$$\begin{bmatrix} H_+(-i\nabla) - E & W \\ W^\dagger & H_-(i\nabla) - E \end{bmatrix} \cdot \begin{bmatrix} F_1' \\ F_2' \end{bmatrix} = 0,\tag{4.49}$$

where

$$H_\pm(-i\nabla) = E_{v0} - (\hbar^2/2m_0)[(\gamma_1 \pm \gamma_2)(k_x^2 + k_y^2) - (\gamma_1 \mp 2\gamma_2)\nabla_z^2],\tag{4.50}$$

$$W = \sqrt{3}(\hbar^2/2m_0)\{[\gamma_2^2(k_x^2 - k_y^2)^2 + 4\gamma_3^2 k_x^2 k_y^2]^{1/2} - 2\gamma_3(k_x^2 + k_y^2)^{1/2}\nabla_z\}.\tag{4.51}$$

$W^\dagger$ is the hermitian conjugate of W. The perturbed wave function $\psi_p(\mathbf{r})$ is given by

$$\psi_p(\mathbf{r}) = \{[\alpha u_1 - \alpha^* u_4)]F_1'(z) + [(\beta u_2 - \beta^* u_3)]F_1'(z)\}\exp[i\mathbf{k}_t.\rho]\tag{4.52}$$

It should be noted that the unitary transformation matrix, used above, depends on both the magnitude and the direction of $\mathbf{k}$. The quantum wells being grown along the crystallographic directions, k_z is fixed for each quantized level. The transformation relation, therefore, depends on the magnitude and direction of the transverse wave vector $\mathbf{k}_t$. Detailed calculations may be done by using different transformation matrices for different $\mathbf{k}_t$. For example, for the <100> and <110> directions of $\mathbf{k}_t$, the operator W is given respectively by,

$$W = \sqrt{3}(\hbar^2/2m_0)k_t(\gamma_2 k_t - 2\gamma_3\nabla_z),\text{(for the} < 100 > \text{direction)}\tag{4.53}$$

$$W = \sqrt{3}(\hbar^2/2m_0)k_t(\gamma_3 k_t - 2\gamma_3\nabla_z),\text{(for the} < 110 > \text{direction)}\tag{4.54}$$

However, the values of γ_2 and γ_3 being nearly the same, the problem is often simplified by assuming that,

$$\gamma_2 = \gamma_3 = (\gamma_2 + \gamma_3)/2 = \bar{\gamma}\tag{4.55}$$

and writing W as

$$W = \sqrt{3}\bar{\gamma}(\hbar^2/2m_0)k_t(k_t - 2\nabla_z)\tag{4.56}$$

4.1.4. ENVELOPE-FUNCTION EQUATION FOR HIGH-ENERGY ELECTRONS

Envelope-function equation for the electrons has been derived in Section 4.1.2 by assuming that the energy of the electron is close to a minimum and that the difference in the energy of this minimum and of other extrema (including the valence band maxima) is large. This assumption may be justified generally only for electron kinetic energies of a few meV, as the band gap in some compounds (i.e., InSb) may be as low as 230 meV. The quantized energies in quantum wells may, however, be a few hundred meV, particularly in narrow wells. The effect of the coupling of the nearby valence bands with the conduction band cannot be neglected for such energies.

Envelope-function equation for the high-energy electrons are obtained by assuming that the contribution of the valence-band extrema are also significant. We illustrate the equation by combining the four-fold degenerate valence-band maxima with the conduction-band minimum. The coupling between the nearby bands is caused by the term $(\hbar/m_0)\mathbf{k}.\mathbf{p}_{lm}$. This term was not included in Section 4.1.2 as the contribution of all extrema, except the conduction-band minimum was considered to be of the second order and included in the analysis by replacing the free electron mass with the effective mass. The $\mathbf{k}.\mathbf{p}_{lm}$ term is, however, required to be included when the contribution of the valence band is also significant.

The cell-periodic part of the Bloch function for the Γ - point conduction-band minimum is of the s-type atomic orbital. Taking the functions for the two spins as $S \uparrow$ and $S \downarrow$ and using the functions for the cell-periodic parts of the four degenerate valence bands we get the following matrix for the Hamiltonian.

$$
\mathbf{H}(\mathbf{k}) = \begin{bmatrix}
E_{c0} - \frac{\hbar^2 k^2}{2m_c^*} & P_+ & \sqrt{\frac{2}{3}} P_{sp} k_z & -P_- & 0 & 0 \\
 & (P' + Q') & R' & -S' & 0 & 0 \\
 & & (P' - Q') & 0 & S' & P_+ \\
 & & & (P' + Q') & R' & \sqrt{\frac{2}{3}} P_{sp} k_z \\
 & & & & (P' + Q') & -P_- \\
 & & & & & E_{c0} - \frac{\hbar^2 k^2}{2m_c^*}
\end{bmatrix}
$$

$$(4.57)$$

where

$$P_{sp} = (\hbar/m_0)(\hbar/i) < S \mid \nabla_z \mid Z >, \qquad (4.58)$$

$$P_\pm = iP_{sp}(k_x \pm ik_y)/\sqrt{2} \qquad (4.59)$$

The diagonal elements and the elements of the upper half are given. Elements of the lower half are the hermitian conjugates of the upper-half elements. The effective mass m_c^* differs from m_c^*, introduced in Section 4.1.2, as in the present formulation contribution of the valence bands are introduced through the coupling terms and are therefore excluded in the definition of m_c^*. The operators P', Q', R' and S' are given by the same expression as P, Q, R and S (See Sect.4.1.3) but $\gamma_1', \gamma_2', \gamma_3'$ differ from γ_1, γ_2 and γ_3 as in their definition also the contribution of the conduction band is excluded.

Energy eigenvalues for the high energy electrons and the dispersion relations are required to be calculated by solving the equation obtained by using the above Hamiltonian. A more complete description requires also the inclusion of the third valence band as in some materials the spin-orbit splitting may be only 40 meV[4.8]. The operator is given by a 8×8 matrix when the coupling to the split-off band is included[4.9]. The analysis of quantum well devices, is rarely based on the solution of these equations, all taken together. Further, the problem is complicated by the presence of the k^2-dependent terms in the diagonal elements which cause the appearance of spurious solutions.

The analysis is simplified by substituting for m_c^* in the envelope function equations, the energy dependent mass $m_c^*(E)$, which is applicable to bulk materials for some simplifying assumptions. It is assumed that the k^2-dependent terms may be neglected in the diagonal elements. The z-direction is also selected in the direction of k and a new set of orthogonal X', Y', Z' functions are chosen for these new coordinates. For this assumption $k_z = k$, and $k_x = k_y = 0$. The matrix then simplifies to [4.4]

$$H(k) = \begin{bmatrix} E_{c0} & \sqrt{2/3}P_{sp}k \\ \sqrt{2/3}P_{sp}k & E_{v0} \end{bmatrix} \quad (4.60)$$

All the elements involving the transverse component of k are zero. In effect, only the conduction band minimum and the valence band corresponding to $J = 3/2, J_z = 1/2$ are coupled.

The above Hamiltonian gives the following dispersion relation.

$$(E - E_{c0})(E - E_{v0}) = (2/3)P_{sp}^2 k^2, \quad (4.61)$$

The matrix element, P_{sp} may be eliminated by noting that the above relation should reduce for $k \to 0$ to

$$(E - E_{c0}) = \hbar^2 k^2 / 2m_c^*. \quad (4.62)$$

We get accordingly,

$$P_{sp}^2 = (3\hbar^2 E_g / 4m_c^*), \quad (4.63)$$

$E_g = E_{c0} - E_{v0}$ or the band gap. Equation (61) may now be rewritten as,

$$(\hbar^2 k^2 / 2m_c^*) = (E - E_{c0})[1 + \alpha(E - E_{c0})], \quad (4.64)$$

where $\alpha = 1/E_g$. The parameter α is referred as the nonparabolicity parameter as for $\alpha \neq 0$, the $E - k$ relation is nonparabolic.

A third order dispersion relation is obtained by including the split-off valence band which is coupled to both the conduction band and the $J = 3/2, J_z = 1/2$ band even for the above simplifying assumptions. The expressions is

$$(E - E_{c0})(E - E_{v0})(E - E_{v0} - \Delta) = P_{sp}^2 k^2 (E - E_0 - 2\Delta/3) \quad (4.65)$$

This expression may, however, be approximated to the form of (64), but α is now given by[4.10]

$$\alpha = (1/E_g)(1 - m_c^*/m_0)^2 (1 + 4\Delta/E_g + 2\Delta^2/E_g^2)(1 + 5\Delta/E_g + 2\Delta^2/E_g^2)^{-1}. \quad (4.66)$$

The free electron contribution, $\hbar^2 k^2/2m_0$, in the diagonal terms is taken into account while deriving the above expression. It should, however, be noted that the diagonal terms include in addition to the free electron term, other terms involving the momentum matrix elements, $p_{cn'}$ etc., with bands other than the lowest conduction band and the three upper most valence bands, effect of which are included in $m_{c'}^*, \gamma_1$ etc. The net effect of these terms is to alter the value of α. To account for these effects, α is treated as an empirical parameter which may be determined from experiments, viz., Burstein shift.

The effect of nonparabolicity has been mostly included in the analysis of the electron states in quantum well devices by using the simplified expression (64), discussed above.

4.1.5. BOUNDARY CONDITIONS

Electron states in heterostructures are obtained by solving the envelope-function equation, subject to the proper boundary conditions. We assume that the perturbing potential varies only in the z direction. The envelope function may then be expressed as,

$$\psi_p(r) = F(z) \exp(i\mathbf{k}_t.\rho), \tag{4.67}$$

where $\mathbf{k}_t$ and ρ represent respectively the wave vector and the position vector. The z-component of the wave function now satisfies the equation,

$$[(-\hbar^2/2m_i^*)\nabla_z^2 + V_p(z)]F_i(z) = E'F_i(z), \tag{4.68}$$

where $E' = E_0 - \hbar^2 k_t^2/2m_i^* - E_{ci}$, E_0 being the energy eigenvalue and E_{ci} the band-edge energy. In heterostructures, two such equations give the electron wave function in the two regions. Boundary conditions are introduced to match the functions at the interfaces and at infinity.

The first boundary condition is obtained by considering that for a quantum well and a heterostructure the wave function should be normalizable and therefore should become zero as $|z|$ goes to infinity.

The second boundary condition is obtained from the requirement that the probability density of the electron is single-valued. To ensure single-valuedness the wave function should be continuous across an interface.

The two conditions are obeyed when it is ensured that,

$$F(z) \to 0 \quad \text{as} \quad |z| \to \infty, \tag{4.69}$$

$$U_A F_A(z_i) = U_B F_B(z_i), \tag{4.70}$$

$$\mathbf{k}_{tA} = \mathbf{k}_{tB}, \tag{4.71}$$

where $U_A F_A(z_i)$ and $U_B F_B(z_i)$ represent the envelope functions respectively for the A and the B material at the interface located at $z = z_i$. It is usually assumed that the cell-periodic parts of the wave functions are identical in the constituent materials, i.e., $U_A = U_B$. This assumption is a limitation of the analysis, using the effective mass approximation. Apparently, deviation from this assumption does not affect the final results in any significant manner, as is indicated by the agreement between the theory and the experiment. This assumption simplifies Eq. (4.70) to

$$F_A(z_i) = F_B(z_i). \qquad (4.72)$$

The third boundary condition is related to the derivatives of the wave function. This boundary condition is obtained by integrating Eq. (4.68) from an infinitesimal distance $-\epsilon$ on one side of the interface to a distance $+\epsilon$ on the other side. However, before carrying out this integration we note that Eq. (4.68) should remain hermitian even when m^* varies with z as in quantum wells. This condition is ensured by formulating the equation as[4.11,12],

$$\{(-\hbar^2/2)\nabla_z[1/m_i^*(z)]\nabla_z + V_p(z)\}F_i(z) = EF_i(z). \qquad (4.73)$$

Integrating the function from $-\epsilon$ to ϵ we get

$$(-\hbar^2/2)\{[1/m_i^*(z)]\nabla_z F_i(z)\}_{-\epsilon}^{\epsilon} + \int_{-\epsilon}^{\epsilon} V_p(z)F_i(z)\,dz = \int_{-\epsilon}^{\epsilon} EF_i(z)\,dz. \qquad (4.74)$$

The second term and the right-hand side term are zero since $F_i(z)$ is continuous. It then follows that $[1/m^*(z)]F_i'(z)$ is continuous across the interface, or

$$(1/m_A^*)F_A'(z_i) = (1/m_B^*)F_B'(z) \qquad (4.75)$$

The prime indicates the z-derivative.

This condition ensures also the continuity of probability current density at the interface and is introduced in some studies from such consideration[4.13]. There has been discussion in the literature[4.9] about the necessity of introducing the condition of the continuity of probability current density since the current density is zero for real $F(z)$ in quantum wells. It is argued, on the other hand, that for states in quantum wells, which are in the continuum, i.e., above the band-edge energy of the barrier layer or when an electron tunnels through a barrier, the probability current density is not zero and the continuity of current density is required to be ensured. As the boundary condition cannot depend on the nature of the problem, the same condition should also apply even when the probability current density is zero.

For high-energy electrons it has been discussed in Section 4.1.4 that the mixing between the different bands is required to be considered. However, in most of the analyses, simplification is done by using the energy dependent effective mass $m^*(E)$ in place of $m^*(0)$. Use of this mass requires that in Eq. (4.75) m_i^* should be replaced by the so-called velocity effective mass, given by[4.14,15],

$$1/m_v = (1/\hbar^2 k)\nabla_k E. \qquad (4.76)$$

The eigenfunction and the energy eigenvalues are, however, required to be obtained by solving the envelope-function equation with the assumed nonparabolic $E(k)$ to give the operator $E(-i\nabla)$. Controversy, however, exists in the literature about the use of m_v in the boundary condition and the use of the so-called energy effective mass, defined as[4.16-18],

$$m_E = \hbar^2 k^2 / 2E, \tag{4.77}$$

has been suggested.

Boundary conditions for the holes are also obtained from the same considerations as for electrons. The condition of the continuity and normalizability of the wave functions give

$$F_B(z) \to 0 \text{ as } |z| \to \infty \tag{4.78}$$

$$\mathbf{k}_{tA} = \mathbf{k}_{tB} \tag{4.79}$$

$$\sum_m F_{mA}(z_i) = \sum_m F_{mB}(z_i) \tag{4.80}$$

$$\text{or } F_{mA}(z_i) = F_{mB}(z_i), \tag{4.81}$$

for each value of m, since the functions are orthogonal.

The boundary condition for the derivative is obtained by integrating the envelope function as for the electrons. We get,

$$\sum_{m=1}^{4} [\sum_{\alpha=x,y} (D_{lm}^{z\alpha} + D_{lm}^{\alpha z}) k_\alpha - 2i D_{lm}^{zz} \nabla_z] F_m \text{ continuous.} \tag{4.82}$$

The cell-periodic parts of the Bloch functions being assumed to be equal in the two materials, the momentum matrix elements, $\mathbf{p}_{lm}$'s, are also equal. Terms involving $\mathbf{p}_{lm}$ are, therefore, omitted in the above condition. On using the expression in the transformed four-fold degenerate model of the valence band [see Eq. (41)] we get for the above condition,

$$\begin{bmatrix} (\gamma_1 - 2\gamma_2)\nabla_z & \sqrt{3}\gamma_3 k_t \\ -\sqrt{3}\gamma_3 k_t & (\gamma_1 + 2\gamma_2)\nabla_z \end{bmatrix} \cdot \begin{bmatrix} F_1' \\ F_2' \end{bmatrix} \text{ continuous} \tag{4.83}$$

4.2. Energy Levels of Electrons

Heterostructures for quantum well devices are so constructed that the potentials produced in them confine the electrons. The confining potentials are due to the band offsets (discussed in Chapter 3) at the interfaces in double-junction quantum wells and superlattices. In single-junction heterostructures, the confining potential is due to the band offset on the one side and due to the accumulated charges on the other side. The determination of energy levels is comparatively simpler for double-junction quantum wells and superlattices since the charge accumulation may be considered negligible. Energy levels in these structures are first worked out.

4.2.1. QUANTUM WELL

The potential distribution in the structure is as shown in Fig. 4.1(b). The bending of the band edges is due to charge redistribution at the interfaces. However, the width and doping of the well is such that the bending may be neglected in the first approximation and the potential may be approximated as being of rectangular shape as shown by the dashed line.

Equation (4.19) is required to be solved subject to the boundary conditions discussed in Section 4.1.5. Choosing $z = 0$ at the left interface between the well and the barrier the solution may be written as,

$$F_w(z) = A \sin(k_w z) + B \cos(k_w z), \tag{4.84}$$

$$F_b(z) = \Theta(-z)C \exp(k_b z) + \Theta(z - L)D \exp[-k_b(z - L)], \tag{4.85}$$

where L is the width of the well and

$$k_i^2 = (2m_i^*/\hbar^2) \mid E_i \mid, \tag{4.86}$$

$$E_i = E_0 - \hbar^2 k_t^2/2m_i^* - E_{ci}, \tag{4.87}$$

$\Theta(x)$ is the Heaviside unit function. The subscript i indicates the well layer for w and the barrier layer for b. E_{ci} is the band-edge energy, k_t is the transverse wave vector which is identical in the well and the barrier because of the requirement of the continuity of the envelope function. In the solution for the barrier layer only one solution has been taken from the two possible solutions on each side to ensure that $F_b(z) \to 0$ as $\mid z \mid \to \infty$. It may also be noted that for $(E_{cw} + \hbar^2 k_t^2/2m_w < E < E_{cb} + \hbar^2 k_t^2/2m_w)$, k_w and k_b are both real.

On applying the boundary conditions at $z = 0$ and at $z = L$ we get the following equation,

$$\begin{bmatrix} 1 & 0 & -r & 0 \\ 0 & 1 & -1 & 0 \\ \sin(k_w L) & \cos(k_w L) & 0 & -1 \\ \cos(k_w L) & \sin(k_w L) & 0 & r \end{bmatrix} = 0, \tag{4.88}$$

where $r = (k_b/k_w)(m_w^*/m_b^*)$. The equation for the eigenvalues of energy, obtained from Eq. (88) may be written as,

$$\tan[k_w L/2 + (1 \pm 1)\pi/4] = r. \tag{4.89}$$

When $E_{cb} \gg E_{cw}$, k_b is much larger than k_w and $k_w L/2 = n\pi/2$. The energy eigenvalues are then given by

$$E_n = E_{cw} + (\hbar^2/2m_w^*)(n\pi/L)^2 + \hbar^2 k_t^2/2m_w. \tag{4.90}$$

This approximation is referred as the infinite-barrier approximation, and is often used in the analysis of quantum well devices.

We also note that there are two series of solutions corresponding to the plus and the minus sign and the wave functions corresponding to them have even and odd symmetry about the centre of the well. This symmetry plays a significant role in the opto-electronic interactions.

Effect of finite barrier potential

The properties of the barrier layer, however, become important, when the energy eigenvalue, E_n , is comparable in magnitude to the barrier potential $V_0[=(E_{cb} - E_{cw})]$. It is then found that E_n's have lower values than those given by the infinite-barrier-potential model. The lowering depends on the magnitude of V_0 and L. These features are illustrated in Fig. 4.3 by computing the energy levels in wells of the $GaAs/Ga_{0.7}Al_{0.3}As$ system with the following values of the physical constants.

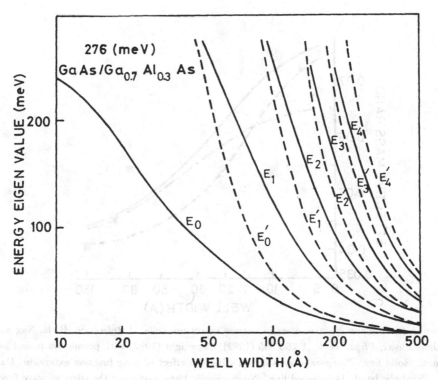

Figure 4.3. Energy eigenvalues for quantum wells. Solid line- Finite Barrier well. Dashed line - Infinite-barrier well.

$m_w^* = 0.067m_0, m_b^* = 0.09m - 0, V_0 = 276\text{meV}, m_0 = 9.1 \times 10^{-31}\text{kg}.$

The calculations show that the values for the finite-barrier model are not much different from those for the infinite-barrier model for well widths larger than about 50 nm, but for smaller widths, the eigenvalues are much lower. The difference becomes more distinct for very narrow wells when the eigenvalue approaches the barrier potential. Further, the electron spends part of the time in the barrier layer when the barrier potential is finite. The in-plane effective mass is therefore a weighted average of the masses in the well and in the barrier layer. As $m_b^* > m_w^*$ the average mass is larger than m_w^*. It may be evaluated for a particular value of n by calculating E for different values of k_t and using the formula $1/m_i^* = (1/\hbar^2 k_t)(\partial E/\partial k_t)$. Such calculations have been done[4.19] for the first subband level in $\text{Ga}_{0.47}\text{In}_{0.53}\text{As/InP}$ wells. The following values of the physical constants were used.

$m_w^* = 0.042m_0, m_b^* = 0.075, V_0 = 240$ meV.

The results are presented in Fig. 4.4. The value of m_i^* is found to increase from the mass for $\text{Ga}_{0.47}\text{In}_{0.53}\text{As}$ to that for InP as the well width is reduced and consequently, the energy eigenvalue approaches the barrier potential and causes

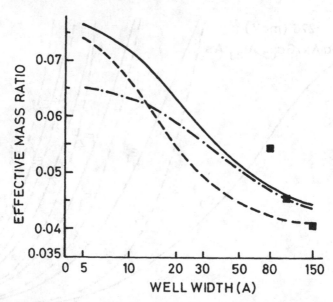

Figure 4.4. In-plane effective mass in narrow quantum wells. [Reference: B. R. Nag and S. Mukhopadhyay, *Phys. Lett. A* **166** 395 (1992), Copyright (1992) with permission from Elsevier Science]. Solid line - Nonparabolic band including the effect of wave function extension. Dashed line - Parabolic band. Dot-dashed line - Nonparabolic band excluding the effect of wave function extension.

Table 4.1 Energy levels in quantum wells(meV).
L= well width(nm).(a) - Parabolic. (b)- Nonparabolic.
I - $GaAs/Ga_{0.3}Al_{0.7}As$.II - $InP/Ga_{0.47}In_{0.53}As$.
III - $Ga_{0.47}In_{0.53}As/Al_{0.48}In_{0.52}As$.

L	I(a)	I(b)	II(a)	II(b)	III(a)	III(b)
20	10	11	14	16	16	18
10	32	34	39	44	49	59
5	80	89	87	99	123	156
2	180	211	172	186	291	345
1	239	266	216	223	-	-
0.5	265	275	233	235	-	-

large penetration of the wave function into the barrier layer. Recent cyclotron resonance experiment has given data confirming this theoretical prediction[4.20].

The extension of the wave function into the barrier layer for finite barrier potential affects also the probabilities of scattering of the electrons. This aspect is discussed in Section 7.3.

Effect of energy band nonparabolicity
The effects of nonparabolicity are incorporated as explained in Section 4.1.4 by using the energy effective mass $[(\hbar^2 k^2/2(E - E_{c0})]$ in Eq. (4.86) and the velocity effective mass, $(\hbar^2 k/\nabla_k E)$, in r of Eq. (4.88). Such analyses have been carried out for $GaAs/Ga_{0.7}Al_{0.3}As$, $Ga_{0.47}In_{0.53}As/InP$ and $Ga_{0.47}In_{0.53}As/Al_{0.48}In_{0.52}As$ systems. The nonparabolic relation used for the analysis is the simplified[4.21-24] Kane relation, which is assumed to apply to both the conduction band and the forbidden band[4.25,26].

Results of the analysis are given in Table. 4.1 along with those for the parabolic band. The nonparabolicity causes significant changes in the values of the energy levels. The change is maximum near about the well width of 2 nm and may be as large as 17 %. Although such changes may not be important in the design of electron devices, these have to be taken into account in the design of optoelectronic devices in which the signal frequency resonates with differences in energy levels.

4.2.2. SUPERLATTICE

Energy levels in superlattices are evaluated by following the same procedure[4.27] as outlined above for the double-junction quantum wells. The effect of the charge redistribution is ignored and it is assumed that the potential distribution is periodic with a uniform period.

We assume that the superlattice is grown along the z direction, the width of the well layers is uniformly L_w and that of the barrier layers L_b .

We choose $z = 0$ at the left interface between a well layer and a barrier layer. The equation for the envelope function $\psi_i(r)$ in the two layers is written as

$$- (\hbar^2/2m_i^*)\nabla^2\psi_i(r) = (E - E_{ci})\psi_i(r), \qquad (4.91)$$

where the subscript i indicates the well layer for w and the barrier layer for b, m_i^* and E_{ci} are respectively the conduction-band-edge effective mass and the energy of the i-layer, E is the energy of the electron.

The equations are to be solved with the boundary conditions at the interfaces as in the double-junction quantum well. The potential variation being periodic,the envelope functions are required to satisfy the additional condition,

$$\psi_i(\mathbf{r} + \hat{\mathbf{z}}L) = \psi_i(\mathbf{r})\exp(ik_zL), \qquad (4.92)$$

where k_z is the component of the wave vector in the z-direction.

The solution for ψ_i may be written as

$$\psi_i(r) = F_i(z)\exp(i\mathbf{k}_t.\rho), \qquad (4.93)$$

since the potential varies only in the z direction; $\mathbf{k}_t$ and ρ are respectively the in-plane wave vector and the position vector. Substituting (4.93) in (4.91) we get,

$$- (\hbar^2/2m_i^*)\nabla_z^2 F_i(z) = (E - \hbar^2 k_{ti}^2/2m_i^* - E_{ci}). \qquad (4.94)$$

Solution of Eq. (4.94) is

$$F_i(z) = A_i\exp(ik_iz) + B_i\exp(-ik_iz), \qquad (4.95)$$

where

$$k_i^2 = (2m_i^*/\hbar^2)(E - \hbar^2 k_{ti}^2/2m_i^* - E_{ci}) \qquad (4.96)$$

The boundary conditions give,

$$\mathbf{k}_{tw} = \mathbf{k}_{tb}, F_w(0) = F_b(0), (1/m_b^*)F_b' = (1/m_w^*)F_w'(0), \qquad (4.97)$$

$$F_w(L_w) = F_b(-L_b)\exp(ik_zL), \qquad (4.98)$$

$$(1/m_w^*)F_w'(L) = (1/m_b^*)F_w'(-L_b)\exp(ik_zL). \qquad (4.99)$$

The prime indicates the derivative with respect to z.

On applying the above conditions to the solutions of Eq. (4.91) we get the following equation,

$$\begin{vmatrix} 1 & 1 & -1 & -1 \\ 1 & -1 & -r & -r \\ a & 1/a & -c/b & -bc \\ a & -1/a & -rc/b & rbc \end{vmatrix} = 0, \qquad (4.100)$$

where

$$a = \exp{(ik_w L_w)}, b = \exp{(ik_b L_b)} \qquad (4.101)$$

$$c = \exp{(ik_z L)}, r = (k_b/k_w)(m_w^*/m_b^*) \qquad (4.102)$$

Simplifying Eq. (4.100) we get for the dispersion relation of superlattices,

$$\cos{(k_z L)} = \cos{(k_w L_w)} \cos{(k_b L_b)} - (1/2)(r + 1/r) \sin{(k_w L_w)} \sin{(k_b L_b)}. \quad (4.103)$$

The variation of the eigenvalues with k_z is illustrated in Fig. 4.5. It may be noted that $(E - E_{cw})$ varies as the square of k_z in the bulk material. For the superlattice, on the other hand, the band breaks into minibands, with boundaries given by $\pm n\pi/L$. Since, L is a few times the lattice constant, a number of such bands may be produced and those are often called minibands. The width of the minibands and their numbers are determined by the barrier potential, the period of the superlattice, L, and the relative values of the widths of the well and the barrier layer i.e L_w and L_b .

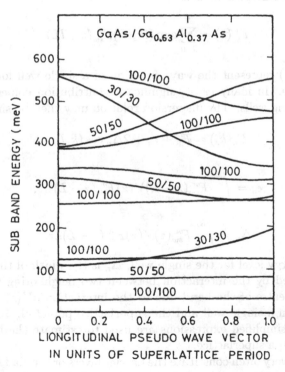

Figure 4.5. Energy eigenevalues in superlattice structures. Figures on the curves give well width/barrier width.

The widths of the bands become very small for barrier potentials of the order of 0.2 - 0.5 eV, when the barrier width is larger than 10 nm[4.28]. The system may then be considered as a combination of single wells, rather than a superlattice. Due to the weak coupling between the wells, the wave function corresponding to one well is not much modified by the proximity of the other wells. Such superlattice systems are referred as multiple quantum well, (MQW) , and are used extensively in the optic quantum well devices discussed in Chapter 10 and 11. Energy levels in such systems are degenerate and may be worked out as discussed in the preceding section. The effects of the extension of the wave function into the barrier layer and the energy band nonparabolicity may also be incorporated into the calculations by following the procedure, already discussed.

For smaller widths, however, Eq. (4.103) is required to be used for obtaining the energy dispersion relation. The effects of nonparabolicity may be readily incorporated by using the energy effective masses to obtain, k_w and k_b for the energy E, and by using the velocity effective mass in Eq. (4.102).

The dispersion relation may also be obtained in the tight-binding approximation by expressing the wave function for the superlattice in terms of the wave functions for the individual wells as

$$F_{k_z}^n(z) = \sum_{l=0}^{N} \exp{(ik_z lL)} F_{sw}^n(z - lL) \tag{4.104}$$

where $F_{sw}^n(z - lL)$ represent the wave function of a single well located at $z = lL$ for the nth mode. In many cases, significant contribution comes only from the nearest neighboring wells. The dispersion relation may then be simplified to

$$E_n(k_z) = E_n + \epsilon_n + 2\Delta_n \cos{(k_z L)}, \tag{4.105}$$

where

$$\epsilon_n = \int_{-\infty}^{\infty} F_{sw}^n(z - L)V(z)F_{sw}^n(z - L)\, dz, \tag{4.106}$$

and

$$\Delta_n = \int_{-\infty}^{\infty} F_{sw}^n(z)V(z)F_{sw}^n(z - L)dz. \tag{4.107}$$

E_n is the nth energy level for the single well, Δ_n is the width of the energy bands, mainly determined by the interaction between two neighboring wells, and ϵ_n is the shift in the centre of the band due to the interaction. $V(z)$ is the potential difference between a single well and the superlattice potential. In many applications, the above simplified expressions are used to estimate the band width and the energy shift in a superlattice.

Superlattices very often constitute the barrier layers or the cladding layers[4.29]. The characteristics of tunneling of electrons through these structure are of importance in such applications, and aspects of this phenomenon are discussed in Section

9.1. Superlattices have also been used in quantum well modulators and switches, which are discussed in Chapter 10.

4.2.3. SINGLE HETEROJUNCTION

The potential variation, causing the quantum well, is produced in single heterojunctions by the redistribution of the charge carriers. Electrons are transferred from the higher-lying conduction band to the lower conduction band or the holes are transferred from the lower-lying valence band to the higher valence band, till the bending of the bands due to this transfer aligns the Fermi levels of the two constituents. Such alignment is illustrated in Fig. 4.6 for the $n-n, n-i$ and $n-p$ junctions in the GaAs/Ga$_{0.7}$Al$_{0.3}$As system. The band offset is determined by the composition of the two constituents as discussed in Chapter 3. The bending of the bands is essentially determined by the doping of the layers. The doping is modulated so that the well layer is undoped and has only the background impurities. A portion of the barrier layer adjacent to the junction is also undoped. The carriers are generated by the doping and the symbols, used in the analysis, are shown in Fig. 4.7. Analysis of the potential profile forming the well is discussed below by considering a $n-p$ junction.

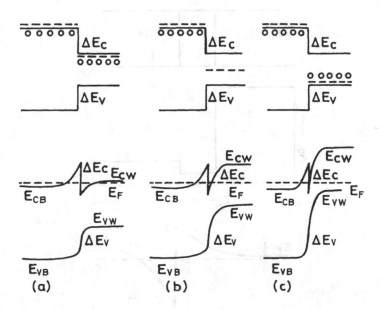

Figure 4.6. Band alignment in single-junction heterostructures. (a) n-n junction. (b) n-i junction. (c) n-p junction. $\Delta E_c (\Delta E_v)$- Conduction band edge energy for well (barrier) layer. $E_{vw}(E_{vb})$- Valence band edge energy for well(barrier) layer. E_F- Fermi level.

Total positive areal charge density, Q_{sb}, in the barrier layer is given by

$$Q_{sb} = |e| \int_{-\infty}^{0} [N_{db}(z) - n_b(z)] \, dz, \qquad (4.108)$$

where $N_{db}(z)$ and $n_b(z)$ represent respectively the donor density and the electron density in the barrier layer.

The charge density Q_{sb} may be expressed in terms of ΔE_c, E_{Fi}, N_{db}, and the thickness of the undoped layer d, by using the following equations which relate the charge density to the local potential and the carrier concentration to the potential and the Fermi level.

$$d^2U/dz^2 = (|e|/\epsilon_b)[N_{db}(z) - n_b(z)], \qquad (4.109)$$

$$n_b(z) = N_{cb}F_{1/2}(\eta_b - |e|U/k_BT), \qquad (4.110)$$

$$N_{cb} = 2.51 \times 10^{25}(m_b^*/m_0)^{3/2}(T/300)^{3/2}, \qquad (4.111)$$

$$F_j(\eta_b) = [\Gamma(j+1)]^{-1} \int_0^\infty W^j \, dW \, [1 + \exp(W - \eta_b)]^{-1}, \qquad (4.112)$$

$$n_b(-\infty) = N_{db} = N_{cb}F_{1/2}(\eta_b), \qquad (4.113)$$

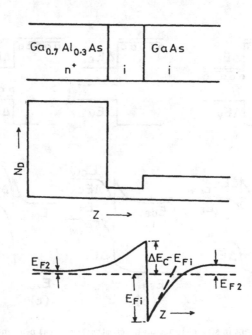

Figure 4.7. Modulation doping. N_d - Donor density. E_{F1}, E_{FI}, E_{F2} - Fermi level in different regions as indicated.

where U is the local electrostatic potential, ϵ_b is the permittivity, m_b^*, m_0 are the effective mass and free-electron mass respectively. T is the lattice temperature. $\Gamma(j)$ is the gamma function and $\eta_b = E_{Fb}/k_B T$, E_{Fb} being the Fermi energy and k_B is the Boltzmann constant. It may be noted that N_{cb} is the same as the prefactor in the Maxwell-Boltzmann distribution function. The subscript cb, db indicate the barrier layer. It is assumed that the donors are fully ionized. The barrier height is of the order of a few hundred millivolts, whereas $k_B T$ is a few millivolts. The contribution of the electron concentration $n_b(z)$ may be shown to be about $(k_B T/|e|U)$ times N_{db} and may be neglected, since $(k_B T/|e|U)$ is small.

Equation (4.108) may now be integrated to relate Q_{sb} with the potential, $(\Delta E_c - E_{Fi} - E_{Fb})$, at the junction. The relation is

$$Q_{sb} \approx [(2\epsilon_b N_{db})(\Delta E_c - E_{Fi} - E_{Fb})]^{1/2} \qquad (4.114)$$

The Fermi level of the well layer, E_{Fi}, at the interface, is treated in the above expression as an unknown. Values of all the other physical constants being known, E_{Fi} may be obtained, by applying the condition that the total structure must be electrically neutral, i.e., the negative areal charge density in the well layer must be equal to Q_{sb}.

The potential distribution in the well layer is controlled ,on the other hand, more by the charge carriers, accumulated near the interface, than by the ionized acceptors. In quantum well devices, the surface charge density is made so large that the potential reaches the bulk value in a distance comparable to the electron wavelength. The charge distribution in the well is, therefore, required to be determined by solving the envelope-function equation, including the potential energy $V_w(z)$ due to the charges, given below:

$$[(-\hbar^2/2m_w)\nabla^2 + V_w(z)]F_w(z) = [E - E_{cw}(o)]F_w(z) \qquad (4.115)$$

The potential $V_w(z)$ is mainly the Hartree potential i.e., the potential due to the ionized impurities and all the electrons except the one under consideration. Electron concentration being large, a second component arising from the charge and correlation effects should also be considered. However, detailed calculations indicate that the effect of this component is comparatively small and it may be neglected in the first approximation[4.30].

The Hartree potential is required to be obtained by solving the Poisson equation,

$$\nabla_z^2 V_w(z) = |e|^2 [n_w(z) + N_{aw}]/\epsilon_w, \qquad (4.116)$$

where N_{aw} is the acceptor density, assumed uniform in the well layer, ϵ_w is the well-layer permittivity and $n_w(z)$ is the electron density at z. It should be noted that the electrons occupy different quantized energy levels according to the Fermi function and they are distributed in space in accordance with the eigenfunction for the particular quantum level. The electron density $n_w(z)$ may be expressed as,

$$n_w(z) = \sum_i n_i |F_{wi}(z)|^2, \tag{4.117}$$

where n_i is the number of electrons in the ith level corresponding to which the eigenfunction is $F_{wi}(z)$. The concentration n_i is related to the Fermi level and the quantized energy levels by the Fermi function, as given below,

$$n_i = (1/2\pi^2) \int \{ \exp[(E_i + \hbar^2 k_t^2/2m_w^* - E_{Fi})/k_B T] + 1 \}^{-1} \, dk_t, \tag{4.118}$$

where E_{Fi} is the Fermi level measured from the conduction-band edge at the interface, and the energy of the electron from the same reference is

$$E(k_t) = E_i + \hbar^2 k_t^2/2m_w^*, \tag{4.119}$$

k_t being the in-plane component of the wave vector, k.

The total charge on the well layer per unit area is

$$Q_{sw} = -|e| \int_0^\infty (n_w(z) + N_{aw}) \, dz, \tag{4.120}$$

and E_{Fi} would be obtained by equating $|Q_{sw}|$ to $|Q_{sb}|$.

The evaluation of $n_w(z)$, however, requires solution of Eq. (4.115) which again involves $V_w(z)$.

Equation (4.116) is required to be solved subject to the boundary conditions,

$$V_w(0) = 0 \tag{4.121}$$

$$\text{and} \ \ \epsilon_w \partial_z V_w(z)|_0 = -Q_{sw} = Q_{sb}. \tag{4.122}$$

The second condition follows from the condition of the continuity of the normal component of the displacement.

For finite values of the barrier potential ΔE_c, the wave functions for the quantized levels will partially extend also into the barrier layer and a second Schrödinger equation is required to be solved for the barrier layer. It is given by

$$[-(\hbar^2/2m_b^*)\nabla^2 + V_b(z)]F_b(z) = (E - E_{cb})F_b(z), \tag{4.123}$$

where the subscript b indicates values for the barrier layer and $V_b(z)$ is the potential energy due to the charge distribution. The potential $V_b(z)$ is considered to be due to only the ionized surface charge density. However, a significant contribution to this density may also come from the electrons in the well for finite barriers, as they spend part of the time in the barrier layer when the wave function extends into it. Equation (4.123) is, therefore, required to be solved after obtaining $V_b(z)$ by including this additional contribution to the charge density in the barrier layer.

The solutions, $F_w(z)$ and $F_b(z)$, are finally required to be matched at the heterojunction interface by applying the boundary conditions, already discussed in Sect.4.2.1 in connection with the double-junction quantum wells.

It should be evident that a complete solution of the equation for evaluating the energy eigenvalues and the wave function is very involved for the single junction quantum wells. The potential, $V_w(z)$ in Eq. (4.115) is required to be obtained by using the solution of the same equation or the equation is to be solved self-consistently[4.31]. It may, therefore, be tackled only iteratively by using numerical techniques. Such solutions have been obtained in a few cases[4.32]. It is, however, found that the problem may be simplified by introducing some approximations, and solutions may be obtained analytically. Such analytic solutions are close to those obtained by applying the rigorous numerical methods and mostly, these are used in the analysis of quantum well devices, using single junctions. These solutions are discussed below.

Infinite-barrier triangular well
The most common and popular approximation is the triangular-well approximation. The potential in the well layer may be expressed in the Hartree approximation as

$$V_w(z) = (eQ_{sw}/\epsilon_w)z + (e^2/\epsilon_w) \int_0^z dz' \int_0^{z'} (N_{aw} + n_w(z'')) \, dz''. \qquad (4.124)$$

Variation of the potential is dominated by the field term for small values of z. The potential, therefore, increases initially linearly with the distance, but as the second term increases faster than the first term the rate of increase gradually decreases till the potential levels off to the bulk value. In a simple approximation, called the triangular-well approximation, the field $\mathcal{E}$ is taken to be equal to

$$\mathcal{E} = (|e|/\epsilon_w) \int_0^\infty [n_w(z)/2 + N_{aw}] \, dz = (|e|/\epsilon_w)(N_{dep} + N_s/2), \qquad (4.125)$$

where N_{dep} and N_s represent respectively the concentration of the ionized acceptors and the accumulated electrons per unit area. The factor of $(1/2)$ is introduced to account for the variation of $\mathcal{E}$ with z. Solutions may be obtained analytically, if it is assumed further that ΔE_c is so large that $F_w(z) = 0$ at the interface.

Equation (4.115) has then the form of the well-studied Airy equation and the wave functions are the Airy functions[4.33,34]. The wave function is given by

$$\psi_w(\mathbf{r}) = F_w(z) \exp(i\mathbf{k}.\rho), \qquad (4.126)$$

$$F_w(z) = A_i[(2m_w^* e\mathcal{E}/\hbar^2)^{1/3}(z - E_i/|e|\mathcal{E})]. \qquad (4.127)$$

E_i 's are the energy eigenvalues, given by

$$E_i = (\hbar^2/2m_w^*)^{1/3}[(3\pi|e|\mathcal{E}/2)(l + \theta_i)]^{2/3}. \qquad (4.128)$$

θ_i have the values of 0.7587, 0.7540, 0.7575 for $l = 0$, 1 and 2 and may be approximated as 0.75 for higher values of l.

The approximation will be more appropriate for the lower energy levels, for which the wave function does not extend to large values of z. However, because of the simplicity of analysis, it is frequently used to derive the characteristics of the quantum well devices using single heterojunctions.

Improved analytic solutions are obtained by using the variational functions for $F_w(z)$. The variational function used for barriers with infinite height is[4.35]

$$F_w(z) = H(z)(b_0^3/2)^{1/2} z \exp(-b_0 z/2). \tag{4.129}$$

The variational parameter b_0 is then determined by minimizing the total energy of the electron, which may now be obtained by solving Eq. (4.115) with the Hartree approximation.

The value of b_0 may be expressed in terms of the surface electron concentration, N_s and the density, N_{dep}, of donor atoms integrated over the distance in which the potential varies in the well layer. The relation is[4.36],

$$b_0 = (33 m_w^* e^2 N_s^*/8\epsilon_w \hbar^2)^{1/3}, N_s^* = N_s + (32/11)N_{dep}. \tag{4.130}$$

More elaborate trial functions have been used for the analysis of wells with finite barriers, which are more realistic[4.37,38]. The function is

$$F_w(z) = Bb^{1/2}(bz + \beta) \exp(-bz/2), \text{ for } z > 0, \tag{4.131}$$

$$= B'b'^{1/2} \exp(b'z/2), \text{ for } z < 0, \tag{4.132}$$

where B, B', b, b' and β are the variational parameters. The constants B, β and B' are expressed in terms of b and b' by using the boundary conditions at the interface and the normalization condition for the wave function. The total energy is calculated by using these functions and then minimizing it by varying b and b' to obtain their values.

The discussion above has been based on the assumption that the electrons occupy only the lowest subband. The following variational function[4.39] has been used for the second subband, to account for its occupancy, in case the well is broad enough.

$$F_w(z) = A(2/b_1^3)^{1/2} z(1 - Bz) \exp(-b_2 z/2). \tag{4.133}$$

It is found after minimizing that

$$A = [(3b_1^3/b_2^5)/4(b_1^2 - b_1 b_2 + b_2^2)]^{1/2}, B = (b_1 + b_2)/6, \tag{4.134}$$

$$b_2 = b_1, A = 0.866b_1, B = b_1/3, b_1 = (77m_w^* e^2 N_s^*/16\epsilon_w \hbar^2)^{1/3}. \tag{4.135}$$

The variational method has been used to GaAs/Al$_x$Ga$_{1-x}$As wells, along with the complete numerical method[4.37] to estimate its accuracy. The energy levels are almost indistinguishable. The space-dependence of the wave functions are slightly different and the average electron position is found to be shifted towards the interface, as expected when the barrier is considered to be finite.

It should be evident from the discussions of this section that an exact evaluation of the wave function or the energy levels is rather involved for single-junction wells. On the other hand, quantum well transistors mostly use these structures and the wave functions, if not the energy levels, determine the scattering probabilities, and are required to be known for any theoretical evaluation of the electron mobility.

It has been the practice to use Eq. (4.128) for this purpose. The effect of the extension of the wave function into the barrier layer may not be important for low electron concentration. For larger electron concentration, however, the effect may not be negligible. The penetration affects mobility in two ways. The alloy scattering in $Al_xGa_{1-x}As$ starts playing a role. Also, the scattering probability decreases, since effectively the well width increases. This point is discussed in greater details in Chapter 7. Equation (4.132) is therefore, required to be used when the electron concentration in the device is large. It should also be mentioned that for such concentrations, the energy band nonparabolicity is also important. However, analysis including the nonparabolicity becomes rather difficult as $E - E_{cw}(z)$ varies with z due to the bending of the conduction-band edge and the variation of E_{cw}. Such analysis has not yet been attempted.

4.2.4. QUANTUM WIRES AND QUANTUM DOTS

It has been mentioned in Chapter 2 that though the major current interest for quantum well electronics is in quantum well systems, quantum wires and quantum dots promise some improved characteristics and are being studied[4.40]. The progress is, however, mostly held back because of technical problems in the realization of a structure with acceptable perfection. Energy levels in these systems are hence only briefly discussed.

Quantum wire or 1D electron gas system
The simplest 1D system is a rectangular two dimensional well with dimensions L_x (in the x direction), L_y (in the y direction) and infinite barrier height. Energy levels in this system are obtained by solving the equation,

$$- (\hbar^2/2m_w)\nabla^2 F_w(\mathbf{r}) + V(x,y)\psi_w(\mathbf{r}) = (E - E_{cw})\psi_w(\mathbf{r}) \qquad (4.136)$$

with $V(x,y) = 0$ for $0 \leq x \leq L_x$ and for $0 \leq y \leq L_y$. Solutions for ψ_w is

$$\psi_w(\mathbf{r}) = A \sin{(m\pi x/L_x)} \sin{(n\pi y/L_y)} \exp{(ik_z z)}, \qquad (4.137)$$

where m, n are integers, A is the normalization constant, k_z is the wave vector in the z direction, in which the electrons have freedom of motion. The x-dependent and y-dependent components of F_w are chosen to fit the boundary condition, that it is zero at the interfaces. The corresponding eigenvalues of energy are

$$E_{mn} = E_{cw} + (\hbar^2\pi^2/2m_w^*)(m^2/L_x^2 + n^2/L_y^2). \qquad (4.138)$$

For the lowest subband energy, $m = n = 1$, the energy is two times higher than the lowest subband energy for the 2D system with similar quantizing dimensions. The higher value of the confinement energy makes it important to consider the effect of the finite barrier height.

Analytical solutions for finite barrier potentials may be obtained if the geometry is assumed to be cylindrical or elliptical. For the cylindrical geometry, assuming the radius of the well to be a, the wave functions may be written as

$$\psi_w = AJ_n(k_w r)\exp{(ik_z z)}, \text{ for the well,} \qquad (4.139)$$

$$\text{and } \psi_b = BK_n(k_b r)\exp(ik_z z), \text{ for the barrier,} \qquad (4.140)$$

where

$$k_i^2 = (2m_i^*/\hbar^2)|(E - E_{ci})|. \qquad (4.141)$$

J_n and K_n are respectively the Bessel and the modified Bessel function of order n.

The eigenvalues of energy are obtained from the solution of the following equation, which is derived by putting the boundary conditions as discussed for the quantum wells.

$$\frac{K_n(k_b a)}{J_n(k_w a)} \cdot \frac{[J_{n+1}(k_w a) - J_{n-1}(k_w a)]}{[K_{n-1}(k_b a) + K_{n+1}(k_b a)]} \cdot \frac{k_w}{k_b} = \frac{m_w}{m_b}. \qquad (4.142)$$

These solutions may be used to estimate the modification in the energy eigenvalues for wells with square cross section by choosing a to be such that the cross sectional areas are the same, i.e., using the following equivalence relation[4.41],

$$a = L/\pi^{1/2}, \text{ where } L_x = L_y = L. \qquad (4.143)$$

For a check, the eigenvalues were compared for the infinite-barrier cylindrical and square geometry. The values were found to agree to within 8%.

Numerical techniques are required to be applied for rectangular quantum wires with finite barriers. The finite- element method has been used[4.42] to evaluate numerically the energy eigenvalues of GaAs/GaAlAs rectangular quantum wires. A convenient method has also been proposed by Gershoni et al[4.40], in which the envelope function is expressed as a sum of two-dimensional orthogonal trigonometric functions, as follows,

$$F(x, y) = \sum_{m,n} A_{mn} \sin{[m(\pi/2 + x/L_1)]} \sin{[n(\pi/2 + y/L_2)]}, \qquad (4.144)$$

where m, n are integers and A_{mn}'s are the corresponding coefficients. L_1 and L_2 are arbitrarily chosen. But, L_1 and L_2 are required to be much larger than L_x and L_y so that $F(x,y)$ has negligible value for $x = L_1$ and $y = L_2$.

Equation (4.136) is modified so as to include the x,y dependence of the effective mass and written as,

$$\{-(\hbar^2/2)[1/m^*(x,y)]\nabla[1/m^*(x,y)]\nabla + V(x,y)\}F(x,y) = EF(x,y). \qquad (4.145)$$

The function (4.144) is substituted in this equation and then the equation is reduced to a matrix equation by using the orthogonal property of the trigonometric functions. Eigenvalues of the matrix equation are obtained by using a software like IMSL. The method has been used to obtain the electron energy eigenvalues in InP/GaInAs quantum wire[4.43]. The method has also been used to calculate the energy shift in InP/GaInAs quantum wires for comparison with the available experimental data. These results are presented in Fig. 4.8. Close agreement has been found between the calculations and the experiment.

The method may be used conveniently to quantum wires of any arbitrary shape, e.g., triangular or arrowheads, discussed in Chapter 2. Calculations for such wires

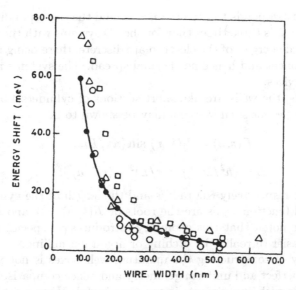

Figure 4.8. Energy eigenvalues in 1D wells with breadth = 5 nm. [After S. Gangopadhyay and B. R. Nag , *Phys. Stat. Sol.* **195**, 123 (1996)]. Solid line - Theory. Open circles - Experiment.

wires have also been reported[4.44].

It should be mentioned that the application of quantum wires to photonic devices requires good estimated values of the energy levels, so that the required dimensions of the well may be worked out with suitable design formulæ. The energy levels may, however,be significantly altered due to the finite height of the barrier. Energy band nonparabolicity is also likely to have significant contribution. The effect of nonparabolicity may be estimated by using the method discussed in Section 4.2.1 for energy levels in double-junction quantum wells. However, the technology of realizing quantum wires is not yet perfect enough to produce wells with well-defined geometries. Evidence for definite two-dimensional quantizations has appeared recently[4.45-49]. Numerical methods discussed above may be fruitfully employed to examine the experimental results obtained for these structures.

Quantum dots or 0D systems

Not many 0D systems have been reported so far in the literature. Some have been produced by etching[4.50] and the confinement is essentially by the semiconductor vacuum interface potential. The value of this potential is of the order of a few eV and it may be assumed justly that the wave function is zero on the surfaces, except for the one in continuity with the substrate. Even if it is assumed that it is zero also on that surface, the error will not be large.

The energy eigenvalues for these conditions are given for a prallelopiped by,

$$E = [(\hbar\pi)^2/2m_w^*](l^2/L_x^2 + m^2/L_y^2 + n^2/L_z^2). \tag{4.146}$$

The lowest mode corresponds to $l = m = n = 1$, and the corresponding eigenvalue for $L_x = L_y = L_z = L$ is three times that for the 2D systems with the well width L. Further, the allowed energies of the electron are discrete, there being no continuous wave vector component and hence no thermal spread. The systems may therefore give good optic devices.

In some reports, the wells are like short sections of cylinders[4.50]. The wave functions and energies for such systems may be shown to be

$$F(r, z) = J_1(k_w r) \, \sin\,(n\pi/L)z. \tag{4.147}$$

$$E = (\hbar^2/2m_w^*)[(n\pi/L)^2 + (r_{lm}/a)^2], \tag{4.148}$$

where a and L are respectively the radius and the length of the cylindrical wells, $J_1(x)$ is the Bessel function, r_{lm}'s are the roots of $J_1(k_w a) = 0$, and l is a positive integer. It may be noted that this geometry introduces no special new features, except that it is easier to realize by etching or argon ion milling.

The technology of constructing the quantum well boxes is not yet developed enough to ensure perfect and uniform geometry and hence comparison of the theoretical energy levels with experiments is not meaningful. However, as the technology improves and systems are realized with a second semiconductor surrounding the structures on all sides, some of the refinements required for estimating the effects of finite barriers and energy band nonparabolicity will have to be considered, as discussed in Section 4.2.1(a) for the 2D system. A few reports[4.47,51] of the realization of such structures have appeared, but comparable calculated values of energy levels are not available.

4.3. Energy Levels of Holes

Electronic quantum well devices are constructed with modulation-doped heterostructures using electrons as the charge carriers, since the electrons have higher mobilities than the holes. However, devices using holes as the charge carriers are also required for complementary logic and attempts are being made to exploit the low in-plane effective mass of holes for this purpose by new techniques, which are discussed in Section 4.4. Photonic devices, on the other hand, involve the holes integrally, as light interaction in many devices causes electrons to be transferred from the quantized electron levels to hole levels or the vice versa. Quantization of hole levels in heterostructures, therefore, plays an important role in photonic devices.

Energy levels of holes corresponding to the degenerate levels are worked out for heterostructures by following the same procedure as discussed for the electrons. However, two coupled effective mass equations are required to be solved, subject to the boundary conditions at the two interfaces. The equations have been derived in Section 4.2. We discuss below the dispersion relations obtained from these equations.

The quantum well is assumed to be grown along a <100> direction, which is also chosen as the z axis. The effective mass equations for the well and the barrier layers are written below assuming that $\gamma_2 = \gamma_3 = \bar{\gamma}$ in order to simplify the expressions.

$$H_{+i}(k_t, \nabla_z^2)F_{1i}(z) + W_i(k_t, \nabla_z)F_{2i}(z) = EF_{1i}(z) \qquad (4.149)$$

$$W_i^\dagger(k_t, \nabla_z)F_{1i}(z) + H_{-i}(k_t, \nabla_z^2)F_{2i}(z) = EF_{2i}(z), \qquad (4.150)$$

where

$$H_{\pm i}(k_t, \nabla_z^2) = E_{vi} - a[(\gamma_{1i} \pm \bar{\gamma}_i)k_t^2 - (\gamma_{1i} \mp 2\bar{\gamma}_i)\nabla_z^2] \qquad (4.151)$$

$$W_i(k_t, i\nabla_z) = \sqrt{3}\bar{\gamma}_i a k_t(k_t - 2\nabla_z), a = (\hbar^2/2m_0) \qquad (4.152)$$

The subscript i is to be replaced by w for the well and by b for the barrier layer.

Before discussing the dispersion relations we note some important characteristics of the holes. The coupling term W is zero for $k_t = 0$ and the two equations are decoupled. The solutions given for E in Section 4.2 are applicable and the eigenvalues of energy may be obtained from Eq. (4.89), by replacing E_{ci} with E_{vi}

and m_{ci} with $m_{\pm i}$. The effective mass of holes, $m_{\pm i}$ is given for the assumption $k_t = 0$, by

$$m_{\pm} = (\gamma_{1i} \mp 2\bar{\gamma}_i)^{-1} m_0 \qquad (4.153)$$

The mass for H_{+i} being larger than that for H_{-i}, a hole corresponding to the plus sign is referred as heavy hole and that corresponding to the minus sign, light holes.

The quantized energy eigenvalues for heavy holes are smaller in magnitude than those of light holes as $m_{+i} > m_{-i}$. The energies of heavy holes are therefore higher. The expression for the Hamiltonians also indicate that the in-plane effective mass of holes is given by

$$m_{\pm i} = (\gamma_{1i} \pm \bar{\gamma}_i)^{-1} m_0 \qquad (4.154)$$

The in-plane effective mass of heavy holes is thus smaller than that of the light holes. As the heavy hole energy levels are higher, these are occupied before the light hole levels. The small in-plane effective mass of heavy holes may be expected to cause a reduction in the density of states of the holes in quantum wells, which should reduce the value of the threshold current density of lasers (see Chapter 10). The expected reduction is, however, much modified by the mixing of the two hole states for $k_t \neq 0$ as discussed below.

For $k_t \neq 0$, the two valence bands are mixed and the coupled equations are required to be solved[4.52] with the proper boundary conditions. The wave vector component k_z for the envelope functions in each layer, corresponding to a particular hole energy E, are the roots of the characteristic equation,

$$[E_{v0i} - a(\gamma_{1i} + \bar{\gamma}_i)k_t^2 + (\gamma_{1i} - 2\bar{\gamma}_i)k_z^2 - E]$$
$$\times [E_{v0i} - a(\gamma_{1i} - \bar{\gamma}_i)k_t^2 + (\gamma_{1i} + 2\bar{\gamma}_i)k_z^2 - E]$$
$$- 3\bar{\gamma}_i^2 a^2 k_t^2 (k_t^2 + 4k_z^2) = 0. \qquad (4.155)$$

The roots are,

$$k_{\pm i}^2 = [(E_{v0i} - E)/(\gamma_{1i} \mp 2\gamma_i)](2m_0/\hbar^2) - k_t^2. \qquad (4.156)$$

The solution for the envelope functions may now be written as

$$F_{1i} = A_1 \sin(k_{+i}z) + B_1 \cos(k_{+i}z) + C_1 \sin(k_{-i}z) + D_1 \cos(k_{-i}z) \qquad (4.157)$$

$$F_{2i} = A_2 \sin(k_{+i}z) + B_2 \cos(k_{+i}z) + C_2 \sin(k_{-i}z) + D_2 \cos(k_{-i}z). \qquad (4.158)$$

The energy eigenvalues are obtained by using these functions in (4.140) and (4.141) and applying the boundary conditions.

We consider first quantum wells with the so-called infinite barrier potentials, i.e., $E_{v0b} \ll E_{v0w}$. In this case, the envelope functions are required to be zero at the well boundaries i.e., at $z = 0$ and at $z = L$. The subscript i may also be dropped as the envelope function does not exist for the barrier layer.

Substituting (4.157) and (4.158) in (4.149) and (4.150) and equating to zero the coefficients of the sine and cosine terms we get the relations between A_2, B_2 and A_1, B_1 and between C_2, D_2 and C_1, D_1. Using these relations and equating F_1 and F_2 to zero for $z = 0$ and for $z = L$, .we get

$$
\begin{vmatrix}
0 & 1 & 0 & 1 \\
c_1 & b_1 & c_2 & b_2 \\
\sin(k_+ L) & \cos(k_+ L) & \sin(k_- L) & \cos(k_- L) \\
-[b_1 \sin(k_+ L) & [c_1 \sin(k_+ L) & -[b_2 \sin(k_- L) & [c_2 \sin(k_- L) \\
+c_1 \cos(k_+ L)] & -b_1 \cos(k_+ L)] & +c_2 \cos(k_- L)] & -b_2 \cos(k_- L)]
\end{vmatrix} = 0.
$$

(4.159)

where

$$b_1 = \sqrt{3} k_t^2 \gamma_2 / e_1, b_2 = \sqrt{3} k_t^2 \gamma_2 / e_2, \tag{4.160}$$

$$c_1 = 2\sqrt{3} \bar{\gamma} k_t k_+ / e_1, c_2 = 2\sqrt{3} \bar{\gamma} k_t k_- / e_2, \tag{4.161}$$

$$e_1 = -[(\gamma_1 - \gamma_2) k_t^2 + (\gamma_1 + 2\gamma_2) k_+^2 - E], \tag{4.162}$$

$$e_2 = -[(\gamma_1 - \gamma_2) k_t^2 + (\gamma_1 + \gamma_2 k_-^2) - E]. \tag{4.163}$$

On simplifying the above equation we get the following dispersion relation for the holes in a quantum well with infinite barrier and width, L,

$$[4k_+^2 k_-^2 + k_t^2(k_+^2 + k_-^2) + 4k_t^4] \sin(k_+ L) \sin(k_- L)$$
$$+ 6k_t^2 k_+ k_- [1 - \cos(k_+ L) \cos(k_- L)] = 0. \tag{4.164}$$

In Fig. 4.9 is illustrated the dispersion relation, $E - k_t$, as obtained from this equation for a GaAs well of thickness 10 nm. Difference of the energy eigenvalues from $E_n(k_t = 0)$ have been plotted against k_t^2, to illustrate the strong nonparabolic nature of the dispersion relation. We note that for $k_t = 0$, the eigenvalues are

$$E_n = E_{v0} - (n\pi)^2 (\gamma_1 \mp 2\bar{\gamma})(\hbar^2 / 2m_0 L^2). \tag{4.165}$$

We get two series of values for the two kinds of holes. The variation of the energy with k_t is found to be quite distinct for the different branches. The strong nonparabolicity of the dispersion relation is a distinctive character of the holes in heterostructures, which affects the opto-electronic properties significantly. One interesting feature of the curves is that the in-plane effective mass is a complex function of k_t and E_n. It varies strongly with k_t and may even be negative for some values of k_t in some bands.

The analysis may be extended to include the effects of the finite barrier height. However, we have to consider the functions in the two barrier layers in addition to those in the well layer. The characteristic equation is obtained by applying the boundary conditions at the two interfaces. Consequently we get a 8×8 matrix for the equation, which is required to be solved numerically for different values of k_t.

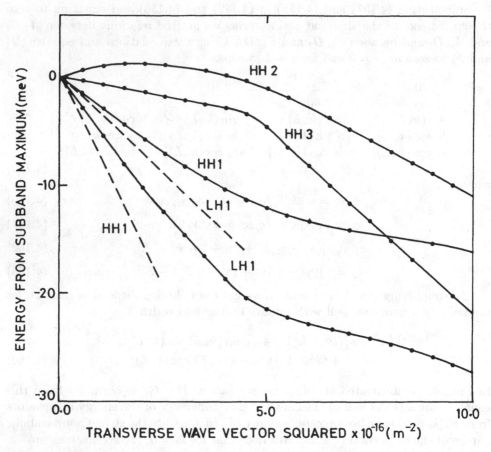

Figure 4.9. Energy-wave vector dispersion curves for holes in a GaAs 10 nm wide quantum well with infinite barrier height. Calculations were done assuming that $\gamma_1 = 2.8, \bar{\gamma} = 2.5$. HH1, HH2 and HH3 are respectively the first, second and third heavy-hole bands. LH1 is the first light-hole band. Dashed curves correspond to the parabolic bands with the effective-mass ratio taken as $(\gamma_1 + \bar{\gamma})^{-1}$ and $(\gamma_1 - \bar{\gamma})^{-1}$ respectively for the heavy holes and the light holes.

The exact dispersion relation changes with the barrier height, but the nature of the $E - k_t$ curves are not much altered. The curves are nonparabolic and the in-plane effective mass becomes zero or negative for some values of k_t in some subbands as for wells with infinite barrier[4.52]

The details of the dispersion relations may not be important for phenomena such as photoluminescence in which only holes near $k_t = 0$ are involved. But, for all other phenomena e.g., transport and light absorption, these details are important.

These details affect the nature of the voltage-current and light absorption characteristics. Some analyses have been reported for the absorption characteristics by using these dispersion relations[4.53,54]. In most other studies, however, the analysis is based on approximately equivalent parabolic and isotropic dispersion relations.

4.4. Energy levels in strained-layer wells

Heterostructures have been discussed so far, by assuming that the constituent materials are lattice-matched, so that there is no distortion or extra potential at the heterojunction interface. Heterostructures may be grown also with materials, the lattice constants of which are different; the difference may be as large as 6.8% as in Si/SiGe system[4.55-58]. For materials with different lattice constants, the difference is accommodated by misfit dislocations when the layers are thicker than about 100 nm. If, however, the layers are thin, and the thickness is less than a critical value, the mismatch is accommodated by generating uniform elastic strains, without producing misfit dislocations. Such heterostructures are referred as strained-layer structures. These layers have crystalline purity, mostly unaffected by the mismatch. But, the mismatch allows realization of hetrostructures with wide-ranging band offsets and band gaps in the well, than is possible by using lattice-matched constituents. Energy levels in such strained- layer wells are discussed in this section.

It is the thin epitaxial layer grown on thick substrates, which gets strained and depending on the difference in the lattice constants, the strain may be compressive or tensile. The in-plane lattice constant of the strained layer becomes the same as that of the substrate and hence when the bulk lattice constant of the former is larger than that of the latter, the strain is compressive and in the reverse case, the strain is tensile. The biaxial in-plane strain causes also a consequential strain in the direction normal to the interface, a compressive in-plane strain causes an extension while a tensile strain causes a compression. The epitaxial layer is, as a result, tetragonally distorted as shown in Fig. 4.10. Energy bands are altered by the strain and these altertions are evaluated by using the phenomenological deformation potential theory.

A discussion of the deformation potential theory is beyond the scope of this book. The interested reader may consult Reference 4.59. We shall only discuss the results which allow us to compute the changes in the band structure of the strained layer. We note that the strain has a hydrostatic and a shear component. The biaxial strain is given by

$$\epsilon_{xx} = \epsilon_{yy} = \epsilon = \Delta a / a, \qquad (4.166)$$

where $\Delta a = a_{\text{sub}} - a_0$, a_{sub} and a_0 being respectively the lattice constant of the

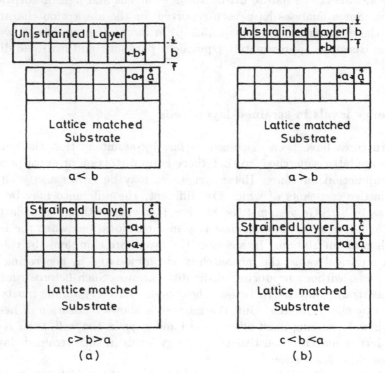

Figure 4.10. Lattice distortion in a strained-layer structure. (a) Compressive strain. (b) Tensile strain.

substrate and of the unstrained epitaxial layer. The strain in the z direction ϵ_{zz} is obtained by considering that the stress in that direction is zero in the equilibrium condition. Consequently[4.60],

$$C_{13}\epsilon_{xx} + C_{23}\epsilon_{yy} + C_{33}\epsilon_{zz} = 0, \qquad (4.167)$$

where C_{13}, C_{23}, C_{33} are the elastic constants, We get from (167),

$$\epsilon_{zz} = [-(C_{13} + C_{23})/C_{33}]\epsilon = (-2C_{12}/C_{11})\epsilon, \qquad (4.168)$$

where use has been made of the equality of the diagonal and non- diagonal components of the elasticity tensor of a cubic crystal.

The hydrostatic or dilational strain is given by

$$\epsilon_{xx} + \epsilon_{yy} + \epsilon_{zz} = 2(1 - C_{12}/C_{11})\epsilon. \qquad (4.169)$$

The shear strain is, on the other hand, given by

$$(\epsilon_{zz} - \epsilon_{xx}) + (\epsilon_{zz} - \epsilon_{yy}) = -2(1 + 2C_{12}/C_{11})\epsilon. \qquad (4.170)$$

It is assumed that the strains, resulting from the lattice mismatch with the substrate, alters the band structure of the thin layer like the external stresses affecting the band structure of the corresponding bulk material. The strain causes changes in the position of the different bands by different amounts.

4.4.1. EFFECT OF STRAIN ON THE CONDUCTION BAND

The conduction band is shifted by an amount proportional to the strain ϵ. The direct determination of the proportionality constant, the so-called deformation potential constant, is difficult to determine for bulk materials as the absolute position of the energy levels are not known. Its value has been calculated for some materials from first principles[4.61]. In most other cases, the value has been inferred from the analysis of the mobility or of the free-carrier absorption[4.22]. For heterostructures, however, the shift in the conduction band changes the band offset, and the shift may therefore be determined by using the methods, which have been used to measure the band offset, as described in Chapter 3[4.62].

4.4.2. EFFECT OF STRAIN ON THE VALENCE BAND

The strain shifts the valence bands relative to the conduction band. The hydrostatic component of the strain given by Eq. (4.169) shifts[4.63] equally the heavy hole and the light hole band by an amount δE_{hy}, where

$$\delta E_{hy} = 2E_h(1 - 2C_{12}/C_{11})\epsilon, \qquad (4.171)$$

E_h being the dialational deformation potential constant.

The shear strain, on the other hand, removes the degeneracy of the valence band maximum and shifts the heavy hole band and the light hole band by equal amounts in the opposite directions. Using the expression for the shearing strain, the shift due to shearing strain may be written as,

$$\delta E_{sh} = E_{sh}(1 + 2C_{12}/C_{11})\epsilon, \qquad (4.172)$$

where E_{sh} is the deformation potential constant for shear strain. Thus, the band-edge energies in the strained layer are,

$$E_c = E_{c0} + E_1\epsilon, \qquad (4.173)$$
$$E_{vH} = E_{v0} + E_1\epsilon + \delta E_{hy} + (1/2)\delta E_{sh}, \qquad (4.174)$$
$$E_{vL} = E_{v0} + E_1\epsilon + \delta E_{hy} - (1/2)\delta E_{sh} \qquad (4.175)$$

where E_c, E_{vH}, E_{vL} represent respectively the band edge energies of the conduction band, heavy hole band and the light hole band, E_1 is the total deformation potential constant which gives the shift of the conduction band.

We note that for a tensile biaxial strain, ϵ is positive and since both E_h and E_{sh} are negative, δE_{hy} and δE_{sh} are negative. The band gap, $(E_{c0} - E_{vH})$, therefore increases from $(E_{c0} - E_{vL})$ by $|\delta E_{hy} + (1/2)\delta E_{sh}|$. We also note that $|E - E_{vL}|$ is smaller than $|E_{c0} - E_{vH}|$. Therefore for a tensile biaxial strain, the light hole band maximum is higher than the heavy hole band maximum in the strained bulk material.

On the other hand, for a compressive biaxial strain, the heavy hole maximum is higher than the light-hole maximum. The relative position of the two maxima are, however, altered further in quantum wells. The space quantization in wells pushes downwards the valence bands, and the light holes have lower energy than the heavy holes because of their lower longitudinal mass. The quantization, therefore acts in the opposite direction to the biaxial tensile strain and in the same direction as the biaxial compressive strain. The difference in energy between the heavy-hole and the light-hole quantized levels is therefore accentuated by compressive biaxial strain and reduced and may even be made negative by the biaxial tensile strain.

The accentuation of the difference in the energy levels by compressive biaxial

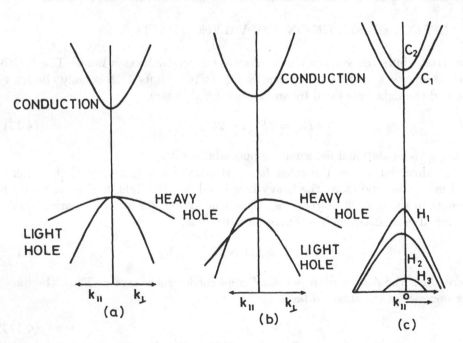

*Figure 4.11.*Schematic diagram showing hole dispersion relation in a strained-layer system. (a) Unstrained bulk material. (b) Strained bulk material. (c) Quantum wells.

strain reduces the coupling between the two hole bands and makes the in-plane effective mass of heavy holes closer to $\gamma_1 - \bar{\gamma}$. In addition, the occupation of the heavy hole band also increases as the separation between the two bands is increased. The density of valence-band states is therefore reduced over a wider energy range by compressive biaxial strain. This reduction offers advantages in the operation of lasers as explained in Chapter 10.

Electron and hole states in strained layer quantum wells are worked out by using the equations discussed in the earlier sections for unstrained layers. The effective mass constants are likely to be affected by strain as the band gaps between the bands change [see Eq. (4.15)]. This likely change is assumed to be negligible. Consequently, we are only required to solve Eq. (4.19) and Eq. (4.155) respectively for finding the states of electrons and holes, replacing E_{ci} and E_{vi} by the values given in Eq. (4.173-4.175). As an illustration, the calculated hole dispersion relation for a compressively strained quantum well system is shown in Fig. 4.11. It should be noted that no significant change is expected in the electron dispersion relations. The hole dispersion relations are not also qualitatively altered, but the light-hole band is pushed further downwards with respect to the heavy-hole bands and the exact shape of the dispersion curves is slightly changed.

The conduction band minimum has been considered in the above discussion to

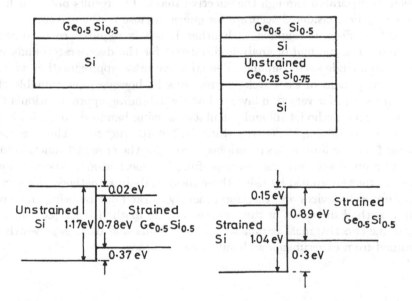

Figure 4.12. Schematic diagram of energy extrema in a $Si/Si_x Ge_{1-x}$ system. (a) Strained $Si_{0.5}Ge_{0.5}$ on unstrained silicon. (b) Strained $Si_{0.5}Ge_{0.5}$ on strained silicon.

lie at the Γ point, which is mostly true for the III-V compounds used in the quantum well devices. Intensive studies are also being done on Si/SiGe systems[4.64-68], in which silicon or $Si_x Ge_{1-x}$ may work as the strained layer depending on the growth sequence. Conduction band minima in such layers are in the Δ directions and are six-fold degenerate. Strain changes the degeneracy in such layers and produces a doublet and a set of four-fold degenerate conduction band minima. Energy levels in Si/Ge systems are worked out following the same procedure as outlined above, by using the envelope function method, but the ordering of the minima have to be properly determined by using the shear deformation potentials. The splitting for the $Ge_x Si_{1-x}$ alloys on Si (001) substrate is illustrated in Fig. 4.12. These data together with the changes in the band gap and the valence band splittings provide the input for the calculation of the electron states in Si/SiGe quantum well systems.

4.5. Conclusion

Energy levels and dispersion relations have been discussed for heterostructures by using the envelope-functions, which are based on the k.p perturbation theory and the effective- mass approximation. Electron states have been obtained by considering only the lowest conduction-band minimum and the hole states by considering only the four-fold degenerate valence bands. The effect of coupling to other bands has been incorporated through the effective mass. The results obtained from the analysis may be considered accurate for energies close to the extrema. For larger energies, the effect of coupling with other bands is required to be directly incorporated. The method of analysis discussed for the degenerate bands may be extended to include such coupling. The effective mass approximation which is an essential component of the envelope functions is, however, questionable close to the interfaces and for very thin layers. The tight-binding approximation[4.69], the pseudopotential method[4.70] and calculations using band-orbitals[4.71] have also been used for obtaining the energy states in heterostructures. These methods do not suffer from the limitations mentioned above for the envelope-function method. The results obtained from the envelope-function method are, however, found to be close to those obtained by using these more elaborate methods. Energy states in quantum well devices are, therefore, mostly worked out by using the envelope-function method discussed in this chapter as the method is simple and requires as input the effective mass parameters, band offsets and band gaps which can be determined from experiments with sufficient accuracy.

CHAPTER 5

OPTICAL INTERACTION PHENOMENA

Quantum well structures have been used to realize various opto-electronic devices with much improved characteristics. The confinement of the electrons and holes and the changed density of states (DOS) have caused the improvement. The charge carriers are concentrated near the band edge energies due to the staircase DOS and the opto-electronic interaction is thereby much enhanced. This has led to a lower threshold current and a higher characteristic temperature in lasers, higher nonlinear coefficients for the realization of optical bistable devices (OBD)[5.1] and degenerate four-wave mixers (DFWM)[5.2]. The quantization of the energy levels have also made it possible to realize a few new opto-electronic devices, e.g., Stark effect modulator[5.3] and intersubband absorption detectors[5.4-6]. The basis of all these devices are the enhanced values of the opto- elctronic interaction parameters of the quantum well structures. The optical interaction phenomena in quantum wells and the underlying theory are discussed in this chapter.

5.1. Optical Interaction in Bulk Materials

The basic theory of the interaction of electrons with the radiation field is first presented by considering bulk materials. The theory is then extended to quantum systems.

In the presence of a radiation field the wave equation[5.7] for the electrons in a solid is as follows

$$(1/2m_0)(-i\hbar\nabla + |e|\mathbf{A}_{op})^2\psi + V_c(\mathbf{r})\psi = E\psi, \tag{5.1}$$

where m_0 is the free-electron mass, $V_c(\mathbf{r})$ is the crystal potential, E is the electron energy, $-i\hbar\nabla$ is the momentum operator and $|e|\mathbf{A}_{op}$ is the additional term arising from the vector potential $\mathbf{A}$ associated with the radiation. It may be noted that all the variables are expressed in SI units.

The vector potential $\mathbf{A}$ is related to the electric field intensity of the radiation by the relation,

$$\mathcal{E} = -\partial\mathbf{A}/\partial t, \tag{5.2}$$

It may be expressed as

$$\mathbf{A} = (1/2)\hat{\mathbf{a}}A_0 \sum_{+,-} \exp[\pm i(\kappa.\mathbf{r} - \omega t)], \tag{5.3}$$

where $\hat{a}$ is a unit vector giving the polarization of the radiation field. A_0 is the magnitude of the vector potential, κ and ω are respectively the wave vector and the frequency of the radiation. Also,

$$\kappa = n_r \omega/c, \tag{5.4}$$

where c is the velocity of light in vacuum and n_r is the refractive index of the material.

A_0 is expressed in terms of the photon energy and the crystal volume, V_c , by equating the energy densities corresponding to one photon in the sample and the density given by the field description , as follows

$$A_0 = (2\hbar/V_c\epsilon_0 n_r^2\omega)^{1/2} \tag{5.5}$$

where ϵ_0 and ϵ are respectively the permittivity of free space and the material, and $n_r^2 = \epsilon/\epsilon_0$.

The operator, $\mathbf{A}_{op}$ is written as

$$\mathbf{A}_{op} = (1/2)\hat{a}(2\hbar/V_c\epsilon_0 n_r^2\omega)^{1/2}[a_\kappa \exp(i\kappa.\mathbf{r}) + a_\kappa^\dagger \exp(-i\kappa.\mathbf{r})], \tag{5.6}$$

where a_κ and $a_\kappa^\dagger$ are respectively the annihilation and the creation operators.

Equation (5.1) is simplified by assuming that the intensity of the radiation is small, so that terms upto the first order in $\mathbf{A}_{op}$ need be retained.

The simplified equation is

$$-(\hbar^2/2m_0)\nabla^2\psi - (i\hbar/m_0)|e|(\mathbf{A}_{op}.\nabla)\psi + V_c(\mathbf{r})\psi = E\psi. \tag{5.7}$$

The effect of the additional term arising from the presence of the radiation is treated as a perturbation. The transition rate from a state $(E_\mathbf{k}, \mathbf{k})$ to $(E_{\mathbf{k}'}, \mathbf{k}')$ is given according to the first order perturbation theory by[5.7,8]

$$T(\mathbf{k}, \mathbf{k}') = (2\pi/\hbar)|M(\mathbf{k}, \mathbf{k}')|^2\delta(E_{\mathbf{k}'} - E_\mathbf{k} \pm \hbar\omega), \tag{5.8}$$

where $\hbar\omega$ is the energy of a photon and the plus and minus signs respectively indicate the emission and the absorption of a photon.

The matrix element $M(\mathbf{k}, \mathbf{k}')$ for the transition is obtained by considering that these are associated with simultaneous changes in the photon population. It is required that the combined wave function for the electron and the photon system be considered. The matrix element for the combined wave function is given by,

$$M(\mathbf{k}, \mathbf{k}') = -(i\hbar/m_0)|e|\sum_\kappa \int \psi_{\mathbf{k}'}^* U_f^* (\mathbf{A}_{op}.\nabla)\psi_\mathbf{k} U_i \, d^3r d\alpha, \tag{5.9}$$

where $\psi_{\mathbf{k}'}U_f$ and $\psi_\mathbf{k}U_i$ represent the wave function for the final and the initial electron-photon system respectively, U_f, U_i being the components representing the photon system. The coordinates of the electron and the photon wave functions are $\mathbf{r}$ and α respectively. The integral is transformed by considering that the annihilation and the creation operators associated with $\mathbf{A}_{op}$ act on U_i , while ∇ operates on $\psi_{\mathbf{k}}$.

The matrix element is expressed in the braket notation as

$$M(\mathbf{k}, \mathbf{k}') = (|e|A_0/2m_0) < \psi_{\mathbf{k}'} | \hat{\mathbf{a}} \exp [\mp i(\kappa.\mathbf{r})].\mathbf{p}) | \psi_{\mathbf{k}} > (n_\kappa + 1/2 \pm 1/2)^{1/2}, \quad (5.10)$$

where $\mathbf{p}$ is the momentum operator $(\hbar/i)\nabla$, n_κ is the occupation number of states with the photon wave vector κ.

The matrix element is expressed also as,

$$M(\mathbf{k}, \mathbf{k}') = i(|e|\omega A_0/2) < \psi_{\kappa'} | \hat{\mathbf{a}} \exp [(\mp i\kappa.\mathbf{r})].\mathbf{r} | \psi_\kappa > (n_\kappa + 1/2 \pm 1/2)^{1/2}, \quad (5.11)$$

by using the identity

$$(-i\hbar/m_0)\mathbf{p} = [H, \mathbf{r}], \quad (5.12)$$

or the equivalence

$$\mathbf{p} = m_0\dot{\mathbf{r}} = i\omega m_0\mathbf{r}, \quad (5.13)$$

where $\mathbf{p}$ is the momentum operator, H is the Hamiltonian operator in the Schrödinger equation and $\mathbf{r}$ is the position vector operator.

Expression (5.11) is simplified for photon-atom interactions, by putting[5.9] $\exp(i\kappa.\mathbf{r}) \approx 1$. It is often referred as the dipole approximation, as $|e|\mathbf{r}$ gives the dipole moment of the atom. The expression helps in envisaging the polarization and direction dependence of absorption and emission from the knowledge of the characteristics of dipole radiation.

On using the electron wave function $N^{-1/2}U_{\mathbf{k}}(\mathbf{r}) \exp(i\mathbf{k}.\mathbf{r})$ the matrix element is found to be,

$$M(\mathbf{k}, \mathbf{k}') = M_0(\mathbf{k}, \mathbf{k}')(n_\kappa + 1/2 \pm 1/2)^{1/2}, \text{where,} \quad (5.14)$$

$$M_0(\mathbf{k}, \mathbf{k}') = (-i|e|\hbar A_0/2m_0)N^{-1} \int U_{\mathbf{k}'}^* \hat{\mathbf{a}} \exp [-i(\mathbf{k}' \mp \kappa).\mathbf{r}].\nabla U_{\mathbf{k}} \exp(i\mathbf{k}.\mathbf{r}) \, d^3r.$$
$$(5.15)$$

The integral in the matrix element may be written also as

$$I = -iN^{-1} \int U_{\mathbf{k}'}^* \exp [-i(\mathbf{k}' - \mathbf{k} \pm \kappa).\mathbf{r}]\hat{\mathbf{a}}.(i\mathbf{k} + \nabla)U_{\mathbf{k}} \, d^3r. \quad (5.16)$$

The integral has a nonzero value only if

$$\mathbf{k}' = \mathbf{k} \mp \kappa, \quad (5.17)$$

since $\exp [-i(\mathbf{k}' - \mathbf{k} \pm \kappa).\mathbf{r}]$ is otherwise a periodic function. The condition implies the conservation of pseudomomentum. On the other hand, the presence of the delta function in the expression for the transition probability implies conservation of energy, since it has a nonzero value only for

$$E_{\mathbf{k}'} = E_{\mathbf{k}} \pm \hbar\omega. \tag{5.18}$$

The two conditions cannot be simultaneously satisfied in bulk material if the electron states $(E_{\mathbf{k}'}, \mathbf{k}')$ and $(E_{\mathbf{k}}, \mathbf{k})$ correspond to the same extremum in the same band. If, however, the states belong to two separate bands with extrema located at the same point in the Brillouin zone, then the energy conservation condition modifies to

$$E_{\mathbf{k}'n} = E_{\mathbf{k}m} \mp E_{gmn} \pm \hbar\omega, \tag{5.19}$$

where $E_{\mathbf{k}'n}$ indicates that the $(E_{\mathbf{k}'}, \mathbf{k}')$ state is in the nth band, $E_{\mathbf{k}m}$ indicates that the $(E_{\mathbf{k}}, \mathbf{k})$ state is in the mth band and E_{gmn} is the energy gap between the bands at $\mathbf{k} = \mathbf{k}' = \mathbf{0}$. Evidently, this condition may be satisfied along with the condition of conservation of momentum if $\hbar\omega = E_{gmn}$.

Electron transitions may be caused directly by the radiation in bulk materials with the valence band maximum and the conduction band minimum located at the same $\mathbf{k}$ in the Brillouin zone. In many compound semiconductors, this condition applies, as the extrema are located at the Γ point. Direct transitions are, therefore, very strong in these materials. The condition applies in quantum wells also between subbands in the same band. Hence, direct intersubband transitions within a band may also be strong in quantum wells, in addition to interband transitions.

It should be mentioned that transitions may also occur between two extrema located at different points in the Brillouin zone but such transitions require the participation of a third particle, e.g., a phonon[5.8] for the conservation of pseudo-momentum.

The integral I has two components, of which the first component,

$$I_1 = \hbar\hat{\mathbf{a}}.\mathbf{k} \int_\Omega U_{\mathbf{c}\mathbf{k}'}^* U_{v\mathbf{k}} d^3 r = \hat{\mathbf{a}}.\hbar\mathbf{k} f', \tag{5.20}$$

would be small, as the cell-periodic functions $U_{\mathbf{c}\mathbf{k}'}^*$ and $U_{v\mathbf{k}}$, corresponding respectively to the conduction band and the valence band are orthogonal for $\mathbf{k}' = \mathbf{k}$. However, as $\mathbf{k}'$ is not exactly equal to $\mathbf{k}$, f' has a small finite value and accounts for part of the transitions.

The second component of the integral,

$$I_2 = \hat{\mathbf{a}}.\mathbf{p}_{cv}, \text{where,} \tag{5.21}$$

$$\mathbf{p}_{cv} = -i\hbar \int_\Omega U_{c0}^* \nabla U_{v0} d^3 r, \tag{5.22}$$

is the momentum matrix element between the states at the extrema. An exact expression for this matrix element is not available, but its value may be estimated by using the expression for the effective mass near the extrema, as obtained from the k.p perturbation analysis given by Eq. (4.15) . The expression may be written in terms of the f-sum as,

$$1/m_{ni}^* = (1/m_0)(1 + \sum_{n'} f_{nn'}), \quad f_{nn'} = (2/m_0)(|\hat{\mathbf{i}}.\mathbf{p}_{nn'}|)^2/(E_n - E_{n'}) \qquad (5.23)$$

where m_{ni}^* is the effective mass at the extremum along the ith direction of the principal axis of the constant energy ellipsoid, E_n and $E_{n'}$ are the energies at the extremum, which are coupled through the k.p perturbation. The summation is required to be extended over all values of n', which are separated by small energies from the band under consideration. In the special case, where transition occurs between two bands, which are very much close to each other than all other bands, the expression may be simplified to

$$1/m_{ni}^* = (1/m_0) + (2/m_0^2)(|\hat{\mathbf{i}}.\mathbf{p}_{cv}|^2)/(E_c - E_v) \qquad (5.24)$$

This expression is often used to estimate the value of $\mathbf{p}_{cv}$ from the knowledge of the conduction-band effective mass m_{ci}^*.

Alternately, $\mathbf{p}_{cv}$ may be estimated by assuming that the conduction-band wave function is like s-type atomic orbitals and the valence-band wave functions are like p-type atomic orbitals, as discussed in Section 4.1.3. The wave functions may be written[5.9-12] in general, as,

$$\psi_{c1} = iS'\alpha'_1, \quad \psi_{c2} = iS'\alpha'_2, \qquad (5.25)$$

$$\psi_{vH1} = (1/\sqrt{2})(X' + iY')\alpha'_1, \psi_{vH2} = (1\sqrt{2})(X' - iY')\alpha'_2 \qquad (5.26)$$

$$\psi_{vL1} = -(1/\sqrt{6})(X' - iY')\alpha'_1 - \sqrt{2/3}Z'\alpha'_2, \qquad (5.27)$$

$$\psi_{vL2} = (1/\sqrt{6})(X' + iY')\alpha'_2 - \sqrt{2/3}Z'\alpha'_1, \qquad (5.28)$$

$$\psi_{vS1} = -(1/\sqrt{3})(X' - iY')\alpha'_1 + (1/\sqrt{3})Z'\alpha'_2, \qquad (5.29)$$

$$\psi_{vS2} = (1/\sqrt{3})(X' + iY')\alpha'_2 + (1/\sqrt{3})Z'\alpha'_1, \qquad (5.30)$$

where S', X', Y', Z' indicate respectively the functions corresponding to the s-type and the p-type atomic orbitals, α'_1 and α'_2 are the spinors. The subscripts indicate the functions for the conduction band ψ_{c1}, ψ_{c2} , heavy hole band, ψ_{vH1}, ψ_{vH2} light hole band, ψ_{vL1}, ψ_{vL2} and the split-off bands ψ_{vS1}, ψ_{vS2} for the different spins. The valence band wave functions are found to be polarized in the direction of the wave vector k and the primed- functions indicate the values transformed from the crystallographic directions of symmetry.

Let z' be chosen in the direction of the wave vector and x', y' two other suitably chosen orthogonal directions, x' being in the plane of z' and the crystallographic z

direction. The transformation relations between the coordinates may be obtained from Fig. 5.1 as given below.

$$\begin{bmatrix} X' \\ Y' \\ Z' \end{bmatrix} = \begin{bmatrix} \cos\theta\cos\phi & \cos\theta\sin\phi & -\sin\theta \\ -\sin\phi & \cos\phi & 0 \\ -\sin\theta\cos\phi & \sin\theta\sin\phi & \cos\theta \end{bmatrix} \cdot \begin{bmatrix} X \\ Y \\ Z \end{bmatrix} \tag{5.31}$$

The corresponding transformation relation for the spinors is [5.13]

$$\begin{bmatrix} \alpha'_1 \\ \alpha'_2 \end{bmatrix} = \begin{bmatrix} \exp(-i\phi/2)\cos(\theta/2) & \exp(i\phi/2)\sin(\theta/2) \\ -\exp(-i\phi/2)\sin(\theta/2) & \exp(i\phi/2)\cos(\theta/2) \end{bmatrix} \cdot \begin{bmatrix} \alpha_1 \\ \alpha_2 \end{bmatrix} \tag{5.32}$$

The wave function for the heavy hole valence band may be written as

$$\psi_{vH} = \alpha_x X + \alpha_y Y + \alpha_z Z, \text{where,} \tag{5.33}$$

$$\alpha_x = (\cos\theta\cos\phi - i\sin\phi)/\sqrt{2}, \tag{5.34}$$

$$\alpha_y = (\cos\theta\sin\phi + i\cos\phi)/\sqrt{2}, \tag{5.35}$$

$$\alpha_z = -\sin\theta/\sqrt{2}. \tag{5.36}$$

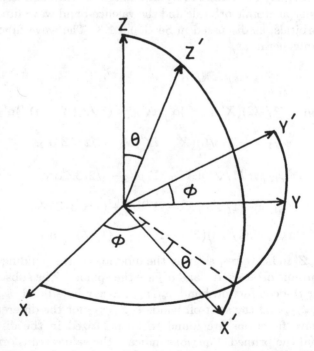

Figure 5.1. Unprimed and primed coordinates used in the transformation. The unprimed coordinates are in the crystallographic directions, while the primed coordinates are chosen with the z' direction of the wave vector.

The matrix element $\mathbf{p}_{cvH}$ is, therefore, given by

$$\mathbf{p}_{cvH} = \hat{\mathbf{i}} < S|p_x|X > \alpha_x + \hat{\mathbf{j}} < S|p_y|Y > \alpha_y + \hat{\mathbf{k}} < S|p_z|Z > \alpha_z, \qquad (5.37)$$

which may be simplified to

$$\mathbf{p}_{cvH} = (m_0/\hbar)P\alpha \qquad (5.38)$$

where

$$< S|p_x|X >=< S|p_y|Y >=< S|p_z|Z >= (m_0/\hbar)P, \qquad (5.39)$$

$$\alpha = \hat{\mathbf{i}}\alpha_x + \hat{\mathbf{j}}\alpha_y + \hat{\mathbf{k}}\alpha_z. \qquad (5.40)$$

It should be noted that in obtaining the above expression it has been assumed that the spin is conserved by using the relations.

$$\alpha_1 \cdot \alpha_1 = \alpha_2 \cdot \alpha_2 = 1; \alpha_1 \cdot \alpha_2 = 0. \qquad (5.41)$$

Similarly, the momentum matrix element between the electrons and the light holes is

$$< \psi_c|\mathbf{p}|\psi_{vL} >= (m_0/\hbar)(P/\sqrt{3})[\hat{\mathbf{i}}(\alpha_x^2 + 2\beta_x^2)^{1/2} + \hat{\mathbf{j}}(\alpha_y^2 + 2\beta_y^2)^{1/2} + \hat{\mathbf{k}}(\alpha_z^2 + 2\beta_z^2)^{1/2}], \qquad (5.42)$$

where

$$\beta_x = \sin\theta \cos\phi, \beta_y = \sin\theta \sin\phi, \beta_z = \cos\theta. \qquad (5.43)$$

The detailed expressions for the matrix element in terms of the components is not important for bulk materials, as the wave vector may have any direction and the matrix element is averaged out. In quantum wells, on the other hand, the direction of the wave vector is not random, but limited to be on the surface of a cone[5.14] with vertical angle $\theta/2$ around z (if it is taken as the direction of quantization) as the direction may be random only for the in-plane component of the wave vector, while the magnitude of the z component is fixed. The detailed expressions introduce an anisotropy factor and will be used to evaluate it when optical transition is discussed for quantum wells.

The magnitude of P may also be obtained in terms of the electron effective mass, the energy band gap and the spin orbit splitting by using the energy dispersion relation obtained from the **k.p** perturbation theory, discussed in Section 4.1.4. In the simplified model, in which only the conduction band, the light hole band and the split-off band are combined, the dispersion relation including the free electron term is found to be[5.11]

$$(E' - E_c)(E' - E_v)(E' - E_v + \Delta) - P^2k^2(E' - E_v + 2\Delta/3) = 0, \qquad (5.44)$$

where $E' = E - \hbar^2k^2/2m_0$, E being the energy eigenvalue, E_c, E_v and Δ are respectively the conduction-band edge energy, valence-band edge energy and the spin-orbit splitting.

The matrix element P may be expressed in terms of the conduction-band edge effective mass m_c^* by noting that for small values of k,

$$E - E_c = \hbar^2 k^2 / 2m_c^*. \tag{5.45}$$

Also $(E' - E_c) \ll E_g$, where E_g is the energy band gap. Using the above expression, P^2 is found to be given by

$$P^2 = \hbar^2(m_0/m_c^* - 1)(E_g/2m_0)(E_g + \Delta)(E_g + 2\Delta/3)^{-1}. \tag{5.46}$$

P may hence be obtained by using the experimental values of E_g, m_c^* and Δ.

In the above formulation, the wave function for the conduction band and the light hole band are taken as,

$$\psi_{\alpha 1} = a_\alpha iS\alpha_1 + (1/\sqrt{2})b_\alpha(X + iY)\alpha/2 + c_\alpha Z\alpha. \tag{5.47}$$

$$\psi_{\alpha 2} = a_\alpha iS\alpha_2 + (1/\sqrt{2})b_\alpha(X - iY) + c_\alpha Z\alpha_2. \tag{5.48}$$

The averaged matrix element, therefore, involves for the bulk material a multiplying constant, γ, given by[5.12]

$$\gamma^2 = (a_c c_j + c_c a_j)^2 + (a_c b_j - b_c a_j)^2, \tag{5.49}$$

where a_c, b_c, c_c are the constants for the conduction band while a_j, b_j and c_j , are the corresponding constants for the valence band. There is some confusion in the literature about the expression for $|\mathbf{p}_{cv}|^2$ in terms of E_g, Δ and m_c^* . An additional factor of $1/2$ is introduced in Eq. (5.9) presumably to account for the fact that spin is conserved in transitions and hence only half the transitions are allowed[5.15]. However, a more clear approach is to use Eq. (5.9), and half the density of states for the final states of the electron. Equation (5.23) has also been misinterpreted[5.16] in some cases by assuming $\mathbf{p}_{nn'}$ to be the same for the heavy hole, the light hole and the split off bands and thereby obtaining the relation,

$$1/m_c* = 1/m_0 + (2P^2/\hbar^2)[1/E_g + 1/E_g + 1/(E_g+\Delta)] \approx 1/m_0 + (6P^2/\hbar^2 E_g). \tag{5.50}$$

But, it may be shown by using expressions (5.25) through (5.30) that $|\mathbf{P}_{cvHH}| = (m_0/\hbar)P/\sqrt{2}, |\mathbf{p}_{cvLH}| = (m_0/\hbar)P/\sqrt{6}, |\mathbf{p}_{cvS}| = (m_0/\hbar)P/\sqrt{3}$ where the subscripts vHH, vLH, vS indicate the heavy hole, the light hole and the split-off hole. On substitution of these values for $|\hat{\mathbf{i}}.\mathbf{p}_{nn'}|$ it is found that

$$1/m_c^* = 1/m_0 + (2P^2/\hbar^2)[(1/2E_g) + (1/6E_G) + 1/3(E_g + \Delta)] \approx 1/m_0 + 2P^2/\hbar^2 E_g. \tag{5.51}$$

This relation is the same as obtained from Eq. (5.46), whereas Eq. (5.50) gives a different value for P.

The matrix elements for transitions from the heavy hole, the light hole and the split-off hole are also different. The values would depend also on the direction relative to the crystallographic directions.

Often an average value[5.15] is used for the matrix element, $\mathbf{p}_{cv}$, given by,

$$|\mathbf{p}_{cv}|^2 = (m_0^2/6m_c)E_g, \quad 1/m_c = 1/m_c^* - 1/m_0, \quad (5.52)$$

but, for quantum wells it is more appropriate to use the matrix elements for the individual bands for calculating the absorption since the energy levels corresponding to the particular heavy and light hole are separated due to quantization. The relevant matrix elements for the heavy hole, the light hole and the split-off bands are given respectively by $(m_0^2/\hbar^2)P^2/2$, $(m_0^2/\hbar^2)P^2/6$ and $(m_0^2/\hbar^2)P^2/3$, for quantum wells grown along a crystallographic direction and light propagating along the same direction.

It should be noted that the value of P obtained from Eq. (5.46) by using the experimental value of m_c^* should not be considered exact. The experimental value of m_c^* is determined by interaction with all the bands, whereas Eq. (5.46) is derived by considering the interaction between only the conduction band and the three valence bands. It is in fact found that the value of P obtained from the electron spin resonance experiment[5.17], which may be considered to be its correct value differs significantly from the value given by Eq. (5.46). For example, the direct experimental value of $2|\mathbf{p}_{cv}|^2/m_0$ for GaAs is 28.8 eV, whereas Eq. (5.46) gives a value of 22.5 eV. The discrepancy may be understood by considering that the contribution of the higher- lying bands is negative [see Eq. (4.15)] and hence when these contributions are neglected a lower value of P is required to get the experimental value of m_c^*. To correct this discrepancy, the matrix element is often generalized as,

$$|\mathbf{p}_{cv}|^2 = \xi(m_0/m_c^*)(m_0 E_g), \quad (5.53)$$

where ξ may be treated as an empirical constant for a particular transition for a particular material.

The matrix element may now be expressed as follows to evaluate the transition probability $T(\mathbf{k}, \mathbf{k}')$.

$$M(\mathbf{k}, \mathbf{k}') = (|e|A_0/2m_0)\hat{\mathbf{a}}.(\hbar k_v f' + \mathbf{p}_{cv})(n_\kappa + 1/2 \pm 1/2)^{1/2}\delta_{\mathbf{k}_c \mathbf{k}_v}. \quad (5.54)$$

The delta function $\delta_{\mathbf{k}_c \mathbf{k}_v}$ is included to indicate that $\mathbf{k}_c = \mathbf{k}_v$, since in relation (17), κ, the photon wave vector, may be neglected in comparison to $\mathbf{k}_c$ and $\mathbf{k}_v$.

It may be mentioned in this connection that the dimensionless quantity, f, is defined as the oscillator strength. It is given by,

$$f = (2/m_0\hbar\omega)|\hat{\mathbf{a}}.\mathbf{p}_{cv}|^2 = 2(|\hat{\mathbf{a}}.\mathbf{r}|^2)(m_0\omega/\hbar), \quad (5.55)$$

as it gives the strength of the equivalent oscillating dipole. The free electron mass m_0 is replaced by m_c^* in the above definition in some publications. Then m_0 in

Eq. (5.55) is also to be replaced by m_c^*. The oscillator strength, however, turns out to be very different for the two definitions as (m_0/m_c^*) is a factor larger than 10.

5.2. Interaction in Quantum Wells

The formulæ of Section 5.1 remain valid for quantum wells, but the expression for the electron wave function is modified due to quantization. As a result of this modification, transitions are allowed in quantum wells between subbands in the same band in addition to the interband transitions. Expressions for the transition probabilities are also altered. Modifications in the interband transitions are first discussed.

5.2.1. INTERSUBBAND TRANSITIONS

Electron wave functions may be taken for the quantized states as (see Section 4.1),

$$\psi_{kvi} = F_{vi}(z)U_{v0}(\mathbf{r}) \exp(i\mathbf{k}_{ti}.\rho), \tag{5.56}$$

$$\psi_{k'cf} = F_{cf}(z)U_{c0}(\mathbf{r}) \exp(i\mathbf{k}'_{tf}.\rho), \tag{5.57}$$

where $\psi_{kvi}, \psi_{k'cf}$ represent respectively the normalized wave functions for the ith subband in the valence band of wave vector $\mathbf{k}$ and the fth subband in the conduction band of wave vector $\mathbf{k}'$, $F_{vi}(z), F_{cf}(z)$ are the corresponding envelope functions, U_{v0}, U_{c0} are the corresponding cell periodic functions at the band edges, $\mathbf{k}_{ti}, \mathbf{k}'_{tf}$ are the in-plane components of the wave vectors and $\mathbf{r}$ is the in-plane position vector.

The matrix element $\mathbf{p}_{cv}$ has now the form,

$$\mathbf{p}_{cv} = -i\hbar \int F_{cf}^*(z)U_{c0}(\mathbf{r}) \exp(-i\mathbf{k}'_{tf}.\rho)\nabla[F_{vi}(z)U_{v0} \exp(i\mathbf{k}_{ti}.\rho)]d^3r. \tag{5.58}$$

The integrals may be simplified by considering that the envelope functions $F_{vi}(z)$ and $F_{cf}(z)$ vary slowly in comparison to the cell periodic function for not too narrow wells. The analysis presented in this treatise, being based on the effective mass approximation, is not likely to be applicable to very narrow wells, for which more elaborate basic formulations would be required. Assuming that the approximation is valid, the integral may be simplified to

$$I = -i\hbar[< F_{cf}(z)|F_{vi}(z) >< U_{c0}|\nabla|U_{v0} > + < U_{c0}|U_{v0} >< F_c(z)|\nabla|F_v(z) >], \tag{5.59}$$

where

$$\mathbf{k}'_{tf} = \mathbf{k}_{ti}. \tag{5.60}$$

The second component of the integral is zero as the functions U_{c0} and U_{v0} are orthogonal. Also, the first component has a non-zero value only if the envelope functions are non- orthogonal. This condition implies that the transitions are allowed only for $f = i$, since $F_{cf}(z)$ and $F_{vi}(z)$ have an overlap only for this condition. However, this restrictive selection rule is not strictly valid as has been found from more elaborate analysis. It is invalidated when the states arise from the mixing of the different basic states, e.g., the second and the third quantized valence band states, which are admixtures of the heavy-hole and the light-hole bands. Also, the one-electron picture from which the selection rule, $f = i$, is derived, is altered by the electron-electron interaction for large electron concentrations as in modulation- doped structures or lasers. These features add fine structures in the absorption spectrum, but are relatively unimportant in the design of the devices, which are based on the main allowed transitions. This aspect is not, therefore, discussed any further. Some results are, however, presented when discussing the experiment in the light of this theory.

The integral for the allowed transitions may also be simplified by considering that $F_{cf}(z)$ and $F_{vi}(z)$ are very similar functions. For wells with infinite barriers, the functions are identical. In that case, the integral I, has the same value as for bulk materials. For wells with finite barriers or for narrow wells, the functions are not identical and the integral differs from the bulk value by a numerical factor. But, this factor is close to unity and interband transitions in quantum wells may be considered to be the same as in bulk material. However, there appears an anisotropy factor, when averaging is done over different directions of k_t , which is different from $(1/3)$ and depends on the direction of the electric field associated with the radiation field.

5.2.2. INTERSUBBAND ABSORPTION

The momentum matrix element for the intersubband transitions is obtained from the formula,

$$\mathbf{p}_{if} = -i\hbar < \psi_{f\mathbf{k}'}|\nabla|\psi_{i\mathbf{k}} >, \tag{5.61}$$

where $\psi_{i\mathbf{k}}$ and $\psi_{f\mathbf{k}'}$ represent respectively the wave functions for the initial and the final states. For electrons in the conduction band, these may be written as,

$$\psi_{i\mathbf{k}} = U_{c\mathbf{k}}(\mathbf{r})F(\mathbf{r}), \psi_{f\mathbf{k}'} = U_{c\mathbf{k}'}(\mathbf{r})F_f(\mathbf{r}) \tag{5.62}$$

It should be noted that as the two subbands belong to the same band the cell-periodic parts of the wave-functions are identical. $F_i(\mathbf{r})$ and $F_f(\mathbf{r})$ denote the envelope functions given by Eq. (4.17). The functions are assumed to be normalized.

On substituting the above expressions in Eq. (5.61), the momentum matrix element is found to be given by

$$\mathbf{p}_{if} = -i\hbar < U_{c\mathbf{k}'}F_f|\nabla|U_{c\mathbf{k}}F_i > \tag{5.63}$$

It may be recalled that the wave function has been approximated so far as

$$\psi_p(\mathbf{r}) = U_{c0}F(\mathbf{r}) \tag{5.64}$$

by considering that the contribution of the second term of $\psi_p(\mathbf{r})$ in the expression (4.20) is negligible. In the case of intersubband scattering, however, this term contributes more than the first term and is required to be retained. On using the full expression,

$$\psi_p(\mathbf{r}) = U_{c0}F(\mathbf{r}) + (\hbar/m_0)\sum_n -i\nabla F(\mathbf{r}).\mathbf{p}_{cn}(E_{c0} - E_n)^{-1}U_{n0}, \tag{5.65}$$

we obtain

$$\mathbf{p}_{if} = < F_f(\mathbf{r})|\mathbf{p}|F_i(\mathbf{r}) > (m_0/m_c^*) \tag{5.66}$$

since

$$1 + (2/m_0)\sum_{n,m\neq n} p_{cn}^2/(E_c - E_n) = m_0/m_c^*, \tag{5.67}$$

m_c^* being the band-edge effective mass.

The matrix element for intersubband scattering is hence given by

$$M(i,f) = (i\hbar|e|A_0/2m_c^*)(n_\kappa + 1/2 \pm 1/2)\hat{a}. < F_f(\mathbf{r})|\mathbf{p}|F_i(\mathbf{r}) > . \tag{5.68}$$

For a well, grown along the z direction, the envelope functions may be written as,

$$F_i(\mathbf{r}) = F_i(z)\exp(i\mathbf{k}_{ti}.\rho), F_f(\mathbf{r}) = F_f(z)\exp(i\mathbf{k}_{tf}.\rho), \tag{5.69}$$

where $\mathbf{k}_{ti}$ and $\mathbf{k}_{tf}$ are the in-plane wave vectors of the initial and the final state belonging respectively to the ith and the fth subband.

$F_i(z)$ and $F_f(z)$ are the z components of the corresponding envelope functions, evaluation of which has been explained in Chapter 4. Non-zero value of the matrix element is obtained only if $\mathbf{k}_{tf} = \mathbf{k}_{ti}$, the photon wave vector being considered to be negligible. Also, F_i and F_f are required to have opposite parity to yield a non-zero value for the matrix element. Thus, intersubband transitions may occur only between neighboring subbands of opposite symmetry obeying the condition of conservation of pseudomomentum. Also, the matrix element has only a z component, the direction in which the quantum well is grown. Consequently, intersubband absorption may occur only if the electric field associated with the radiation has a component perpendicular to the interfaces.

Analytic expressions may be given for the matrix elements for wells with infinite barrier height by using the functions $\sqrt{(2/L)}\cos(i\pi/L)z$ and $\sqrt{(2/L)}\sin(f\pi/L)z$ respectively for $F_i(z)$ and $F_f(z)$ for a well of width L. The expression is

$$M(i,f) = [|e|A_0\hbar^2 a_z 4if/(f^2 - i^2)Lm_c^*](n_p + 1/2 \pm 1/2)^{1/2}, \tag{5.70}$$

where a_z is the z component of the polarization vector $\hat{a}$. It should be noted that the values of i and f are such that $(i + f)$ is odd.

The oscillator strength for the intersubband absorption is defined in the effective-mass approximation as,

$$f = (2/m_c^* \omega \hbar) < F_f(z)|\mathbf{p}|F_i(z) >^2 . \qquad (5.71)$$

It may be expressed for the wells of infinite barrier height as,

$$f = (64/\pi^2)i^2 f^2/(f^2 - i^2)^3. \qquad (5.72)$$

Expressions have been obtained so far by assuming that the $E - k$ relation is isotropic and it has been concluded that intersubband interaction may happen only if the radiation field has an electric field component along the z direction. In materials with anisotropic mass, however, the matrix element has the form[5.18]

$$M(\mathbf{k}_{tm}, \mathbf{k}'_{tn}) = (|e|A_0/2)(n_\kappa + 1/2 \pm 1/2)^{1/2}\hat{a}.\mathbf{M}^{-1}. < F_n^*(z)|\mathbf{p}|F_m(z) >, \qquad (5.73)$$

where $\mathbf{M}^{-1}$ is the inverse-effective mass tensor, obtained from the energy wave-vector relation, assumed to be given by

$$E - E_c = (\hbar^2/2)\mathbf{k}.\mathbf{M}^{-1} \cdot \mathbf{k}. \qquad (5.74)$$

It should be mentioned also that the intersubband interaction involves higher levels of energy. The energy-band nonparabolicity has often significant effect for interactions involving such energy levels. Effects of nonparabolicity may be incorporated in the same formalism by using the nonparabolic dispersion relation of Kane[5.11] or Rössler[5.19].

Expressions for the matrix element have been derived above by using the wave equation including the radiation field in the Schrödinger equation. The expression may also be derived by using the equivalence given by Eq. (5.13). The matrix element may be written by using the equivalence as

$$M(i, f) = i(|e|A_0/2\hbar)(E_f - E_i)\hat{a} < \psi_f|\mathbf{r}|\psi_i > (n_p + 1/2 \pm 1/2)^{1/2}. \qquad (5.75)$$

This expression is applicable, however, only when the wave function becomes zero for large values of z. It should not, therefore be applied to unbounded superlattice systems, as large errors may arise[5.20].

5.3 Excitons

Photon-electron interaction has been discussed in Section 5.1 and 5.2 for transitions of the electron from a quasi-free state (e.g., the valence band or the ground state in the conduction band) to another quasi-free state (e.g., states in the conduction band). The electron and the hole created by such transitions are assumed to be completely decoupled and described by the wave functions for the one-electron approximation. However, the electron-hole pair created by the interaction may be bound to each other by the Coulomb force and move about freely as a bound-pair quasi-particle like an electron or a hole. The bound pairs are referred as excitons and the states so created by the transfer of an electron from an occupied level to an unoccupied level are called exciton states. The energy of the states may be discrete and located below the conduction-band edge or form a continuum within the conduction band. The excitonic effects dominate the light absorption property near the band gap and form the basis of photonic quantum well devices[5.21,22].

It should, however, be noted that the excitonic effects are important only for low carrier and low impurity concentration. For large carrier concentration, the excitonic effects are much reduced. Large impurity concentration may also destroy the excitons, even if the electron concentration be small as in compensated materials. Also, the binding energy of the excitons being small these are mostly deexcited at room temperature in bulk materials. In quantum wells, the binding energy is larger and the carrier concentration and the impurity concentration are also small and hence the excitonic effects are more important in quantum wells.

5.3.1. EXCITED-STATE WAVE FUNCTIONS

The wave function for the excited state is obtained from the one-electron Bloch functions by considering that for an N-electron system the ground-state wave function may be written in the Hartree-Fock approximation[5.12] as

$$\psi_0 = \frac{1}{N!^{1/2}} \begin{vmatrix} \psi_{\mathbf{k}_1}(\mathbf{r}_1) & \psi_{\mathbf{k}_1}(\mathbf{r}_2) & \cdots & \psi_{\mathbf{k}_1}(\mathbf{r}_h) & \cdots & \psi_{\mathbf{k}_1}(\mathbf{r}_N) \\ \psi_{\mathbf{k}_2}(\mathbf{r}_1) & \psi_{\mathbf{k}_2}(\mathbf{r}_2) & \cdots & \psi_{\mathbf{k}_2}(\mathbf{r}_h) & \cdots & \psi_{\mathbf{k}_2}(\mathbf{r}_N) \\ \cdots & \cdots & \cdots & \cdots & \cdots & \cdots \\ \psi_{\mathbf{k}_h}(\mathbf{r}_1) & \psi_{\mathbf{k}_h}(\mathbf{r}_2) & \cdots & \psi_{\mathbf{k}_h}(\mathbf{r}_h) & \cdots & \psi_{\mathbf{k}_h}(\mathbf{r}_N) \\ \cdots & \cdots & \cdots & \cdots & \cdots & \cdots \\ \psi_{\mathbf{k}_N}(\mathbf{r}_1 & \psi_{\mathbf{k}_N}(\mathbf{r}_2) & \cdots & \psi_{\mathbf{k}_N}(\mathbf{r}_h) & \cdots & \psi_{\mathbf{k}_N}(\mathbf{r}_N) \end{vmatrix} \tag{5.76}$$

where the electrons (total number N) with the position vectors $\mathbf{r}_1, \mathbf{r}_2, \ldots \mathbf{r}_h, \ldots, \mathbf{r}_N$ are assumed to be in the uppermost fully filled valence band. The subscript $\mathbf{k}_h$ indicates an electron in the valence band with the wave vector $\mathbf{k}_h$.

In the excited state, an electron from the valence band is excited to an empty conduction band state. Only the uppermost valence band and the lowest conduction band are assumed to be involved in such transitions. Wave function for the excited state is obtained by replacing one of the valence-band Bloch function $\psi_{\mathbf{k}_h}$ with wave vector $\mathbf{k}_h$, by a conduction band Bloch function with wave vector $\psi_{\mathbf{k}_s}$

as follows,

$$\phi_{\mathbf{k}_e,\mathbf{k}_h} = \frac{1}{N!^{1/2}} \begin{vmatrix} \psi_{\mathbf{k}_1}(\mathbf{r}_1) & \psi_{\mathbf{k}_1}(\mathbf{r}_2) & \cdots & \psi_{\mathbf{k}_1}(\mathbf{r}_e) & \cdots & \psi_{\mathbf{k}_1}(\mathbf{r}_N) \\ \psi_{\mathbf{k}_2}(\mathbf{r}_1) & \psi_{\mathbf{k}_2}(\mathbf{r}_2) & \cdots & \psi_{\mathbf{k}_2}(\mathbf{r}_e) & \cdots & \psi_{\mathbf{k}_2}(\mathbf{r}_N) \\ \cdots & \cdots & \cdots & \cdots & \cdots & \cdots \\ \psi_{\mathbf{k}_e}(\mathbf{r}_1) & \psi_{\mathbf{k}_e}(\mathbf{r}_2) & \cdots & \psi_{\mathbf{k}_e}(\mathbf{r}_e) & \cdots & \psi_{\mathbf{k}_e}(\mathbf{r}_N) \\ \cdots & \cdots & \cdots & \cdots & \cdots & \cdots \\ \psi_{\mathbf{k}_N}(\mathbf{r}_1) & \psi_{\mathbf{k}_N}(\mathbf{r}_2) & \cdots & \psi_{\mathbf{k}_N}(\mathbf{r}_e) & \cdots & \psi_{\mathbf{k}_N}(\mathbf{r}_N) \end{vmatrix} \qquad (5.77)$$

Wave functions for the various excited stated are obtained by permuting the states with wave vectors $\mathbf{k}_h$ and $\mathbf{k}_e$, $\psi_{\mathbf{k}_h}$ and $\psi_{\mathbf{k}_e}$, amongst the available states. The excitonic states are obtained when the effect of the Coulomb force between the excited electron and the hole created in the valence band is taken into account. The corresponding wave function may be obtained by linearly combining the excited-state wave functions, $\phi_{\mathbf{k}_e,\mathbf{k}'_h}$'s. The wave function for the excitonic state $\psi_{\mathrm{ex}}(\mathbf{r}_1, \mathbf{r}_2 \ldots \mathbf{r}_N)$ may hence be written as

$$\psi_{\mathrm{ex}}(\mathbf{r}_1, \mathbf{r}_2, \mathbf{r}_e, \mathbf{r}_N) = \sum_{\mathbf{k}_e, \mathbf{k}_h} A(\mathbf{k}_e, \mathbf{k}_h) \phi_{\mathbf{k}_e,\mathbf{k}_h}, \qquad (5.78)$$

where $A(\mathbf{k}_e, \mathbf{k}_h)$'s are the unknown coefficients, which should be suitably chosen. These coefficients are evaluated in the Wannier model[5.23] by using the effective-mass equation. The assumption underlying the model is that the excitons have large spatial extent such that the electron and the hole states are given by the band structure and the Coulomb interaction may be treated as a perturbation. It is also required that the interaction potential may be evaluated by using the macroscopic dielectric constant.

The effective-mass equation is obtained by using the result that the combined electron states in the valence band in the absence of an electron in the state $\psi_{\mathbf{k}_h}$, may be represented by a positive charge and a positive mass or a hole with the wave functions $\psi^*_{-\mathbf{k}_h}$ or $\theta \psi_{\mathbf{k}_h}$, where θ is the time reversal operator. The wave function $\phi_{\mathbf{k}_e,\mathbf{k}_h}$ may then be replaced by $\psi_{\mathbf{k}_e}(\mathbf{r}_e)\psi_{\mathbf{k}_h}(\mathbf{r}_h)$ when the position vectors of the excited electron and the created hole are given respectively by $\mathbf{r}_e$ and $\mathbf{r}_h$. The wave equation may be written by considering only the electron-hole pair as

$$\{\sum[-(\hbar^2/2m_0)\nabla_i^2 + V_c(\mathbf{r}_i, \mathbf{R}_i)] + V(\mathbf{r}_e, \mathbf{r}_h)\}\psi_{\mathrm{ex}} = E\psi_{\mathrm{ex}}, \qquad (5.79)$$

where ∇_i is the operator for the variable $\mathbf{r}_i$, which is either $\mathbf{r}_e$ or $\mathbf{r}_h$. $V_c(\mathbf{r}_i, \mathbf{R}_i)$ is the potential energy of the ith particle due to its being in the field of the nuclear charge (with coordinate $\mathbf{R}_i$) and all other electrons. $V(\mathbf{r}_e, \mathbf{r}_h)$ is the potential energy due to the Coulomb interaction between the electron-hole pair, is given by

$$V(\mathbf{r}_e, \mathbf{r}_h) = -(e^2/4\pi K_d\epsilon_0)|\mathbf{r}_e - \mathbf{r}_h|^{-1}, \qquad (5.80)$$

where $\mathbf{r}_e$ and $\mathbf{r}_h$ are respectively the position vectors of the electron and the hole. The constant K_d in the above expression is the static dielectric constant, when electrons and holes are far apart and the relative motion between them is small, so that the polarization is due to both the electrons and the lattice displacement. As the particles approach each other, K_d changes to the high-frequency dielectric constant, which corresponds to the case of polarization being due to only the electron displacement, lattice polarization being unable to follow the field variation. For still smaller values of $|r_e - r_h|$, when the relative motion of the electron and the hole becomes more rapid, K_d is further reduced and other effects, e.g., correlation effect and break down of the effective-mass approximation, become important. Expression (5.80) for the interaction potential is not applicable for such distances. The critical distance is of the order of 50 Å in bulk III-V compounds and the approximation may not be considered a limitation for bulk materials. However, this aspect should be considered, when applying the present theory to quantum well systems, where dimensions are of the same order.

Replacing ψ_{ex} by $\sum_{k_e, k_h} A(k_e, k_h)\phi_{k_e, k_h}$ and taking the scalar product of $\psi_{k_e}^*$ and $\psi_{k_h}^*$ with each term, the equation is transformed to[5.24]

$$[E_c(k_e) - E_v(k_h) - E]A(k_e, k_h) + \sum_{k_e, k_h} F(k_e, k_h, k'_e, k'_h)A(k'_e, k'_h) = 0, \quad (5.81)$$

where

$$F(k_e, k_h, k'_e, k'_h)A(k'_e, k'_h) = \int \psi_{k_e}^* \psi_{k_h}^* V(r_e, r_h)\psi_{k_e}\psi_{k_h} d^3r_e d^3r_h. \quad (5.82)$$

$E_c(k_e)$ and $E_v(k_h)$ represent respectively the energy of an electron with the wave vector k_e in the conduction band and that of a hole with the wave vector k_h in the valence band. The k-dependence of the cell-periodic parts of the wave functions, U_{ck_e} and U_{vk_h} are considered to be weak and are neglected at this stage of approximation. The wave functions may hence be simplified to

$$\psi_{ck_e} = \sqrt{1/N}U_{c0}\exp(ik_e.r) \text{ and } \psi_{vk_h} = \sqrt{1/N}U_{v0}\exp(ik_h.r),) \quad (5.83)$$

N being the total number of unit cells in the crystal over each of which the cell periodic parts are normalized. Then multiplying each term of Eq. (5.81) by $\exp i(k_e.r_e + k_h.r_h)$ and summing over all values of k_e and k_h the effective-mass envelope-function equation is obtained as follows .

$$[E_e(-i\nabla_e) - E_v(-i\nabla_h) - E]F(r_e, r_h) + V(r_e, r_h)F(r_e, r_h) = 0. \quad (5.84)$$

The subscripts e and h indicate respectively the variables corresponding to the electron and the hole. The envelope function $F(r_e, r_h)$ is defined as

$$F(r_e, r_h) = (1/V_c) \sum_{k_e, k_h} A(k_e, k_h)\exp i(k_e.r_e + k_h.r_h), \quad (5.85)$$

V_c being the crystal volume.

It should be noted that the effective-mass equation for the excitons has been obtained by assuming that the cell-periodic parts of the wave functions are weakly dependent on the wave vectors. The dependence causes a mixing of the wave functions from the different bands in the formation of the exciton state. This mixing is, therefore, neglected in the present formulation. It appears from the comparison of the present theory with experiments that this approximation is reasonably valid.

5.3.2. EXCITONIC WAVE FUNCTIONS

The effective-mass equation is solved by assuming an isotropic parabolic band for both the electrons and the holes, so that the equation may be written as

$$[-(\hbar^2/2m_e^*)\nabla_e{}^2 - (\hbar^2/2m_h^*)\nabla_h{}^2 - (e^2/4\pi K_d\epsilon_0)(|\mathbf{r}_e - \mathbf{r}_h|)^{-1}]F(\mathbf{r}_e, \mathbf{r}_h) \quad (5.86)$$
$$= (E - E_g)F(\mathbf{r}_e, \mathbf{r}_h),$$

where m_e^*, m_h^* are respectively the band-edge effective mass of the electron and the hole.

The equation is transformed by using the following relations,

$$(1/2)(\mathbf{r}_e + \mathbf{r}_h) = \mathbf{R}, \mathbf{r}_e - \mathbf{r}_h = \mathbf{r}, \nabla_e + \nabla_h = \nabla_R, (1/2)(\nabla_e - \nabla_h) = \nabla_r. \quad (5.87)$$

and

$$F(\mathbf{r}, \mathbf{R}) = \sqrt{1/V_c}\exp(i\mathbf{K}.\mathbf{R})\phi(\mathbf{r}). \quad (5.88)$$

The function $\phi(\mathbf{r})$ satisfies the equation,

$$-(\hbar^2/2m_e^*)[\nabla_r + (1/2)iK]^2 - (\hbar^2/2m_h^*)[-\nabla_r + (1/2)iK]^2 - e^2/4\pi K_d\epsilon_0 r \quad (5.89)$$
$$= (E - E_g)\phi(\mathbf{r}).$$

The coefficients $A(\mathbf{k}_e, \mathbf{k}_h)$ of Eq. (5.85) are given by,

$$A(\mathbf{k}_e, \mathbf{k}_h) = V_c^{-3/2} \int \exp[i(\mathbf{K} - \mathbf{k}_e - \mathbf{k}_h).\mathbf{R}] \exp(-i\mathbf{k}.\mathbf{r})\phi(\mathbf{r})d^3r d^3R, \quad (5.90)$$

where

$$\mathbf{k} = (1/2)(\mathbf{k}_e - \mathbf{k}_h). \quad (5.91)$$

For non-zero value of the integral, it is required that

$$\mathbf{K} = \mathbf{k}_e + \mathbf{k}_h. \quad (5.92)$$

The exciton wave vector is thus the vector sum of the electron and the hole wave vectors.

The coefficients are given by

$$A(\mathbf{k}_e, \mathbf{k}_h) = A(\mathbf{k}) = V_c^{1/2} \int \phi(\mathbf{r}) \exp(-i\mathbf{k}.\mathbf{r})d^3r. \quad (5.93)$$

The exciton wave functions may now be expressed by neglecting the k-dependence of U_c and U_v as,

$$\psi_{ex} = (1/N)U_{c0}\sum_k A(k)\exp(i\mathbf{k}.\mathbf{r})\exp(i\mathbf{K}.\mathbf{R})$$

$$= U_{v0}\sqrt{V_c}\phi(\mathbf{r})\exp(i\mathbf{K}.\mathbf{R}). \tag{5.94}$$

Explicit expression for $\phi(\mathbf{r})$ may be obtained conveniently by changing the variable $\mathbf{r},\mathbf{R}$ to $\mathbf{r}',\mathbf{R}'$ as given below,

$$\mathbf{r}' = \mathbf{r}; \mathbf{R}' = \mathbf{R} + [(1/2)(m_e^* - m_h^*)/M].\mathbf{r}, M = m_e^* + m_h^*. \tag{5.95}$$

and putting

$$F(\mathbf{r}'',\mathbf{R}'') = V_c^{-1/2}\exp(i\mathbf{K}.\mathbf{R}')\phi'(\mathbf{r}''). \tag{5.96}$$

The equation for $\phi'(\mathbf{r}')$ is,

$$[-(\hbar^2/2\mu)\nabla_r^2 - e^2/4K_d\epsilon_0 r']\phi'(\mathbf{r}') = (E - \hbar^2 K^2/2M - E_g)\phi'(\mathbf{r}'). \tag{5.97}$$

where

$$1/\mu = 1/m_e^* + 1/m_h^*. \tag{5.98}$$

Solutions are of the hydrogenic form and are given by

$$\phi_n'(\mathbf{r}') = \phi'(\mathbf{r}) = R_{nl}(r)Y_l^m(\theta,\phi), \tag{5.99}$$

where $R_{nl}(r)$ and $Y_l^m(\theta,\phi)$ are respectively the radial function and the spherical harmonic; l,m are the orbital and the magnetic quantum numbers and n is the order of the energy level.

The energy of the exciton for discrete states is given by

$$E = E_g + E_n + \hbar^2 K^2/2M. \tag{5.100}$$

E_n is an eigenvalue for and is given by

$$E_n = -(e^2/4\pi K_d\epsilon_0)^2/2n^2(\hbar^2/\mu), \tag{5.101}$$

where n is an integer, and μ is the reduced mass.

The $E - K$ relations for the excitonic states are schematically illustrated in Fig. 5.2 and the energy levels corresponding to the excitonic states in the one-electron $E - k$ diagram in Fig. 5.3. The first few states are discrete and below the conduction band and in the forbidden gap. Equation (5.97) admits also solutions with positive values of E_n, which, however, form a continuum as shown in Fig. 5.3. Energy eigenvalues are given by

$$E(k) = \hbar^2 k^2/2\mu, \tag{5.102}$$

where k is a continuously variable quantum member.

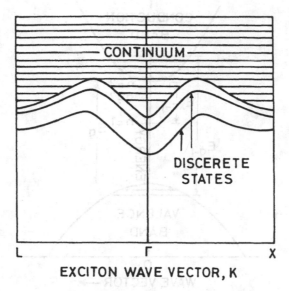

CONTINUUM

DISCRETE
STATES

L   Γ   X

EXCITON WAVE VECTOR, K

Figure 5.2. Schematic diagram showing the dispersion relation for the excitons. The kinetic energy of the excitons varies as the square of the wave vector.

The function $\phi_1'(r)$ for the ground ($n = 1$) excitonic state is

$$\phi_1'(r) = (\pi a_B^3)^{-1/2} \exp(-r/a_B), \tag{5.103}$$

where a_B is the Bohr radius given by, $a_B = 4\pi K_d \epsilon_0 \hbar^2 / \mu e^2$.

The coefficients , $A(\mathbf{k}_e, \mathbf{k}_h)$, are evaluated by using the equivalence, $F(\mathbf{r}', \mathbf{R}') = F(\mathbf{r}, \mathbf{R})$, or

$$\phi_n'(r') \exp(i\mathbf{K}.\mathbf{R}') = \phi_n(r) \exp(i\mathbf{K}.\mathbf{R}). \tag{5.104}$$

The coefficient for the ground state is,

$$A(k') = V_c^{-1/2} (64\pi/a_B^5)^{1/2} [k'^2 + (1/a_B)^2]^{-2}, \tag{5.105}$$

where

$$\mathbf{k}' = (m_e^* \mathbf{k}_h - m_h^* \mathbf{k}_e)/M. \tag{5.106}$$

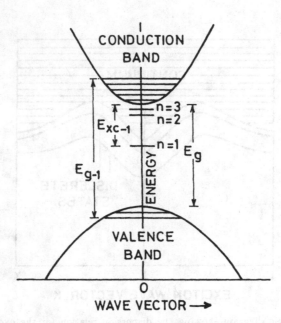

Figure 5.3. Schematic representation of the excited states for the electron wave vector, $\mathbf{K} = \mathbf{0}$. First few states are discrete and are in the forbidden gap below the conduction-band edge. The shaded region shows the electron and the hole states which contribute to the ground exciton state.

Significant values of $\mathbf{k}_e$ and $\mathbf{k}_h$ which form the excitons are of the order of $1/a_B$. The mixing effect of the k-dependence of the cell-periodic parts (neglected in the present analysis) may be estimated from the result that $k \sim 1/a_B$. It is found that the effect is of the order of 10^{-2} -10^{-3} and may be neglected for III-V compounds.

5.3.3. EXCITONIC OPTICAL INTERACTION MATRIX ELEMENT

The Matrix element for optical interactions in which an exciton state is formed may now be worked out by using the wave function of the exciton ψ_{ex} and of the ground state ψ_0 The matrix element is, (see Section 5.1)

$$M(f, i) = (-i\hbar|e|A_0/m_0) < \psi_{ex}| \sum_i \exp(i\kappa.\mathbf{r}_i)\hat{\mathbf{a}}.\nabla|\psi_0 > . \tag{5.107}$$

Replacing ψ_{ex} and ψ_0 by the detailed expressions, the matrix element may be expressed as

$$M(f, i) = (-i\hbar|e|A_0/m_0) \sum_{\mathbf{k}_e,\mathbf{k}_h} A(\mathbf{k}_e, \mathbf{k}_h) < \psi_{\mathbf{k}_e}| \sum_i \exp(i\kappa.\mathbf{r}_i)\hat{\mathbf{a}}.\nabla|\theta\psi_{\mathbf{k}_h} >, \tag{5.108}$$

ψ_{k_h} being the wave function for the hole, created by the interaction.

The significant component of the sum is obtained by neglecting the k-dependence of the cell-periodic parts. The integral has a non-zero value if $\mathbf{k}_e$ and $\mathbf{k}_h$ be such that

$$\mathbf{k}_e + \mathbf{k}_h = \mathbf{K}. \tag{5.109}$$

Since K is small, $\mathbf{k}_e = -\mathbf{k}_h$. This result implies also that $\mathbf{K} = \kappa$, i.e., the exciton wave vector is equal to the photon wave vector. The significant part of the matrix element may be written as,

$$M(f, i) = (|e|A_0/2m_0) \sum_{\mathbf{k}_e} A(\mathbf{k}_e + \mathbf{K}/2, -\mathbf{k}_e + \mathbf{K}/2)\hat{\mathbf{a}}.\mathbf{p}_{cv}. \tag{5.110}$$

The above expression for the matrix element may also be written in terms of the envelope functions as follows,

$$M(f, i) = (|e|A_0/2m_0)\hat{\mathbf{a}}.\mathbf{p}_{cv}V_c^{1/2}\phi'_n(o)\exp i\mathbf{K}.(\mathbf{R}' - \mathbf{R}). \tag{5.111}$$

The magnitude of the matrix element which enters the expression for the transition probability is hence

$$M(f, i) = V_c^{1/2}(|e|A_0/2m_0)\phi'_n(o)\hat{\mathbf{a}}.\mathbf{p}_{cv}. \tag{5.112}$$

The function $|\phi'_n(o)|$ is the magnitude of the eigenfunction $\phi'_n(r')$ of Eq. (5.99). It is given for the discrete states characterized by integers n, by

$$|\phi'_n(o)| = (\pi a_B^3 n^3)^{-1/2}, \tag{5.113}$$

whereas for the continuum state characterized by quantum number k it is

$$|\phi'_n(o)| = [(\pi/ka_B)\exp(\pi/ka_B)/\sinh(\pi/ka_B)]^{1/2}. \tag{5.114}$$

5.3.4. EXCITONS IN QUANTUM WELLS

Analysis of the characteristics of excitons in quantum wells is made difficult by the confining heterojunction potential, which destroy the spherical symmetry of the wave functions. The theory is developed by applying the same procedure as discussed in Section 5.3.1 for excitons in bulk materials, but analytically closed solutions cannot be worked out even for the effective-mass wave equation. Variational techniques have been applied to obtain the energy eigenvalues. Principles of these techniques[5.25-27] and results obtained therefrom are briefly discussed in this section.

The Hamiltonian for the excitons may be written as follows in the effective-mass formalism by applying the procedure discussed in Section 5.1.3 for excitons in bulk materials.

$$H = \sum_i [-(\hbar/2m_i)\nabla_i^2 + V_c(\mathbf{r}_i, \mathbf{R}_j) + V_{zi}(\mathbf{r}_i)] - (e^2/4\pi K_d\epsilon_0)|\mathbf{r}_e - \mathbf{r}_h|^{-1}, \tag{5.115}$$

where the subscript i is either e or h and indicates the electron or the hole. The first term gives the kinetic energy, the second term, the potential energy due to the field of the atoms, the third term, that due to the heterojunction potentials, which are different for the electrons and the holes; the fourth term gives the potential energy due to the Coulomb attraction between the electron and the hole. As for the bulk excitons, the wave function ψ_{ex} is expressed as a linear combination of the wave functions for the excited states, $\phi_{\mathbf{k}_e, \mathbf{k}_h}$, and the effective-mass wave equation for the envelope function is obtained by following the same procedure. But, now the equation includes V_{ze} and V_{zh} as shown below,

$$[-(\hbar^2/2m_e^*)\nabla_e^2 - (\hbar^2/2m_{h\pm})\nabla_h^2 - (e^2/4\pi K_d\epsilon_0)|\mathbf{r}_e - \mathbf{r}_h|^{-1}$$
$$+V_{ze}(\mathbf{r}_e) + V_{zh}(\mathbf{r}_h)]F(\mathbf{r}_e, \mathbf{r}_h) = (E - E_g)F(\mathbf{r}_e, \mathbf{r}_h), \qquad (5.116)$$

where the plus and minus signs in $m_{h\pm}$ indicate the heavy hole or the light hole.

As discussed in Section 4.3, the effective mass of holes is radically altered in quantum wells due to the mixing of the wave functions of the two kinds of holes. The in-plane effective mass is different in different subbands and changes largely, in magnitude and even in sign with the in-plane component of the wave vector. The problem is simplified[5.27,28] by assuming that the hole dispersion relation may be expressed as

$$E_{h\pm} = -(\hbar^2/2)[(1/m_{hz\pm}^*)k_z^2 + (1/m_{ht\pm}^*)k_t^2], \qquad (5.117)$$

where $m_0/m_{hz\pm}^* = \gamma_1 \mp 2\gamma_2, \gamma_1$ and γ_2 being the Luttinger parameters, $k_t^2 = k_x^2 + k_y^2$. The transverse mass $m_{ht\pm}^*$ has been taken in some studies[5.27] to be equal to its value,obtained by neglecting the mixing , i.e., $m_0/(\gamma_1 \pm \gamma_2)$. In some other studies [5.28] it has been taken to be equal to $m_0/(\gamma_1 \pm (3 - \alpha)\gamma_2)$ with α suitably chosen between 2 and 1.

Equation (5.116) is further simplified by considering that the energy due to the Coulomb interaction may be considered small in comparison to the subband energies due to the quantization. The envelope function may then be written as,

$$F(\mathbf{r}_e, \mathbf{r}_h) = f_e(z_e)f_{h\pm}(z_h)G(\rho_e, \rho_h), \qquad (5.118)$$

where $f_e(z_e)$ and $f_{h\pm}(z_h)$ are respectively the solutions for the electron and the hole corresponding to the subband energies.

Equation(5.116) is transformed as in the case of bulk excitons by putting,

$$x = x_e - x_h, y = y_e - y_h, z = z_e - z_h, \qquad (5.119)$$

$$X = (m_e^* x_e + m_{ht\pm}^* x_h)/M_\pm, Y = (m_e^* y_e + m_{ht\pm}^* y_h)/M_\pm, \qquad (5.120)$$

and putting

$$G(\rho_e, \rho_h) = \exp(i\mathbf{K}.\mathbf{P})G(\rho, z), \qquad (5.121)$$

where
$$\mathbf{P} = \hat{\mathbf{i}}X + \hat{\mathbf{j}}Y, \rho^2 = x^2 + y^2. \tag{5.122}$$

The reduced equation is
$$[-(\hbar^2/2\mu_\pm)(\partial^2/\partial x^2 + \partial^2/\partial y^2) - (e^2/4\pi K_d\epsilon_0)(\rho^2 + z^2)^{-1/2}]G(\rho, z)$$
$$= (E_{ex} - \hbar^2K^2/2M_\pm)G(\rho, z), \tag{5.123}$$

where
$$1/\mu_\pm = 1/m_e^* + 1/m_{ht\pm}^*, E_{ex} = E - E_{cn} - E_{vn} - E_g, \quad M_\pm = m_e^* + m_{hz\pm}^*. \tag{5.124}$$

The difference in energy between the hole level and the electron level is represented by $E_{cn} + E_{vn} + E_g$. Equation (5.123) cannot be solved analytically and variational techniques have been used [5.25-27]to evaluate the eigenvalues of energy. Different variational functions have been used for this purpose. Some of the functions are quoted below.

$$G(\rho, z) = N \exp(-\beta\rho). \tag{5.125}$$
$$G(\rho, z) = N \exp -\beta(\rho^2 + z^2). \tag{5.126}$$
$$G(\rho, z) = N \exp[-\alpha_i z^2 - (\alpha_j + \beta)\rho^2]. \tag{5.127}$$

N is a normalizing constant, β is the variational parameter, α_i, β_i are chosen constants. The value of the variational parameter is chosen by minimizing $< \psi_{ex}|H|\psi_{ex} >$, which also gives the eigenvalues of energy.

Results of such analysis[5.27,29], illustrated in Fig. 5.4, clearly indicate that the binding energy of excitons increases with the narrowing of the wells. For example, the energy increases from about 4.5 meV to 9.5 meV as the well width is reduced from 500 Å to 30 Å in the GaAs/Ga$_{0.7}$ Al$_{0.3}$ As system. In the early analysis, $f_{ez}(z_e)$ and $f_{hz}(z_h)$ were taken to correspond to very large values (tending to infinity) of V_{ze} and V_{zh} . However, when the solution for finite values of V_{ze} and V_{zh} are taken, the binding energies are found to be lowered, particularly for very narrow wells. For very narrow wells, the binding energy is found to decrease instead of increasing with the narrowing of the well.

5.3.5. EXCITONIC OPTICAL MATRIX ELEMENT FOR QUANTUM WELLS

The development of the optical matrix element gets complicated for the general case in which $z_e \neq z_h$. The procedure may be much simplified by assuming that $z_e = z_h$ i.e, $z = 0$. The exciton wave function may then be written as

$$\psi_{ex} = U_{c0}U_{v0}f_{ei}(z_e)f_{hj}(z_h) \sum_{\mathbf{k}_{te},\mathbf{k}_{th}} A(\mathbf{k}_{te}, \mathbf{k}_{th}) \exp i(\mathbf{k}_{te}.\boldsymbol{\rho}_e + \mathbf{k}_{th}.\boldsymbol{\rho}_h). \tag{5.128}$$

and the matrix element reduced to

$$M(f, i) = (|e|A_0/2m_0)A_c^{1/2}G(\rho, 0)\hat{\mathbf{a}}.\mathbf{p}_{cv}, \tag{5.129}$$

A_c is the in-plane area of the well.

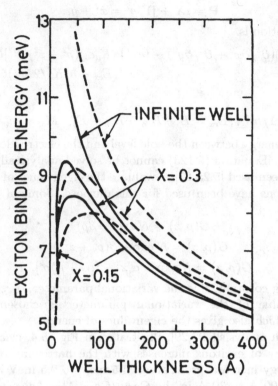

Figure 5.4. Calculated values of the heavy-hole exciton binding energy of the ground state in a GaAs/Al$_{0.3}$Ga$_{0.7}$As quantum well for different well thicknesses. The binding energy decreases with increase in the well thickness, from about 9 meV to 4.5 meV as the well thickness increases to 50 nm. [After R. L. Greene, K. K. Bajaj and D. E. Phelps, *Phys. Rev.* **B 29**, 1807 (1984); Copyright (1984) by the American Physical Society].

$G(\rho, 0)$ is the solution of the equation,

$$[(-\hbar^2/2\mu_\pm)(\partial^2/\partial x^2 + \partial^2/\partial y^2) - (e^2/4\pi K_d \epsilon_0 \rho)]G(\rho, 0) = E'G(\rho, 0), \qquad (5.130)$$

where

$$E' = E_{ex} - \hbar^2 \mathbf{K}_t^2/2M, \mathbf{K}_t = \kappa_t, \kappa_z = 0. \qquad (5.131)$$

It may be shown[5.30] that the eigenvalues of energy for this strictly 2D ($z = 0$) excitons are given by

$$E'_n = \mu_\pm e^4[2(4\pi K_d \epsilon_0)^2 \hbar^2 (n - 1/2)^2]^{-1}. \qquad (5.132)$$

The binding energy for the first excitonic level in the extreme 2D case is four times the value for the bulk excitons. This value, as discussed earlier, is not, however, reached as the well width is reduced because of the finite value of the potential barrier.

The wave function, $G(\rho, 0)$, obtained from Eq. (5.132) is

$$G(\rho, 0) = (8/a_B^2)^{1/2} \exp -(2\rho/a_B). \tag{5.133}$$

On using this function in (5.131) we find that the ratio of the strength of optical excitonic interaction in quantum wells and in bulk material is $(8L/a_B)$. Since $a_B \sim L$, the strength is expected to be 8 times larger. Although this increase is not realized in practice, the excitonic absorption in quantum wells is found to be much stronger and to dominate the optical properties. An understanding of the quantum well optic devices, therefore, require a good acquaintance with the behavior of excitons in the wells.

5.4. Bound and Localized Excitons

We have considered so far the excitons formed by the binding of a free electron and a free hole. These excitons may move about freely and are referred as free excitons. Excitons may, however, be bound to the ionized or the neutral impurity atoms. The excited electron, the hole and the impurity atom are coupled by the Coulomb force. Under conditions in which the total energy of the system (electron + hole + impurity atom) is reduced (corresponding to an increase in the binding energy of the exciton), a bound exciton is formed. Such excitons cannot move about, but remain near the impurity atom. The required conditions for a bound exciton to exist have been studied by Hopfield[5.31] and others[5.32]. It is the ratio of the electron mass and the hole mass which determines whether a bound exciton may be formed. The conditions being exclusive an exciton may be bound either to a donor or to an acceptor. In the case of excitons bound to neutral impurities it was observed empirically[5.33] that the dissociation energy of the exciton-neutral impurity complex is about 10% of the impurity binding energy. This result, usually referred as the 'Haynes rule' is also supported by detailed calculations. Bound excitons have been detected in the photoluminescence spectra of quantum wells, in many experiments, some of which are discussed in Section 5.6.

Excitons may also be localized by potential fluctuations[5.34,35] . The interfaces of the heterostructures have roughnesses extending over distances comparable to the exciton radius and of the height of a monolayer. The width of the wells changes and the energies of the quantized levels change consequently. The energy being therefore different in different regions of the well, the excitons are confined in low-energy regions. In wells of alloy semiconductors, e.g., $Ga_{0.47}In_{0.53}As$, the variation of the alloy potential may also produce local potential minima and cause localization of the excitons.

Excitons bound to impurities and localized by interface defects and alloy potentials are common features of quantum wells [5.36.37]. In fact, luminescence produced due to the recombination of such excitons provide a means of assesing the quality of heterostructures and even in the mapping of the defects. This aspect of the excitons is discussed in Section 5.6.

Optical interaction results in various phenomena, e.g., light absorption, photoluminescence, optical nonlinearity and Stark effect. Quantum well optoelectronic devices are realized by exploiting these phenomena. Theory for exploiting the phenomena and their important characteristics are discussed in the following sections.

5.5. Absorption

The light incident on a quantum well structure may cause excitation of the ground state electrons from the valence band to the conduction band, or from one subband to a higher subband. The excitation may cause the appearance of a free electron at a higher energy or an exciton. Whenever such excitation occurs, the required energy is supplied by the photon and in effect light is absorbed. The phenomenon is characterized by an absorption coefficient. It is defined as the light energy absorbed in unit length per unit incident energy and is given by

$$\alpha = -(dn_p/dt)[(n_p - n_{p0})(c/n_r)]^{-1}, \qquad (5.134)$$

where n_p is the photon density in the sample when light is incident, n_{p0} is its value in the absence of the incident light, c is the velocity of light in vacuum, n_r is the refractive index of the material. Evidently, $(n_p - n_{p0})(c/n_r)$ gives the number of photons incident per unit time per unit area and (dn_p/dt) gives the rate of absorption of these photons. The ratio of the two quantities is equal to α by definition.

The rate of photon absorption is given according to Fermi's Golden rule[5.38-41] by

$$
\begin{aligned}
-dn_p/dt =\ & (V_cD)(V_cD')(2\pi/\hbar)\rho(\omega) \int \{|M_0(\mathbf{k},\mathbf{k}')|^2 n_\kappa f(\mathbf{k})[1 - f(\mathbf{k}')] \\
& \times\delta(E_{\mathbf{k}'} - E_{\mathbf{k}} - \hbar\omega) - |M_0(\mathbf{k}',\mathbf{k})|^2(n_\kappa + 1)f(\mathbf{k}')[1 - f(\mathbf{k})] \\
& \times\delta(E_{\mathbf{k}} - E_{\mathbf{k}'} + \hbar\omega)\}d^3k d^3k',
\end{aligned}
\qquad (5.135)
$$

where $\mathbf{k}$ and $\mathbf{k}'$ are respectively the initial and the final state wave vectors, $E_{\mathbf{k}}$ and $E_{\mathbf{k}'}$ are the corresponding energies. D and D' are the electron density of states for the initial and the final states, V_c is the volume of the crystal. $M_0(\mathbf{k},\mathbf{k}')n_\kappa^{1/2}$ and $M_0(\mathbf{k}',\mathbf{k})(n_\kappa + 1)^{1/2}$ are the matrix elements for transitions from the $\mathbf{k}$ - state to the $\mathbf{k}'$ -state and for the reverse transitions,n_κ being the occupation probability of the photons. The functions $f(\mathbf{k})$ and $f(\mathbf{k}')$ give the distribution of electrons with wave vectors $\mathbf{k}$ and $\mathbf{k}'$ respectively. Frequency of the incident signal is assumed to be ω, so that the energy of a photon is $\hbar\omega$ and the corresponding density of states for the photons is $\rho(\omega)$.

First term in the integral gives the number of electrons excited per unit time from the $\mathbf{k}$-state to the $\mathbf{k}'$-state by absorbing light. The second term gives the rate of reverse transitions from the $\mathbf{k}'$-state to the $\mathbf{k}$-state by emitting light. In the absence of light, the integral must be zero and from the principle of detailed balance each term must be zero. This condition gives

$$|M_0(\mathbf{k}, \mathbf{k}')|^2 n_{\kappa 0} f(\mathbf{k})[1 - f(\mathbf{k}')] = |M_0(\mathbf{k}', \mathbf{k})|^2 (n_{\kappa 0} + 1) f(\mathbf{k}')[1 - f(\mathbf{k})]. \quad (5.136)$$

where $n_{\kappa 0}$ is the occupation probability in the absence of the incident light. It is evident that $E_{\mathbf{k}'}$, and $E_{\mathbf{k}}$ are related because of the delta function as

$$E_{\mathbf{k}'} = E_{\mathbf{k}} + \hbar\omega. \quad (5.137)$$

On using the Fermi distribution function for $f(\mathbf{k})$ and $f(\mathbf{k}')$ and the Bose distribution function for the photon occupation probability it is found that,

$$n_{\kappa 0} f(\mathbf{k})[(1 - f(\mathbf{k}')] = (n_{\kappa 0} + 1) f(\mathbf{k}')[1 - f(\mathbf{k})]. \quad (5.138)$$

This condition when combined with (5.136) gives

$$|M_0(\mathbf{k}, \mathbf{k}')| = |M_0(\mathbf{k}', \mathbf{k})|, \quad (5.139)$$

i.e., the matrix element for transitions from the $\mathbf{k}$-state to the $\mathbf{k}'$- state and from the $\mathbf{k}'$-state to the $\mathbf{k}$-state are equal. This equality is often taken as the starting point in (5.135) which then means in effect that detailed balance is assumed to apply.

Equation (5.135) may now be simplified to

$$
\begin{aligned}
-(dn_p/dt) &= (V_c D)(V_c D')(2\pi/\hbar) \int |M_0(\mathbf{k}, \mathbf{k}')|^2 (n_p - n_{p0}) \\
&\quad \times [f(\mathbf{k}) - f(\mathbf{k}')]\delta(E_{\mathbf{k}'} - E_{\mathbf{k}} - \hbar\omega) d^3 k d^3 k' \quad (5.140)
\end{aligned}
$$

considering that $n_p = \rho(\omega) n_\kappa$ and $n_{p0} = \rho(\omega) n_{\kappa 0}$. The absorption coefficient,

$$
\begin{aligned}
\alpha &= (V_c D)(V_c D')(2\pi/\hbar)(n_r/c) \int |M_0(\mathbf{k}, \mathbf{k}')|^2 \\
&\quad \times [f(\mathbf{k}) - f(\mathbf{k}')]\delta(E_{\mathbf{k}'} - E_{\mathbf{k}} - \hbar\omega) d^3 k d^3 k.' \quad (5.141)
\end{aligned}
$$

Absorption coefficient for the different kinds of interaction may be worked out by putting the corresponding expression for the matrix element, as discussed below.

5.5.1. INTERBAND ABSORPTION

In interband absorption, a free electron from the valence band with wave vector
k is excited to a free state in the conduction band, having a wave vector **k'**. Such
transitions may occur due to the interaction with the photons alone only in direct
band gap semiconductors in which the conduction band minimum and the valence
band maximum are located at the same point of the Brillouin zone. This is because
photons cannot cause a large change of momentum as required for the indirect
band gap semiconductors in which the conduction band minimum is separated by
a large wave vector from the valence band maximum. Interband transitions may
occur in indirect band gap semiconductors if an additional particle, e.g., a phonon,
an impurity atom or a defect, participates in the interaction along with the photon
and such interactions are comparatively infrequent as three particles are involved.
Quantum well electronics is, therefore, based on interband absorption caused by
the interaction of the electrons with the photons alone. The formula for bulk
materials is first given. The matrix element is [see Eq. (5.54)]

$$|M_0(\mathbf{k}, \mathbf{k'})|^2 = (e^2 A_0^2/4m_0^2)[|\hat{\mathbf{a}}.\hbar\mathbf{k}|^2 f'^2 + |\hat{\mathbf{a}}.\mathbf{p}_{cv}|^2]\delta_{\mathbf{kk'}}. \qquad (5.142)$$

The Krönecker delta function is included to indicate that $\mathbf{k} = \mathbf{k'}$ when the photon
wave vector is assumed to be small in comparison to **k** or **k'** or when **k**-selection
rule applies. The density of states D and D' for bulk materials are $2/(2\pi)^3$ and
$1/(2\pi)^3$. The factor of 2 in D accounts for the spin degeneracy and its absence in
D' indicates that spin is conserved in the transitions. The factor DD' is, however,
changed when the **k**-selection rule applies, since it restricts transition from a k-
value to the same k-value. An electron from a particular state in the valence
band may be excited to a single state and not to all other available states. The
number of states involved in the transitions from the phase space volume $V_c d^3k$
to $V_c d^3k'$ is therefore the density of joint states[5.39,40] $2V_c d^3k/(2\pi)^3$ and not
$2V_c^2 d^3k d^3k'/(2\pi)^6$. Equation (5.141) has for **k** -selection rule the form,

$$\begin{aligned}
\alpha &= V_c(2/(2\pi)^3)(2\pi/\hbar)(e^2 A_0^2/4m_0^2)(n_r/c)\int[|\hat{\mathbf{a}}.\mathbf{p}_{cv}|^2 + \hbar^2|\hat{\mathbf{a}}.\mathbf{k}|^2 f'^2] \\
&\quad \times [f_v(E_{v\mathbf{k}}) - f_c(E_{c\mathbf{k}})]\delta(E_{c\mathbf{k}} - E_{v\mathbf{k}} - \hbar\omega)d^3k, \qquad (5.143)
\end{aligned}$$

where $E_{v\mathbf{k}}$ and $E_{c\mathbf{k}}$ are respectively the energy of the electron with the wave vector
k in the valence band and in the conduction band. These are related to k for an
isotropic parabolic band by the following relations.

$$E_{v\mathbf{k}} = E_{v0} - \hbar^2 k^2/2m_h, \quad E_{c\mathbf{k}} = E_{c0} + \hbar^2 k^2/2m_e \qquad (5.144)$$

where m_e and m_h are respectively the effective mass of the conduction-band and
the valence-band electron, E_{c0} and E_{v0} being the corresponding energies for $k = 0$.

The argument of the delta function may be written as

$$E_{ck} - E_{vk} - \hbar\omega = (\hbar^2 k^2/2m_r) + E_g - \hbar\omega, \quad (5.145)$$

where m_r is called the reduced mass and is defined as

$$1/m_r = 1/m_e + 1/m_h, \quad (5.146)$$

E_g is the band gap, being equal to $E_{c0} - E_{v0}$. The distribution functions for the electrons in the valence band and in the conduction band are indicated respectively by $f_v(\mathbf{k})$ and $f_c(\mathbf{k})$.

The integral is evaluated by substituting

$$\hbar^2 k^2/2m_r = E \quad (5.147)$$

and using the property of the delta function. One obtains,

$$\alpha = C_{ab}(\hbar\omega - E_g)^{1/2}[|\hat{\mathbf{a}}.\mathbf{p}_{cv}|^2 + (2m_r f'/3)(\hbar\omega - E_g)][f_v(E_{vk}) - f_c(E_{ck})], \quad (5.148)$$

where

$$C_{ab} = \sqrt{2}e^2 m_r^{3/2}/\pi m_0^2 \epsilon_0 c n_r \hbar^3 \omega, \quad (5.149)$$

$$k = (2m_r/\hbar^2)^{1/2}(\hbar\omega - E_g)^{1/2}. \quad (5.150)$$

The matrix element $\mathbf{p}_{cv}$ is assumed to be independent of k and $|\hat{\mathbf{a}}.\mathbf{k}|^2$ is replaced by its average value $k^2/3$. A_0^2 has been replaced by$(2\hbar/V_c\epsilon_0 n_r^2\omega)$ [see Eq. (5.5)].

On comparing the magnitudes, the second component of the absorption, arising from $|\hat{\mathbf{a}}.\mathbf{k}|$ is found to be negligible in comparison to that due to $|\hat{\mathbf{a}}.\mathbf{p}_{cv}|^2$. Now, $|\hat{\mathbf{a}}.\mathbf{p}_{cv}|^2$ may be approximated as $\xi(m_0^2/m_c^*)E_g$ [see Eq. (5.53)] with $\xi = 1/6$. It may also be assumed that $f_c(E_{ck}) \approx 0$ and $f_v(E_{vk}) \approx 1$.

Expression for the interband absorption coefficient for bulk materials may, now, be written as,

$$\alpha = C_{ab}\xi(m_0^2/m_c^*)E_g(\hbar\omega - E_g)^{1/2}. \quad (5.151)$$

Expression (5.150) is modified for the quantum wells due to the change in the density of states resulting from quantization. The term $[V_c/(2\pi)^3]d^3k$ is replaced by $[A_c/(2\pi)^2]d^2k_t$, where A_c is the in-plane area of the sample and k_t is the in-plane wave vector. After carrying out the integration, following the same procedure as explained above for the bulk material, the absorption coefficient is found to be given by

$$\alpha = C_{ab2D}\xi_{2D}(m_0^2/m_c^*)E_{gn}\sum H[\hbar\omega - (E_{cn} + E_{vm} + E_g)]\delta_{mn}, \quad (5.152)$$

where

$$C_{ab2D} = e^2 m_r/Ln_r\epsilon_0\omega c m_0^2 \hbar^2, \quad (5.153)$$

L is the width of the well, $H(x)$ is the Heaviside unit function, ξ_{2D} is also different from the constant ξ for the bulk material. E_{vm}, E_{cn} are the energy levels in the valence band and in the conduction band corresponding to $k_t = 0$, between which the transitions occur. The delta function indicates that absorption occurs between levels obeying the rule $m = n$.

It is evident from Eq. (5.152) that the absorption in quantum wells increases in steps as the frequency increases and becomes equal to the difference between the quantized levels of different order. The first step occurs at $\hbar\omega = E_g + E_{c1} + E_{v1}$, the second step at $\hbar\omega = E_g + E_{c2} + E_{v2}$ and so on. It is of interest to note that

$$C_{ab2D} = [\hbar\pi/(2m_r)^{1/2}L]C_{ab} \tag{5.154}$$

and the step in the absorption coefficient for quantum wells with infinite barrier height is approximately equal in magnitude to the absorption coefficient in bulk materials at the corresponding frequencies since for quantum wells with infinite barrier height.

$$E_{cn} + E_{vm} = n^2\hbar^2\pi^2/2m_rL^2 \tag{5.155}$$

and

$$(\hbar\omega - E_g)^{1/2} = (E_{cn} + E_{vm})^{1/2} = n\hbar\pi/(2m_r)^{1/2}L \tag{5.156}$$

This equality, however, does not apply for wells with finite barrier height. In fact, the steps occur at lower frequencies due to the reduced values of $E_{cn} + E_{vm}$. It is, also, interesting to find that the absorption coefficient in quantum wells varies as the inverse of the well width and the total absorption is, therefore, independent of the well width. The total absorption above the threshold frequency in a multiquantum well system therefore, varies directly as the number of wells and does not depend on the the width of the individual wells.

Formulae for the absorption coefficients have been obtained by assuming ideal conditions in which k-selection rule applies. Practical conditions are often so different that the k-selection rule is not strictly valid. Absorption coefficient is estimated for such conditions by using the so-called no-k-selection rule[5.41,42]. The matrix element in such transitions is taken to be independent of k and the absorption coefficient is evaluated from Eq. (5.142) by omitting the $\delta_{kk'}$ term. Expression for the absorption coefficient is obtained by substituting,

$$E_k = E_{vm} - \hbar^2k^2/2m_h, E_{k'} = E_g + E_{cn} + \hbar^2k'^2/2m_e. \tag{5.157}$$

The expression is,

$$\alpha = (e^2m_em_h/4\pi L^2m_0^2\epsilon_0n_rc\omega\hbar^4)|M|^2\sum_n H[\hbar\omega - (E_g + E_{cn} + E_{vm})]$$
$$\times[\hbar\omega - (E_g + E_{cn} + E_{vm})]. \tag{5.158}$$

The no-k-selection rule predicts results which are qualitatively different from those obtained by applying the k-selection rule. The absorption, instead of increasing in steps, increases with ω and the slope changes as the frequency crosses the thresholds for quantized levels of different order.Also, it varies as the inverse square of the well width. The magnitude of such absorption is, however, difficult to estimate theoretically. $|M|^2$ has the dimension of $L^3 p^2$ (p - momentum) and is determined by the impurity concentration. It is therefore required to be treated as an experimental parameter.

5.5.2. EXCITONIC ABSORPTION

Light absorption in the interband transition process discussed in Section 5.5.1 takes place when an electron is transferred from the valence band to the conduction band. Total number of transitions depends, therefore, on the number of states in the two bands. On the other hand, in excitonic absorption, an electron in the valence band is annihilated and an exciton is created in its place. The ground state and the excitonic states are mutually exclusive. Excitons obey Bose statistics[5.43] and there is no limit to the number that can be created at a particular energy, unless it is very large when other processes become effective to limit the number. The absorption coefficient is, therefore, determined by the rate of transitions. If the number of transitions per unit volume per unit time be W and the incident radiation intensity be $I(\omega)$, then the absorption coefficient α is given by

$$\alpha = W\hbar\omega/I(\omega) \qquad (5.159)$$

$W\hbar\omega$ being the energy absorbed per unit volume per unit time.

Transitions per unit time per unit volume is

$$W = (1/V_c)(2\pi/\hbar)|M(f,i)|^2\delta(\hbar\omega - E_g - E_{ex_n}), \qquad (5.160)$$

where E_{ex_n} is the excitonic energy.

Replacing $|M(f,i)|^2$ by Eq. (5.112), we get

$$W = (2\pi/\hbar)(e^2 A_0^2/4m_0^2)|\phi'_n(o)|^2|\hat{\mathbf{a}}.\mathbf{p}_{cv}|^2\delta(\hbar\omega - E_g - E_{ex_n}), \qquad (5.161)$$

$I(\omega)$ may be expressed in terms of A_0^2 as

$$I(\omega) = (1/2)cn_r\epsilon_0\omega^2 A_0^2. \qquad (5.162)$$

The absorption coefficient is

$$\alpha(\omega) = C_{ex}|\phi'_n(o)|^2|\hat{\mathbf{a}}.\mathbf{p}_{cv}|^2\delta(\hbar\omega - E_g - E_{ex_n}), \qquad (5.163)$$

where

$$C_{ex} = \pi e^2/cn_r\epsilon_0\omega m_0^2. \qquad (5.164)$$

This expression indicates that the absorption spectrum will be discrete lines at the frequencies,

$$\omega = (E_g + E_{ex_n})/\hbar. \qquad (5.165)$$

The lines are, however, broadened in experimental samples by lattice vibrations, impurities, strain fields and other crystal imperfections. The lines are given for such broadening by a Lorentzian or a Gaussian curve. The delta function is then replaced by $S(\hbar\omega)$, which has either of the two forms given below.

$$S(\hbar\omega) = (\hbar\Gamma/2\pi)[(\hbar\omega - E_g - E_{ex_n})^2 + (\hbar\Gamma/2)^2]^{-1} \qquad (5.166)$$

$$\text{or} \quad S(\hbar\omega) = (1/2\pi)^{1/2}\sigma^{-1}\exp[-(\hbar\omega - E_g - E_{ex_n})^2/2\sigma^2] \qquad (5.167)$$

Γ and σ give respectively the half power band width of the Lorentzian and the Gaussian curve. Values of these constants are determined by the processes which cause the broadening. Some of these processes are discussed in Section 5.4 in relation to photoluminescence.

Excitonic energy levels are crowded close to the band edge for large values of n. The absorption spectrum due to the discrete levels merge together to form bands. Absorption coefficient for this region is expressed as

$$\alpha(\omega) = C_{ex}|\phi'_n(o)|^2|\hat{\mathbf{a}}.\mathbf{p}_{cv}|^2 S_1(\hbar\omega,) \qquad (5.168)$$

where $S_1(\hbar\omega)$ is defined as the density of excitonic states per unit energy interval, including spin degeneracy.

On using Eq. (5.103) for the excitonic energy, we get for $S_1(\hbar\omega)$,

$$S_1(\hbar\omega) = 2/(dE_{ex_n}/dn) = n^3/R, \qquad (5.169)$$

$$\text{where} \quad R = (e^2/4\pi K_d\epsilon_0)^2\mu/2\hbar^2. \qquad (5.170)$$

On replacing $|\phi'_n(0)|^2$ [Eq. (5.113)] and $S_1(\hbar\omega)$, the absorption coefficient is found to be given by,

$$\alpha(\omega) = C_{ex}(1/\pi)(2\mu/\hbar^2)^{3/2}R^{1/2}|\hat{\mathbf{a}}.\mathbf{p}_{cv}|^2 \qquad (5.171)$$

Expression for the absorption coefficient may be extended also to frequencies such that $\hbar\omega > E_g$ and the relevant excitonic states extend into the conduction band. The expression, obtained by using for $|\phi'_n(o)|^2$ the relevant expression [Eq. (5.114)] is as given below.

$$\alpha(\omega) = C_{ex}(1/2\pi)(2\mu/\hbar^2)^{3/2}[\pi\gamma\exp(\pi\gamma)/\sinh(\pi\gamma)](\hbar\omega - E_g)^{1/2}|\hat{\mathbf{a}}.\mathbf{p}_{cv}|^2, \quad (5.172)$$

$$\text{where} \quad \gamma = 1/ka_B = (2\mu/\hbar^2)^{-1/2}(\hbar\omega - E_g)^{-1/2}a_B^{-1}. \qquad (5.173)$$

Equation (5.172) reduces to Eq. (5.171) as k tends to zero. The absorption due to a few discrete states well below the band edge show up as discrete lines in the absorption, but the excitonic absorption below the band edge merges with the absorption due to the free electron states and thereby eliminates the absorption edge.

5.5.3. ABSORPTION SPECTRUM

The absorption spectrum for experimental samples is the combined effect of band-to-band and excitonic transitions. The nature of the spectrum for bulk samples is illustrated schematically in Fig. 5.5(a) and the experimental results are presented in Fig. 5.5(b). The theoretical curve is obtained by using Eq. (5.151) and

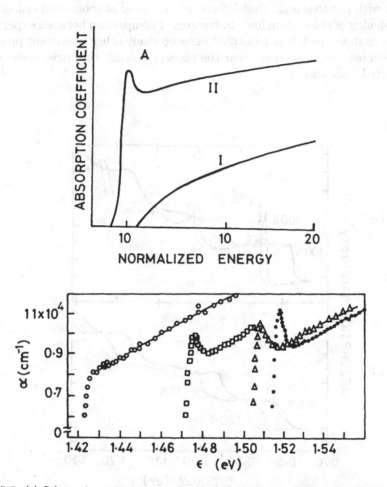

Figure 5.5. (a) Schematic representation of optical absorption coefficient. I - Spectrum without excitonic absorption showing increasing absorption from a threshold wavelength. II - Spectrum with excitonic absorption. A - peak due to first excitonic level, which occurs in the long wavelength end of the spectrum. (b) Experimental excitonic spectra in GaAs: o 294 K, ◇ 186 K, ● 21 K . [After M. D. Sturge, *Phys. Rev.* **127**, 768 (1962); Copyright (1962) by the American Physical Society].

Eq. (5.172) respectively for the band-to-band and the excitonic absorption. The theory gives a temperature-independent spectrum. The experimental curve is, on the other hand, very much dependent on temperature. The absorption near the band edge changes drastically. Excitonic absorption involving the discrete states are clearly visible at low temperatures but at room temperature, such absorption is mostly absent. The binding energy of the excitons [Eq. (5.101)] being small compared to $k_B T$ at room temperature, these are mostly de-excited. In addition, interaction with phonons and other defects are enhanced at room temperature and cause broadening of the exciton line. In fact, exact comparison between experiment and theory, as developed, is not justified because many other important processes are not taken into consideration. Near the absorption edge the curve is less sharp than predicted from theory.

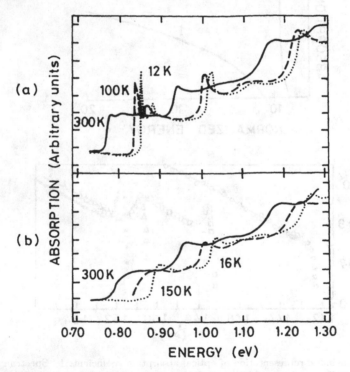

Figure 5.6. Absorption spectra of an InGaAs/AlInAs multiple quantum well structure with well width of 100 Å for different temperatures. Note the steps in the absorption. The well is doped n-type ($N = 4.8 \times 10^{11}$ cm^{-2}). [After G. Livescu, D. A. Miller, D. S. Chemla, M. Ramaswamy, T. Y. Chang, N. Sauer, A. C. Gossard and J. H. English, *IEEE J. Quantum Electron.* **24**, 1677 (1988); Copyright: (=A9 1988=IEEE].

The experimental curve differs also significantly from the theoretical curve at high frequencies at which electrons are transferred deep into the conduction band. This deviation is explained when the nonparabolicity of the energy band and the k-dependence of the cell-periodic functions are taken into account. Absorption for such frequencies are not, however, of much relevance to the operation of quantum well devices. This aspect of the absorption spectrum is not discussed any further.

It has been mentioned earlier (Section 5.3.4) that the binding energy of excitons is much enhanced in quantum wells. This feature is strongly reflected on the absorption spectrum for 2D systems[5.44]. The excitonic absorption peaks are seen even at room temperature as illustrated in Fig. 5.6. More importantly, the absorption spectra exhibit very clearly the excitons formed due to transitions between the higher order quantized energy levels. Such studies, when extended to wells such that the restriction of the selection rule $\Delta n = 0$ (see Section 5.1.2) is relaxed, yield a wealth of data for the identification of the quantized energy levels. Experiments have been done on parabolic wells[5.45] (the composition of the well varies such that the conduction-band edge has a parabolic shape), and absorption

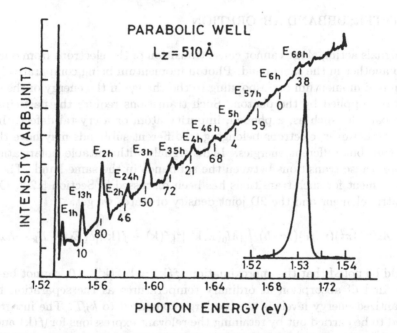

Figure 5.7. Excitonic spectra in a prabolic quantum well. The peaks correspond to a transition from a heavy-hole or a light-hole level to the nth electron level. The subscript nmh or nml indicates transitions from the nth heavy or light hole-level to the nth electron level. Note that the selection rule $\Delta n = 0$ is not obeyed in a parabolic well. [After R. C. Miller, A. C. Gossard, D. A. Kleinman and O. Munteanu, *Phys. Rev. B* **29**, 3740 (1984); Copyright (1984) by the American Physical Society].

spectrum has been obtained as shown in Fig. 5.7., which exhibits a large number
of excitonic peaks. Such spectra have been usefully employed[5.46] to determine
the value of band offset and provide direct evidence of quantization of the energy
levels.

Absorption has been discussed in this section for signals of low amplitude,
such that the density of the carriers or of the excitons, created in the absorption
process, is not large enough to produce any significant effect. The absorption coefficient may be considered independent of the light intensity for such conditions.
As the intensity is increased, other processes become effective to make the absorption coefficient dependent on the magnitude of the light intensity. This aspect of
absorption is discussed in Section 5.5.

It should, however, be mentioned in this section that band-to-band transitions
should show a step increase at the threshold frequency as characteristic of the 2D
behavior. This is usually masked by the excitonic absorption. However, when excitonic absorption is washed out, such step increase is often observed, as illustrated in
Fig. 5.6.

5.5.4. INTERSUBBAND ABSORPTION

Light signals acting alone cannot cause transition of the electrons from one energy
level to another in the same band. Photon momentum being comparatively small,
the required momentum corresponding to the change in the energy of the electron
cannot be supplied by the photon. Such transitions require the participation of
a third particle, such as, a phonon, impurity atom or a crystal defect. In quantum wells, however, electrons belonging to different subbands may have the same
momentum but different energies. Light signals with suitable polarization may,
therefore, cause transitions between the subbands in the same band. The optical
matrix element for such transitions has been discussed in Section 5.2.2. On using
this matrix element and the 2D joint density of states we get,

$$\alpha = A_c(2/4\pi^2)(n_r/c)(2\pi/\hbar) \int |M_0(\mathbf{k}, \mathbf{k}')|^2 [f(\mathbf{k}) - f(\mathbf{k}')] \delta(E_{\mathbf{k}'} - E_{\mathbf{k}} - \hbar\omega) d^3k$$

$$(5.174)$$

It should be noted that the approximation, $f(\mathbf{k}) = 1, f(\mathbf{k}') = 0$, cannot be applied
to this kind of absorption at ordinary temperatures as the separation between
the quantized energy levels is not too large compared to $k_B T$. The integration is
required to be carried out by retaining the relevant expressions for $f(\mathbf{k})$ and $f(\mathbf{k}')$.
We assume a parabolic energy band so that $E_{\mathbf{k}}$ and $E_{\mathbf{k}'}$ may be expressed as,

$$E_{\mathbf{k}} = E_{cm} + \hbar^2 k^2 / 2m_e, \quad E_{\mathbf{k}'} = E_{cn} + \hbar^2 k'^2 / 2m_e, \quad (5.175)$$

where m_e is the electron effective mass, E_{cm} and E_{cn} are the energies of the mth
and the nth subbands between which the transition occurs. The distribution

functions may also be written as,

$$f(\mathbf{k}) = \{\exp[(E_\mathbf{k} - E_F)/k_BT] + 1\}^{-1}, \quad f(\mathbf{k'}) = \{\exp[(E_{\mathbf{k'}} - E_F)/k_BT] + 1\}^{-1},$$
$$(5.176)$$

where E_F is the Fermi energy, k_B is the Boltzmann constant and T is the sample temperature. It may also be recalled that

$$|M(\mathbf{k}, \mathbf{k'})|^2 = (e^2 A_0^2/4m_0^2)|a_z < F_n|\partial/\hbar\partial z|F_m > |^2\delta_{\mathbf{kk'}},\qquad(5.177)$$

where F_n and F_m are the envelope functions for the nth and the mth subband, a_z is the component of polarization vector in the z direction, i.e., the direction of the width of the quantum well. The photon wave vector is assumed to be small enough for being neglected in comparison to $\mathbf{k}$ or $\mathbf{k'}$. Consequently, $\mathbf{k} = \mathbf{k'}$.

On using the above relations we get,

$$\alpha = C_{\text{int}}\cos^2\theta| < F_n|\partial/\partial z|F_m > |^2\ln\langle\{1 + \exp[(E_F - E_{cm})/k_BT]\}$$
$$\times\{1 + \exp[(E_F - E_{cn})/k_BT]\}^{-1}\rangle\delta(E_{cn} - E_{cm} - \hbar\omega),\qquad(5.178)$$

where θ is the angle between the polarization vector of the radiation and the z direction,

$$C_{\text{int}} = e^2 k_BT/L\epsilon_0 n_r\omega c m_e.\qquad(5.179)$$

The intersubband absorption occurs ideally at a single frequency given by

$$\omega_0 = (E_{cn} - E_{cm})/\hbar.\qquad(5.180)$$

In practice, however, the absorption spectrum is modified by various factors, e.g., nonuniformity in well width, electron - electron collisions and electron-lattice collisions. The effect of such broadening mechanism is represented by replacing the delta function by a Lorentzian function[5.47] as in the case of the excitonic absorption. The absorption coefficient is then written as,

$$\alpha = C_{\text{int}}\cos^2\theta < F_n|\partial/\partial z|F_m >^2 (\pi\hbar^2/m_e k_BT)(N_{sm} - N_{sn})$$
$$\times(\Gamma/2\pi)[(\hbar\omega - \hbar\omega_0)^2 + (\Gamma/2)^2]^{-1},\qquad(5.181)$$

where Γ is the broadening parameter, which is equal to the full width at half the maximum of absorption (FWHM). The Fermi functions have also been replaced by the carrier concentrations in the two bands by using the relation,

$$N_{si} = (m_e k_BT/\pi\hbar^2)\ln\{1 + \exp[(E_F - E_{ci})/k_BT]\},\qquad(5.182)$$

where N_{si} is the total carrier concentration in the ith band, the subband energy for which is E_{ci}.

Intersubband absorption has been observed in quantum wells of $Al_{1-x} Ga_x As/GaAs$[5.48-50], $Al_{0.48}In_{0.52}As/Ga_{0.47}In_{0.53}As$[5.51-54], $Ga_{0.47}In_{0.53}As/$

InP[5.55], $In_{0.15}Ga_{0.85}As/Al_{0.35}Ga_{0.65}As$[5.56], Si/Si_xGe_{1-x}[5.57] and more recently in $Ga_{0.5}In_{0.5}P$ /GaAs[5.58] system. The experimental values of the oscillator strength have been found to be close to those given by the theory. Refinements in the theory has also been made by including the effects of the energy band nonparabolicity which is expected to be important [5.59], since in this kind of absorption electrons at higher energies are involved. The effect is found to cause a shift in the peak frequency and some asymmetry in the line shape.

The restriction that the intersubband absorption may occur only with light signals having a polarization component in the direction of the well width, has been attempted to be removed by using gratings[5.60-62], or wells with semiconductors of anisotropic effective mass [5.63] or more recently by using the p-type wells[5.64]. The basic theory of these modifications is the same as presented above. Only the anisotropies of the effective mass have to be properly considered to explain the observed results. These are not discussed any further. It should, however, be mentioned that the study of intersubband absorption is being pursued with different structures[5.65], as it is expected to provide the basis of new kinds of long wavelength infrared detectors.

5.6. Quantum - Confined Stark Effect

The shift of the absorption edge in bulk semiconductors due to an electric field is known as Franz-Keldysh effect[5.66,67]. A significant change, may be produced by utilizing this effect only by applying very large fields. In addition, the excitonic effects are less prominent in bulk materials due to thermal dissociation at room temperature. Effects are further diminished in the presence of an electric field due to decreased overlap between the electrons and the holes. The absorption edge is smeared out as a result and Franz-Keldysh effect has no practical application. In quantum wells, on the other hand, excitonic effects persist strongly at room temperature and at the same time an electric field does not diminish much the coupling between the electrons and the holes as they are confined within the well. The absorption edge, therefore, remains sharp and band gap modification by an electric field has been used successfully for realizing practical devices in quantum-well electronics.

The effect of the electric field appears as a shift in the quantized energy levels of the excitons and also of free carriers and as a result, the excitonic peaks as well as the absorption edge shift. The shift is similar to the shift in the atomic or molecular emission lines in the presence of an electric field, which is known as Stark effect[5.68] after its discoverer. In the case of quantum wells, the shift in the energy levels due to an electric field is often referred as quantum-confined Stark effect. Theory of the effect and some experimental results are presented in this section. Practical application of the effect will be discussed in Section 9.2.

The effect is analyzed by using the envelope functions formalism, discussed in Section 4.1. We consider a single quantum well of width L and barrier potential V_0. The potential distribution in the well in the presence of an electric field is as shown in Fig. 5.8. The equation for the envelope function is

$$[-(\hbar^2/2m_i^*)|\nabla_z^2 + V_{ci} + |e|\mathcal{E}z]F_i(z) = EF_i(z), \qquad (5.183)$$

where m_i^*, V_{ci} and F_i are respectively the effective mass, the conduction-band edge potential and the envelope function for the ith layer; $\mathcal{E}$ is the electric field applied in the positive z direction and $|e|$ is the magnitude of the charge of the carrier. Holes are treated in the same manner as electrons, by assuming an isotropic dispersion relation with an average effective mass. Solution is here illustrated for the electrons only.

It is evident that the electrons remain confined within the well only if the barrier potential is infinitely large. Energy levels may be obtained for this case by applying the method discussed in Section 4.2. For finite barrier, on the other hand, the well is leaky as the electrons escape by tunneling when an electric field is applied. There is, therefore, no bound state. But the electrons remain for a long time at some particular state before escaping. These states correspond to

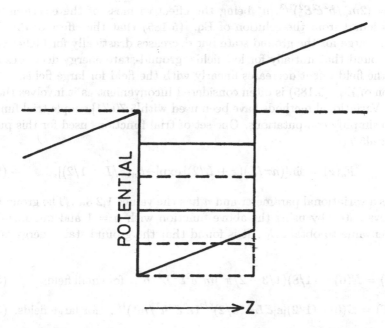

Figure 5.8. Potential distribution and the energy levels in the presence of an external field. Dashed lines give the potential and the energy levels in the field-free condition. Full lines give the same quantities in the presence of a field.

the bound states of the field - free well. These are the so-called quasi-bound states. These states determine the emission lines and absorption edges. The objective of the analysis is to determine the quasi-bound states. It should be mentioned, however, that the finite value of the confinement time of the carriers broaden the emission lines and the absorption edges. The analysis gives also the magnitude of this broadening, although it may not be of practical interest as it is significant at fields, much larger than the operating fields of the devices, utilizing the effect.

Solutions for the infinite barrier wells are obtained by noting that Eq. (5.183) is the Airy equation and $F_i(z)$ may be written in general as[5.69,70].

$$F_i(z) = C_i A_1[-\alpha(|e|\mathcal{E}z - E)] + C_2 B_i[-\alpha(|e|\mathcal{E}z - E)], \qquad (5.184)$$

where C_1 and C_2 are constants, A_i and B_i are the two independent Airy functions. For infinite barrier wells, $F_i(z) = 0$ for $|z| = L/2$ and application of this boundary condition gives the following equation for the eigenvalues of energy.

$$A_i[-\alpha(|e|\mathcal{E}L/2 - E)]B_i[\alpha(|e|\mathcal{E}L/2 + E)]$$
$$= A_i[\alpha(|e|\mathcal{E}L/2 + E)]B_i[-\alpha(|e|\mathcal{E}L/2 - E)], \qquad (5.185)$$

where $\alpha = (2m_e^*/\hbar^2 e^2 \mathcal{E}^2)^{1/3}, m_e^*$ being the effective mass of the electrons in the well. It is found from the solution of Eq. (5.185) that the effect of the field is significantly large for the ground state but decreases drastically for higher energy. Also, it is found that initially for low fields, ground-state energy decreases as the square of the field but it decreases linearly with the field for large fields.

Solution of Eq. (5.185) is often considered inconvenient as it involves the Airy functions. Variational methods have been used with[5.70,71] simple trial functions in order to simplify computations. One set of trial functions used for this purpose has the form[5.71],

$$F_n(z) = \sin[(n\pi/L)(z + L/2)] \exp[-\beta_n(z/L + 1/2)], \qquad (5.186)$$

where β_n is a variational parameter and n has the values 1,2,3.... The ground state energy is evaluated by using the above function with $n = 1$ and minimizing the energy eigenvalue to obtain β_1 . It is found that the ground state energy is given by

$$E(\mathcal{E}) = E(0) - (1/8)(1/3 - 2/\pi^2)m_e^* e^2 \mathcal{E}^2 L^4/\hbar^2, \quad \text{for small fields} \qquad (5.187)$$

and $E(\mathcal{E}) = E(0) - (1/2)|e|\mathcal{E}L + (3/2)^{5/2}(e^2\mathcal{E}^2\hbar^2/m_e^*)^{1/3}, \quad \text{for large fields.}$ (5.188)

The values obtained from the above expressions are very close to those given by the exact solutions obtained from Eq. (5.185). Values of the higher energy levels may be obtained by suitable combinations of these functions to ensure normalization.

However, as infinite barriers idealize the problem and for practical devices the barrier is finite these methods are not discussed any further.

A straightforward application of the usual methods is not possible for the analysis of Stark levels in wells with finite barriers, since there is no bound state. Solutions have been obtained by applying several intuitive approaches[5.71]. In one method, it has been argued that since the electron will behave as a free electron after its escape, the wave function should give a propagating wave for $z < -L/2$. Accordingly, solutions have been taken to be given by,

$$F_1(z) = C_1 A_i[\phi_b(z, E)] + i B_i[\phi_b(z, E)], \quad \text{for } z < -L/2; \quad (5.189)$$

$$F_2(z) = C_2 A_i[\phi_w(z, E)] + C_3 B_i[\phi_w(z, E)], \quad \text{for } -L/2 < z < L/2; \quad (5.190)$$

$$F_3(z) = C_4 A_i[\phi_b(z, E)], \quad \text{for } z > L/2, \quad (5.191)$$

where

$$\phi_b(z, E) = -(2m_b^*/\hbar^2 e^2 \mathcal{E}^2)^{1/3}[|e|\mathcal{E}z + (V - E)], \quad (5.192)$$

$$\phi_w(z, E) = -(2m_w^*/\hbar^2 e^2 \mathcal{E}^2)^{1/3}[|e|\mathcal{E}z - E], \quad (5.193)$$

m_b^* and m_w^* are respectively the effective mass in the barrier and in the well. Equation for the energy eigenvalues is obtained, when the above solutions are used to satisfy the boundary conditions at $|z| = L/2$. Solutions for the GaAs/Ga$_x$ Al$_{1-x}$ As well are given in Fig. 5.9., which show some interesting features in the shift of hole energies. The shift is initially negative, reaches a minimum and then changes rapidly to a positive value, whereas for electrons the shift is negative and the magnitude increases monotonically with the field.

A second method[5.73] used for the purpose is the same as developed by Gamow for the evaluation of the decay time of alpha particles. It is assumed that there exists an infinite potential barrier at $z = -L_\infty$, where L_∞ is a very large quantity. It is then assumed that $F(-L_\infty) = 0$. On applying this condition, $F(z)$ is found to be given for $z < -L/2$, by

$$F(z) = C_1\{A_i[\phi_b(z, E)] - A_i[\phi_b(-L_\infty, E)]B_i[\phi_b(z, E)]/B_i[\phi_b(-L_\infty, E)]\}. \quad (5.194)$$

The other two equations remain unchanged. Solution of the equations may now be obtained by applying the boundary conditions at $z = -L_\infty$ and $|z| = L/2$, as the electron is effectively bound in the negative z direction at $z = -L_\infty$ and decays exponentially in the positive z direction. The solution can also be normalized by using the relation,

$$\int_{-L_\infty}^{L_\infty} F(z)^2 dz = 1. \quad (5.195)$$

The energies of the quasi-bound states are then obtained by using the condition that for these states,

$$\rho \int_{-L/2}^{L/2} F(z)^2 dz = \text{a maximum}, \quad (5.196)$$

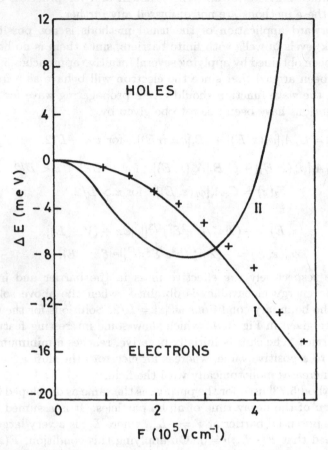

Figure 5.9. Calculated energy shifts of electrons and holes for different applied fields. I - Electrons. II - Holes. [After E. J. Austin and M. Jaros, *Phys. Rev. B* **31**, 5569 (1985); Copyright (1985) by the American Physical Society].

where ρ is the density of states and is given by

$$\rho = (2m_b^*/\hbar^2)^{1/2} L_\infty^{1/2}/(e\mathcal{E})^{1/2}. \tag{5.197}$$

A numerical method[5.69] has also been proposed, in which a free electron is assumed to be incident from the left hand side with a function $\exp(ik_0 z)$, ($k_0^2 = (2m_b^*/\hbar^2)E$. The transmitted amplitude T at the right-hand interface of the well is then evaluated by using the solutions based on Airy functions.

Alternately, the well and the barrier region are divided into sections and the transmission amplitude is obtained by using the formula,

$$\begin{bmatrix} T \\ 0 \end{bmatrix} = \prod_{n=1}^{r} (1/2) \begin{bmatrix} 1+\gamma_n & 1-\gamma_n \\ 1-\gamma_n & 1+\gamma_n \end{bmatrix} \cdot \begin{bmatrix} \exp(i\beta_n) & 0 \\ 0 & \exp(-i\beta_n) \end{bmatrix} \cdot \begin{bmatrix} 1 \\ R \end{bmatrix} \quad (5.198)$$

where $\gamma_n = k_n m_{n+1}^* / k_{n+1} m_n^*$, $\beta_n = k_n(d_n - d_{n-1})$, $\beta_1 = 0$ $k_n = (2m_n^*/\hbar^2)(E - V_n)^{1/2}$.

Equation (5.198) is obtained by considering that solutions for $F(z)$ may be written as

$$F_i(z) = A_i \exp(k_i z) + B_i \exp(-k_i z), \quad (5.199)$$

where the index i indicates the section under consideration. On applying the boundary conditions,

$$F_i(z) = F_{i+1}(z), \quad (1/m_i)\partial_z F_i(z) = (1/m_{i+1})\partial_z F_{i+1}(z) \quad (5.200)$$

at the interface, A_{n+1}, B_{n+1} are found to be related to A_n, B_n by the following relations,

$$\begin{bmatrix} A_{n+1} \\ B_{n+1} \end{bmatrix} = (1/2) \begin{bmatrix} 1+\gamma_n & 1-\gamma_n \\ 1-\gamma_n & 1+\gamma_n \end{bmatrix} \cdot \begin{bmatrix} A_n \\ B_n \end{bmatrix} \quad (5.201)$$

Equation (5.198) is obtained by successively using Eq. (5.201) and introducing the phase shifts suffered by the forward and the reflected wave in travelling from the $n, n+1$ interface at d_n to the $n+1, n+2$ interface at d_{n+1}. Values of energy for which T is maximum gives the quasi-bound states. Results obtained by this method for the same quantum well as studied by the other two methods are found to be nearly the same, particularly for the ground state.

The three methods discussed above are based on different approaches but give similar results. Either of them may be applied when computers are used to obtain the solutions. However, the third method offers some advantages when effects of the energy band nonparabolicity and of the band bending caused by the redistribution of carriers are to be taken account.

It has been discussed in Section 5.5 that absorption near the band edge is caused predominantly by the creation of excitons in quantum wells. The modification of the exciton levels are therefore required to be analyzed in addition to the shift of the free-carrier levels for the interpretation of the quantum-confined Stark effect. Analysis of Stark shifts of excitonic levels is, on the other hand, fairly complex. Important results may, however, be obtained with reasonable accuracy by introducing suitable approximations as discussed below: the equation to be solved may be written as[5.69],

$$\begin{aligned} \{ -(\hbar^2/2\mu)(\partial^2/\partial x^2 + \partial^2/\partial y^2) - (\partial/\partial z_e)(\hbar^2/2m_{ez})\partial/\partial z_e + |e|\mathcal{E} z_e \\ + V_e H(|z| - L/2) - (\partial/\partial z_h)(\hbar^2/2m_{hz})(\partial/\partial z_h) - |e|\mathcal{E} z_h \\ + V_h H(|z| - L/2) + E_g - (e^2/4\pi\epsilon)[x^2 + y^2 + (z_e - z_h)^2]^{-1/2} \} \psi_{\text{ex}} \\ = E_{\text{ex}} \psi_{\text{ex}}, \quad (5.202) \end{aligned}$$

where ψ_{ex} and E_{ex} are the wave functions and energy eigenvalues for the excitons. μ is the reduced mass, x, y are the in-plane relative coordinates, $z_e(z_h)$, $m_{ez}(m_{hz})$, $V_e(V_h)$ are respectively the z coordinate, the effective mass and the barrier potential for electrons (holes). E_g is the energy band gap of the well material assumed to be negligibly affected by the electric field. The well width is taken to be L and $H(x)$ is the Heaviside unit function. The permittivity is taken to be the same in the well and in the barrier material, so that the image effects may be considered negligible. The term involving the centre-of-mass, M, has not been included, as for optical interaction, the wave vector, K, for the motion of the centre of mass may be taken to be zero as explained in Section 5.3.4. The holes are also represented by an isotropic mass or an anisotropic mass with different z and in-plane components. The actual description of holes may, however, be more complex as discussed in Section 4.3. Equation (5.202) has been solved by using variational methods. The following discussion is in accordance with the initial analysis of Miller et al[5.69]. The solution is obtained by using the trial function,

$$\psi_{ex}(\mathbf{r}, z_e, z_h) = \psi_e(z_e)\psi_h(z_h)\phi_{e-h}(\mathbf{r}) \tag{5.203}$$

where $\psi_e(z_e)$ and $\psi_h(z_h)$ are the solutions respectively for the electrons and the holes in the absence of the interaction. These functions satisfy the equations,

$$[-(\partial/\partial z_e)(\hbar^2/2m_{ez})(\partial/\partial z_e) + |e|\mathcal{E}z_e + V_e H(|z| - L/2)]\psi_e(z_e) = E_e\psi_e(z_e), \tag{5.204}$$

$$[-(\partial/\partial z_h)(\hbar^2/2m_{hz})(\partial/\partial z_h) + |e|\mathcal{E}z_h + V_h H(|z| - L/2)]\psi_h(z_h) = E_h\psi_h(z_h), \tag{5.205}$$

where $E_e(E_h)$ are the energy eigenvalues for the electrons (holes) when these are not coupled. The wave functions $\psi_e(z_e)$ and $\psi_h(z_h)$ are evaluated by the methods discussed above. An approximate method has been used for the sake of convenience. Solution for the infinite-barrier well is used instead of that for the finite-barrier well, by taking advantage of the fact that the two solutions may be made to agree by using an effective width for the well to take account of the extension of the wave function into the barrier region for the finite barrier height. The effective width may be chosen from the consideration of only the right energy eigenvalue or may be evaluated by considering the effective penetration of the wave function on the two sides of the well for a chosen energy. For $\phi_{e-h}(\mathbf{r})$, a trial function is chosen; one such function is,

$$\phi_{e-h}(\mathbf{r}) = (2/\pi)^{1/2}(1/\lambda)\exp(-r/\lambda), \tag{5.206}$$

where λ is the variational parameter. The eigenvalues are finally obtained from,

$$E = <\psi_{ex}|H|\psi_{ex}>, \tag{5.207}$$

where H is the complete Hamiltonian in Eq. (5.202) and E is the total eigenvalue. It may, however, be decomposed as follows:

$$E = E_e + E_h + E_{exk} + E_{exV} \tag{5.208}$$

where E_e and E_h are given by Eq. (5.204) and (5.205),

$$H_{\text{exk}} = -(\hbar^2/2\mu)(\partial^2/\partial x^2 + \partial^2/\partial y^2),$$

$$E_{\text{exk}} = <\phi_{e-h}|H_{\text{exk}}|\phi_{e-h}>, \qquad (5.209)$$

$$E_{\text{exV}} = <\psi_{\text{ex}}|V(\mathbf{r}_e, \mathbf{r}_h|)\psi_{\text{ex}}>, \qquad (5.210)$$

$V(\mathbf{r}_e, \mathbf{r}_h)$ represents the Coulomb potential term, E_{ex} is required to be evaluated numerically and for the sake of convenience often variational wave functions are used for ψ_e and ψ_h for this purpose, instead of the exact solutions given by the Airy functions.

An alternative approach has been used to evaluate the exciton binding energy. The Hamiltonian is separated into components representing the 2D electron, the hole and the hydrogen atom. The solution is obtained in terms of some tabulated functions[6.74]. Results of such calculations are presented in Fig. 5.10 for the GaAs/Al$_{1-x}$Ga$_x$As system. Exciton binding energy is found to decrease with increasing field and at a larger rate for wide wells.

The net effect of the electric field as discussed in this section is to shift the electron and the hole subband energies and also the exciton binding energies and thereby shift the peak position and the edge of the absorption curves. The change in the electron energy being faster than in the heavy-hole energy the band gap

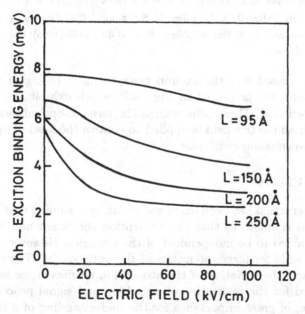

Figure 5.10. Heavy hole exciton binding energy for different applied fields and well widths. Well widths are indicated by the numbers on the curves. [After Der-San Chuu and Yu-Tai Shi, *Phys. Rev.* B **44**, 8054 (1991); Copyright (1991) by the American Physical Society].

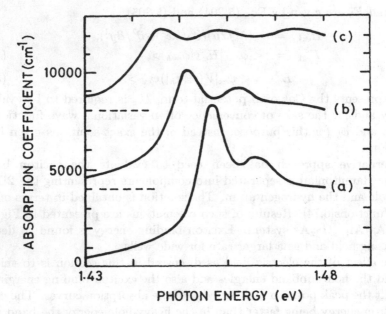

Figure 5.11. Absorption spectra of a quantum well for different applied fields, $\mathcal{E}$. The zeroes for the different fields are displaced for clarity. (a) $\mathcal{E} = 1 \times 10^4$ V/cm, (b) $\mathcal{E} = 4.7 \times 10^4$ V/cm, (c) $\mathcal{E} = 7.3 \times 10^4$ V/cm. [After D. A. B. Miller, D. S. Chemla, T. C. Damen, A. C. Gossard, W. Wiegmann, T. H. Wood and C. A. Burrus, *Phys. Rev. B* **32** , 1043 (1985); Copyright (1985) by the American Physical Society].

is found to be decreased and the exciton peak to move to higher wavelengths. Experimental results are presented in Fig. 6.7, which exhibit the theoretically predicted characteristics. These results form the basis of optical modulators and switches, in which an electric field is applied to control the level of optical signals by varying the transmission coefficient.

5.7. Nonlinear Effects

Electron-optic interaction has been discussed so far by assuming that the intensity of the optic signal is small, so that the absorption coefficient and the refractive index may be assumed to be independent of the intensity. However, this assumption ceases to be valid for large intensities of the optic signal. The change in the refractive index and the variation of the absorption coefficient for large intensity signals are utilized for the realization of switching and signal processing devices and are, therefore, of great importance for the understanding of a class of quantum well devices. The theory of the optic nonlinearity and a detailed discussion of the experimental characteristics are, however, beyond the scope of this book. Only simplified theories of the nonlinearity and phenomenological explanations are presented in this section, which should help in the explanation of the operating principles of photonic quantum well devices, discussed in Chapter 10.

5.7.1. NONRESONANT NONLINEARITY

The physical processes which cause nonlinearity are broadly divided into two classes. the nonresonant and the resonant processes. The nonresonant processes produce nonlinearity through the nonlinear response of the bound electron cloud and the free charges. These are effective for all the frequencies, both below and above the frequency of E_g/h, E_g being the band gap. However, the phenomenon is usually studied for frequencies below E_g/h, so that the results are not mixed with those arising from carrier generation.

The response time for the nonresonant nonlinearity is of the order of the period of the optic signal as the electrons and the charge cloud respond almost instantaneously. The polarization due to the displacement of the charges by the optic signal may, in general, be written as,

$$\mathbf{P} = [\chi^{(1)}].\mathcal{E} + [\chi^{(2)}].\mathcal{E}.\mathcal{E} + [\chi^{(3)}].\mathcal{E}.\mathcal{E}.\mathcal{E} + \dots \tag{5.211}$$

where $[\chi^{(n)}]$ is the susceptibility of the nth order, $\mathcal{E}$ is the electric field associated with the optic signal. $[\chi^{(n)}]$'s are, in general, tensors. However, for isotropic materials these are scalars, and $\chi^{(2)}$ may also be taken as zero. Limiting the expansion to the third order term, and assuming that the field is given by,

$$\mathcal{E} = (1/2)[\mathcal{E}_0 \exp(i\omega t) + c.c], \tag{5.212}$$

the polarization for the signal frequency, ω, may be expressed as,

$$P = [\chi^{(1)} + (3/4)\chi^{(3)}\mathcal{E}_0^2]\mathcal{E}_0 \cos(\omega t). \tag{5.213}$$

The refractive index, n_r, is, hence,

$$n_r = n_{r0} + (3/4)\chi^{(3)}\mathcal{E}_0^2/2\epsilon_0, \tag{5.214}$$

where ϵ_0 is the free-space permittivity and n_{r0} is the small-signal refractive index, given by, $n_{r0}^2 = 1 + \chi^{(1)}/\epsilon_0.\mathcal{E}_0^2$ may be replaced in terms of the light intensity I, by using the relation,

$$I/(c/n_r) = (1/2)\epsilon\mathcal{E}_0^2, \quad \epsilon = n_r^2\epsilon_0, \tag{5.215}$$

to obtain,

$$n_r = n_{r0} + [(3/4)\chi^{(3)}/n_r^2\epsilon_0^2 c]I = n_{r0} + n_2 I, \tag{5.216}$$

$$n_2 = 3\chi^{(3)}/4n_r^2\epsilon_0^2 c, \tag{5.217}$$

where c is the velocity of light in free space.

The third-order coefficient $\chi^{(3)}$ is usually expressed in e.s.u. and n_2 in cm^2/kW, assuming that I is given in kW/cm^2 . On using these units, (5.217) gives,

$$n_2 = 4(\pi/n_r)^2 \chi^{(3)},\tag{5.218}$$

The nonresonant nonlinearity arising from the bound charges are caused by the excitation of the virtual electron-hole, (e-h), pairs. Expression for the susceptibility is obtained by summing over complete sets of such states after taking into account the coupling of the fields of the Maxwell's equations through the nonlinearity. The basic aspects of the phenomenon may, however, be clarified by including an anharmonic force of restitution in the classical theory of polarization.

The polarization due to the bound charges is given by,

$$P = Nex,\tag{5.219}$$

where N is the concentration of the atoms and x is the displacement of the charge cloud having an effective charge e. The equation of motion giving x is,

$$\dot{x} + \gamma\ddot{x} + \omega_0^2 x + \mu x^3 = (e/m)\mathcal{E},\tag{5.220}$$

where $\omega_0^2 x$ and μx are respectively the harmonic and the anharmonic forces of restitution terms, γ is the damping factor, and m is the effective mass of the charge cloud. Considering that $\mu x \ll \omega_0^2 x$, the equation may be solved by the method of successive approximation. We get,

$$\chi^{(3)} = (m\mu/N^3 e^4)[\chi^{(1)}]^4.\tag{5.221}$$

Expression (5.221) may be used to estimate the magnitude of $\chi^{(3)}$ from the knowledge of $\chi^{(1)}$ and μ. The value of μ has been estimated by assuming that the linear and the nonlinear restoring forces are equal when x equals the lattice constant. The estimated values[5.75] of $\chi^{(3)}$ for different semiconductors is found to be of the order of 10^{-10} -10^{-11} e.s.u.

A second source of nonresonant nonlinearity arises from the response of the free electrons in doped semiconductors with strong nonparabolic bands. This contribution may be worked out by using the dispersion relation (see Chapter. 4),

$$\hbar^2 k^2/2m^* = E(1 + \alpha E), \alpha \approx 1/E_g,\tag{5.222}$$

and the equation of motion,

$$\hbar\dot{k} = e\mathcal{E},\tag{5.223}$$

The velocity of the electron, v, is,

$$v = (1/\hbar)\nabla_k E = (\hbar k/m^*)/(1 + 2\alpha E) \approx (\hbar k/m^*) - \alpha m^*(\hbar k/m^*)^3.\tag{5.224}$$

The corresponding displacement at the fundamental frequency for an optic field, $\mathcal{E}_0 \exp(i\omega t)$, is

$$x = [-(e/m^*\omega^2) - (3/4)(e^3/m^*\omega^4)\alpha\mathcal{E}_0^2]\mathcal{E}_0 \exp(i\omega t)/i\omega]\mathcal{E}_0 \exp(i\omega t). \quad (5.225)$$

Using Eq. (5.225) we get for $\chi^{(3)}$,

$$\chi^{(3)} = (\alpha/N)[\chi^{(1)}]^2. \quad (5.226)$$

Nonresonant nonlinearity arising from this process has been measured in low band gap materials, viz., InSb. The magnitude of $\chi^{(3)}$ is found[5.76] to be of the order of 10^{-10} e.s.u. for the electron concentration of 2×10^{16} /cm^3 , which is in agreement with the value given by Eq. (5.226).

5.7.2. RESONANT NONLINEARITY

Optic signals with photon energies larger than the band gap excite electrons from the valence band, which may form excitons or behave as free e-h pairs. The nonlinearity produced by such real excitation of the electrons is referred as resonant nonlinearity. It may be distinguished from the nonresonant nonlinearity by considering that in this case, the nonlinearity is produced indirectly by the photo-excited carriers, whereas the nonresonant nonlinearity is produced directly by the optic field.

The excited carriers cause nonlinearity by changing the refractive index. This change may be evaluated[5.77] from the knowledge of the absorption coefficient by using the Kramers- Kronig relation between the imaginary and the real parts of permittivity, $\epsilon_2(\omega)$ and $\epsilon_1(\omega)$. The relation is ,

$$\epsilon_1(\omega) = \epsilon_0 + (2/\pi)P_r \int \omega'\epsilon_2(\omega')(\omega'^2 - \omega^2)^{-1}d\omega', \quad (5.227)$$

where P_r indicates the principal value of the integral, i.e., the value of the integral excluding the singular point $\omega = \omega'$. We note that the absorption coefficient $\alpha(\omega)$ is given in terms of $\epsilon_2(\omega)$ by the relation,

$$\alpha(\omega) = (\omega/n_r c\epsilon_0)\epsilon_2(\omega) \quad (5.228)$$

The change in the refractive index Δn_r is given by,

$$\Delta n_r = \Delta\epsilon/2n_r\epsilon_0. \quad (5.229)$$

Using the expression for α given by Eq. (5.228) and the relations (5.227), we get

$$\Delta n_r = (\hbar^2 e^2/4\pi^3 m_0^2\epsilon_0 n_r) \int |p_{cv}|^2 (E_{ck'} - E_{vk})^{-1}$$
$$\times [(E_{ck'} - E_{vk})^2 - \hbar^2\omega^2][f_c(E_{k'}) - f_v(E_k)]d^3k, \quad (5.230)$$

where $\mathbf{k}$ and $\mathbf{k}'$ are the initial and the final wave vectors, the suffixes c and v indicate the conduction band and the valence band, $f(E_{\mathbf{k}})$ is the energy dustribution function, ω is the frequency of the optic signal, other symbols have the same meaning as defined earlier. Replacing the momentum matrix element by the value given by the $\mathbf{k}.\mathbf{p}$ perturbation method expression (5.230) has been simplified to,

$$\Delta n = (N_e e^2 / 2n_r m_e^* \epsilon_0 \omega^2)[\omega_g^2/(\omega_g^2 - \omega^2)], \tag{5.231}$$

where $\omega_g = E_g/\hbar$, E_g being the band gap, n_r is the refractive index and N_e is the photo-excited electron density. N_e may be eliminated by using the expression,

$$N_e = (\eta \alpha \tau / \hbar \omega)I, \tag{5.232}$$

where I is the intensity of the incident light, α is the absorption coefficient, η is the quantum efficiency and τ is the lifetime of the carriers or the duration of the light pulse, whichever is smaller.

Substituting (5.231) in (5.230) we get for n_2,

$$n_2 = (\eta \alpha \tau e^2 / 2n_r m_e^* \epsilon_0 \hbar \omega^3)[\omega_g^2/(\omega_g^2 - \omega^2)] \tag{5.233}$$

Values of $\chi^{(3)}c$ have been obtained[5.78] by using this expression for n_2 and relation (5.217), and found to agree with the experiments. Some semiconductors, e.g., $Hg_x Cd_{1-x} Te$, yield a value as large as 10^{-2} e.s.u. However, the response time for this kind of nonlinearity is of the order of 10^{-7} s.

In addition to changing the susceptibility directly, the photo-excited carriers cause significant changes in the band- edge absorption coefficient. This change occurs through four basic processes.

First, the excited carriers loose energy partially through non-radiative recombination and energy exchange with the phonons. The transferred energy increases the lattice temperature. In most cubic semiconductors, the band gap, E_g , decreases with the rise in the lattice temperature following the relation,

$$E_g(T) = E_g(0) - \alpha T/(1 + \beta/T), \tag{5.234}$$

where $\alpha \sim 4 \times 10^{-4}$ eV/K, $\beta \sim 300$. The absorption band edge is red-shifted as a result of the rise in temperature and the red shift increases with the intensity of light.

Second, the energy band gap is altered due to the so-called band gap renormalization. Pauli exclusion principle excludes occupation of the same site by two charges with the same quantum number and produces a forbidden zone for other charges around a charge. This is referred as the exchange hole. Coulomb repulsion also prevents occupation of the same site by two charges of the same kind, and produces the so-called Coulomb hole around a charge. These two effects cause reduction of the electron and the hole energy and reduces the band gap to cause a red-shift of the band edge.

Third, the created charges collectively screen the field of a charge and reduces effectively the Coulomb attraction between the electrons and the holes. Excitons, which are produced by Coulomb binding are therefore de-excited in screened fields. The bleaching of the excitons alters the band edge absorption characteristic very significantly.

Fourth, the excited carriers fill-up the states and reduces the density of the available states which may be occupied by subsequently generated carriers. This effect is referred as phase-space-filling (PSF). The PSF is particularly effective in reducing the excitonic absorption.

All the resonant nonlinear processes, described above, are effective in quantum wells as in bulk materials. However, the changes in the excitonic absorption may be caused by relatively small optic power in quantum wells. In photonic device, therefore, mostly the change in the excitonic absorption is exploited.

Excitonic absorption in quantum wells, in contrast to bulk materials, is not significantly affected by the screening of the Coulomb force. The 2-D screening length λ_{2D} is given by,

$$1/\lambda_{2D} = m^* e^2 / 2\pi \epsilon \hbar^2. \tag{5.235}$$

under degenerate conditions. The effect of the screening therefore becomes saturated as the carrier concentration reaches high values. On the other hand, PSF plays an important role in reducing the excitonic absorption, as explained below by using a simple model[5.79].

We consider a sample of area A and assume that for an incident radiation of intensity I, N_x excitons are in equilibrium with N_e electrons and N_h holes per unit area at a lattice temperature T. The probability of an electron or hole being outside the space occupied by one exciton is,

$$P_1 = 1 - A_x/A, \tag{5.236}$$

where A_x is the area occupied by an exciton. The total probability of N_e electrons and N_h holes being outside A_x is ,

$$P = (1 - A_x/A)^{(N_e + N_h)A} \approx \exp[-(N_e + N_h)A_x]. \tag{5.237}$$

Also, as N_x excitons occupy an area $N_x A_x$, the area available to the free electrons and holes is $(1 - N_x A_x)$.

The excitonic absorption coefficient $\alpha(I)$ for the signal intensity I, is therefore reduced by the factor,

$(1 - N_x A_x) \exp[-(N_e + N_h)A_x]$ and we get

$$\alpha(I) = \alpha(0)(1 - N_x A_x) \exp(-2N_e A_x), \tag{5.238}$$

where $\alpha(0)$ is the absorption coefficient in the absence of light for which the effect of PSF is nonexistent. Also, N_e and N_h are taken to be identical as the sample is neutral.

We may eliminate N_e and N_x from Eq. (5.238) by using the relation,

$$N_0 = N_x + N_e, \tag{5.239}$$

and the Saha equation relating the bound excitons with the free carriers as given below,

$$N_e^2/N_x = C(E_{ex}, T), \tag{5.240}$$

where N_0 is the number of photons absorbed per unit area in the quantum well and $C(x)$ is a function of the exciton binding energy E_{ex} and the lattice temperature T.

Combining the two relations we obtain,

$$N_e \approx N_0 - N_0^2/C, \tag{5.241}$$

$$(1 - N_x A_x) \exp(-2N_e A_x) \approx (1 - N_0^2/C A_x)(1 - 2N_0 A_x + 2N_0^2/CA) \approx (1 - 2N_0 A_x). \tag{5.242}$$

Also,

$$N_0 = [\alpha(I)\tau/\hbar\omega]L, \tag{5.243}$$

where L and τ are respectively the width of the well and the life time of the excited carriers. Substituting,

$$I_s = [\hbar\omega/\alpha(0)L\tau](1/2A_x), \tag{5.244}$$

and putting,

$$(1 - N_x A_x) \exp(-2N_e A_x) \approx (1 - 2N_0 A_x,) \tag{5.245}$$

we get

$$\alpha(I) = \alpha(0)/(1 + I/I_s). \tag{5.246}$$

Absorption coefficient for the nonlinear conditions has been determined[5.79] experimentally by first measuring the transmission coefficient of a probing signal and then measuring the change in the coefficient in the presence of a pump signal of high intensity. The linear transmitted signal and the differential nonlinear signal as obtained in such experiments, are illustrated in Fig. 5.12. It was concluded from a detailed analysis of the absorption spectrum that the coefficient for the excitonic absorption was strongly affected by the pump signal, but that for the intersubband absorption remained almost unaffected. The nonlinearity coefficient $\chi^{(3)}$ was found to have a value of 6×10^2 e.s.u, which is about 6 times the value for silicon and much larger than the coefficient for other nonlinear mechanism. The response time for such nonlinearity was also found to be of the order of 10^{-9}s. The saturation intensity I_s has also been determined from experiments for GaAs/Ga$_x$ Al$_{1-x}$As[5.80],Ga$_{0.47}$In$_{0.53}$As/ InP [5.81] and Ga$_{0.47}$In$_{0.53}$As/Al$_{0.48}$In$_{0.52}$As[5.82] system. Results were found to be in close agreement with Eq. (5.246) . The value

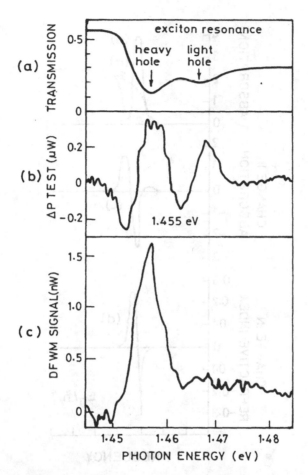

Figure 5.12. Absorption spectra of quantum well illustrating the optical nonlinearity. (a) Transmission without the pump beam. (b) Different nonlinear transmission induced by the pump beam. (c) Intensity of the four-wave degenerate mixing signal. [After D. S. Chemla, D. A. B. Miller, P. W. Smith, A. C. Gossard, and W. Weigmann, *IEEE J. Quantum Electron.* **QE-20**, 265 (1984); Copyright: (=A9 1984= IEEE)].

of the saturation intensity was less than 1 kW/cm² for the first two systems and was about 3 kW/cm² for the third system[5.83].

The expected change in the refractive index associated with the saturation of the excitonic absorption is illustrated schematically in Fig. 5.13. The excitonic peaks are absent under the saturation conditions and the change in the refractive index shows a large negative peak on the lower side and a large positive peak

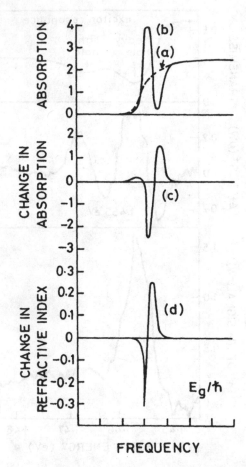

Figure 5.13. Schematic diagram showing the changes in the absorption and the refractive index for high intensity radiation. (a) The absorption spectrum for small amplitude radiation, showing the excitonic peak. (b) The absorption spectrum spectrum for high intensity radiation when the exciton is bleached. (c) Change in the absorption. (d) Consequent change in the refractive index.

on the higher side of the excitonic resonance frequency. These large changes in the refractive index of quantum wells may be caused by relatively low-intensity light signals produced by laser diodes. The optical switches and bistable devices described in Chapter 10 have been realized by utilizing this nonlinearity.

5.8. Photoluminescence

Photoluminescence is the phenomenon of the emission of light due to the recombination of the electrons and the holes created by a pump signal. The sample is excited[5.84] by a strong source of light, often a laser, and the spectrum of the emitted signal is measured by a double grating spectrometer coupled to a cooled photomultiplier tube, which is monitored by a photon counter. The results of such measurement[5.85] on a set of quantum wells with varying widths are shown on Fig. 5.14.

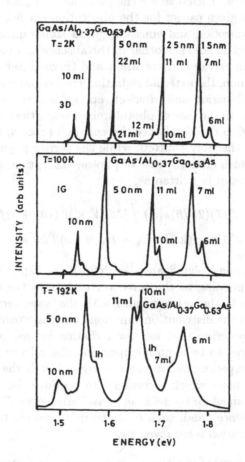

Figure 5.14. Photoluminescence spectrum of quantum wells of different widths at three temperatures. Widths are indicated by the numbers on the curves in nm. The splitting of the lines are ascribed to pseudo-monolayer flat islands. [After H. Hillmer, A. Forchel, C. W. Tu and R. Sauer, *Semicond. Sci. Technol.* **7**, B235 (1985); Copyright: IOP Publishing Ltd.].

Emission peaks are observed at wavelengths, which decrease with the width of the well. The spectrum has a small band width, depending on the temperature and the interface morphology. The characteristic feature of photoluminescence is the appearance of the sharp lines. This is due to the fact that the electrons and the holes, created by the exciting pump signal, relax very rapidly to the end or levels close to the ends of the bands. These then recombine radiatively to produce the luminescence, and thereby produce sharp lines characteristic of the band gap The phenomenon is widely used for the characterization of quantum well structures.

It is also relevant to the understanding of the quantum well lasers. A discussion of the physics of photoluminescence should therefore be useful as the background material for these devices. Formulae for the evaluation of photoluminescence are essentially the same as given earlier for the absorption coefficient in Section 5.5. The two processes, absorption and emission go on simultaneously all the time. These processes counterbalance each other in the absence of an external agency to produce the equilibrium carrier concentration and thermal radiation. In the case of absorption phenomenon, the external radiation creates more electron-hole pairs than are lost by recombination and in effect, energy is absorbed and the carrier population is increased. In the case of photoluminescence, the excess electron-hole pairs created by the pump signal enhance the emission process, which exceeds the absorption and light is therefore emitted, while the carrier population is reduced.

Equation (5.135) which gives the rate of photon absorption, gives also the rate of photon emission, when it is written as,

$$
\begin{aligned}
dn_p/dt = \ & (V_cD)(V_cD')(2\pi/\hbar)\rho(\omega) \int |M_0(\mathbf{k}, \mathbf{k}')|^2 \{(n_\kappa + 1)f(\mathbf{k})[1 - f(\mathbf{k}'] \\
& - n_\kappa f(\mathbf{k}')[1 - f(\mathbf{k})]\} \delta(E_\mathbf{k} - E_{\mathbf{k}'} - \hbar\omega) d^3k d^3k'.
\end{aligned}
\tag{5.247}
$$

The symbols have the same significance as defined for Eq. (5.135). But, now the initial state corresponding to the wave vector $\mathbf{k}$ is in the conduction band, while the final state is in the valence band with the wave vector $\mathbf{k}'$. Further, the factors arising from the distribution functions were approximated to be unity for absorption, by considering that for low radiation intensities the conduction band could be considered to be almost empty and the valence band almost full. For the luminescence experiments, on the other hand, it is the excited electrons and the consequential holes which recombine to produce the radiation. These excess carriers are accounted for by defining quasi-Fermi levels E_{Fc} and E_{Fv} for the conduction and the valence band, such that the increased electron concentration n and the hole concentration p are given by,

$$
n = (V_cD) \int \{\exp[(E_\mathbf{k} - E_{Fc})/k_BT] + 1\}_{-1} d^3k
\tag{5.248}
$$

$$
\text{and} \quad p = (V_cD') \int \{\exp[(E_{Fv} - E\mathbf{k}')/k_BT] + 1\} d^3k'
\tag{5.249}
$$

Equation (5.246) may be written by using the quasi-Fermi levels in the distribution functions as follows.

$$
\begin{aligned}
dn_p/dt = \; & (V_c D)(V_c D')(2\pi/\hbar)\rho(\omega) \int |M_0(\mathbf{k}, \mathbf{k}')|^2 \\
& \times \{\exp[(E_{\mathbf{k}} - E_{\mathbf{k}'} - \Delta E_F)/k_B T]\}^{-1}\{1 + n_\kappa \\
& - n_\kappa \exp[(E_{\mathbf{k}} - E_{\mathbf{k}'} - \Delta E_F)/k_B T]\} \\
& \times \delta(E_{\mathbf{k}} - E_{\mathbf{k}'} - \hbar\omega)d^3k\,d^3k'.
\end{aligned}
\tag{5.250}
$$

where $\Delta E_F = E_{Fc} - E_{Fv}$ and it has been assumed that $(E_{\mathbf{k}} - E_{Fc}), (E_{Fv} - E_{\mathbf{k}'}) \gg k_B T$. The integral may be evaluated by following the same procedure as used for the evaluation of the absorption coefficient. We first consider the photoluminescence due to band-to-band transitions in a bulk material. We get,

$$
\begin{aligned}
dn_p/dt = \; & (c/n_r)C_{\mathrm{ab}}(\hbar\omega - E_g)^{1/2}\exp[-(\hbar\omega - \Delta E_F)/k_B T] \\
& \times \xi(m_0^2/m_c^*)E_g\{1 + n_\kappa - n_\kappa \exp[(h\omega - \Delta E_F)/k_B T]\}.
\end{aligned}
\tag{5.251}
$$

However, the radiation signal emitted by the sample over a solid angle determined by the receiver is measured in the photoluminescence experiment. The photon, emission rate for which is given by Eq. (5.250), is, however, emitted in all directions and with all polarizations. The energy radiated per unit time, per unit solid angle, per unit energy interval, and per unit volume of the crystal is obtained by multiplying (dn_p/dt) by the energy of a photon, i.e., $\hbar\omega$ and by the optical density of states i.e., the number of photons states per unit volume, per unit solid angle and per unit energy interval. The density of states ,taking into account the two states of polarization is given by,

$$
G(\omega) = 2(1/2\pi)^3 d^3k/d(\hbar\omega),
\tag{5.252}
$$

Replacing k by $\omega n_r/c$, we get

$$
G(\omega) = 2(n_r/2\pi c\hbar)^3(\hbar\omega)^2.
\tag{5.253}
$$

Multiplying (dn_p/dt) by $G(\omega)\hbar\omega$, we get for the energy radiated per unit solid angle per unit energy interval per unit time per unit volume of the crystal,

$$
\begin{aligned}
L(\omega) = \; & C_{\mathrm{lum}}\xi(m_0^2/m_c^*)E_g(\hbar\omega)^2(\hbar\omega - E_g)^{1/2}\exp[(\Delta E_F - \hbar\omega)/k_B T] \\
& \times \{1 + n_\kappa - n_\kappa \exp[(\hbar\omega - \Delta E_F)/k_B T]\},
\end{aligned}
\tag{5.254}
$$

where $C_{\mathrm{lum}} = e^2 m_r^{3/2} n_r/2\sqrt{2}\pi^4 m_0^2 c^3 \hbar^5 \epsilon_0$.

Under conditions of thermal equilibrium, $\Delta E_F = 0$ and $n_p = [\exp(\hbar\omega/k_B T) - 1]^{-1}$ and hence $L(\omega) = 0$, and there is no radiation other than the thermal radiation. When, however, excess carriers are generated by the pump signal, $\Delta E_F > 0$

and $L(\omega)$ has a finite value and there is luminescence at the frequency ω. So long as ΔE_F is less than $\hbar\omega$, there is only enhanced luminescence. If, however, ΔE_F is greater than $\hbar\omega$, the second term, representing the stimulated radiation, becomes positive and laser action sets in.

In photoluminescence experiments, the luminescence spectrum is often more important. It is very different from the absorption spectrum. Considering the more rapidly-varying components in the expressions, we find that whereas absorption varies as $(\hbar\omega - E_g)^{1/2}$ luminescence varies as $(\hbar\omega - E_g)^{1/2} \exp[-(\hbar\omega - E_g)/k_BT]$ and hence it has the shape of a resonance curve. The luminescence spectrum[5.85], as illustrated in Fig. 5.14 starts from zero at $\hbar\omega = E_g$, peaks at $\hbar\omega = E_g + k_BT/2$ and then falls off exponentially. The peak moves closer to the band gap as the temperature is reduced. This property of photoluminescence is often used to determine the band gap.

We note that it is not necessary to derive separately the expressions for the photoluminescence spectrum for the excitonic transitions in bulk materials, or for transitions in quantum wells. The luminescence $L(\omega)$ may be written as

$$
\begin{aligned}
L(\omega) = \ & (c/n_r)\alpha(\omega)G(\omega)\hbar\omega \exp[(\Delta E_F - E_g)/k_BT] \\
& \times \exp[-(\hbar\omega - E_g)/k_BT] \\
& \times \{1 + n_\kappa - n_\kappa \exp[(\hbar\omega - \Delta E_F)/k_BT]\}.
\end{aligned}
\tag{5.255}
$$

The absorption spectrum has peaks for excitonic transitions or for inter - and intra - subband transitions in quantum wells; the peaks in the photoluminescence spectrum are,however, much sharper. It should, however, be mentioned that the experimental spectrum is broadened by several mechanisms which are not included in the expression for $\alpha(\omega)$, derived in Section 5.5 for ideal conditions. The broadening is more important for the photoluminescence experiment, as the ideal spectrum is more sharp. Phenomenological discussions were made about this broadening in the absorption spectrum. However, more intensive studies have been made of the broadening of the photoluminescence spectrum, as the data may be used to assess the purity of the quantum well structures.

The photoluminescence spectrum is broadened by several processes[5.86]. Some of these arise from the structural and the compositional variations. The width of the well varies along the plane of the heterointerfaces. Monolayer thickness variation having different extents occur even in very well-grown structures. The potential barrier also varies due to the variation in the composition of the barrier or the well material, when these are mixed compounds. There are also structural defects in the interfaces. All these variations cause broadening due to band-filling, when significant number of carriers migrate to the well. The broadening due to these various causes may be easily shown to be given by the following expressions:

$$
\Delta E = (dE/dL)\Delta L \approx 2E_0(\Delta L/L),
\tag{5.256}
$$

for well width variation of ΔL over a width of L, E_0 being the corresponding subband energy.

$$\Delta E = (dE/dV)\Delta V, \tag{5.257}$$

for barrier potential variation of ΔV over a barrier potential of V. The derivative (dE/dV) may be evaluated using the theory discussed in Section 4.2.

$$\Delta E = 2.36(dE_g/dx)x^{1/2}(1-x)^{1/2}(a_0^3/4L)^{1/2}(K_d\epsilon_0 a_B m_0/m_r)^{-1}, \tag{5.258}$$

for composition variation in the well material, where E_g is the energy band gap, x is the fraction of one compound in the ternary material, a_0 is the lattice constant, K_d is the dielectric constant, a_B is the Bohr radius and m_r/m_0 is the reduced effective mass ratio.

$$\Delta E = n_s\pi\hbar^2/m^*, \tag{5.259}$$

for band-filling effect, where n_s is the areal electron concentration in the well. The broadening due to the above factors may be computed by using the estimated

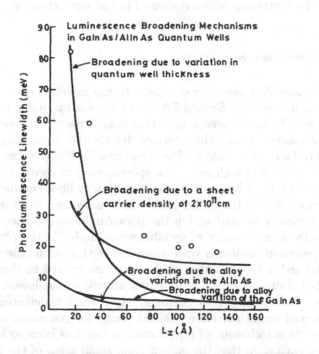

Figure 5.15. Luminescence broadening mechanisms in GaInAs/AlInAs quantum wells. Contribution of different mechanisms to the line width are shown separately. Broadening in thin wells is mainly due to the variation of quantum well thickness while for thick wells, the broadening is due to band-filling by the sheet carrier density, which is taken to be 2×10^{11} cm^{-2}. [After D. F. Weltch, G. W. Wicks and L. F. Eastman, *Appl. Phys. Lett.* **46**, 991 (1985); Copyright: American Institute of Physics].

values of $\Delta L, \Delta V$ and n_s. Such computed values for the $Ga_{0.47} In_{0.53} As/Al_{0.48} In_{0.52}$ As wells are illustrated in Fig. 5.15. The experimentally observed broadening may be explained by using properly chosen values of $\Delta L, \Delta V$ and n_s. The broadening for $L < 5$ nm is mainly due to width variation, while that for $L > 5$ nm is due to band filling. Broadening due to alloy variation may, however, be considered relatively unimportant.

Although the well-width dependence of the broadening may be explained by considering the mechanisms discussed above, the scattering of the excitons by the lattice vibrations and other imperfections also cause broadening.The broadening ΔE resulting from scattering of s unit time is[5.87,5.88],$\Delta E = \hbar s$.

This kind of broadening is also important and is required to be considered particularly to explain the increase of broadening with increase in the lattice temperature. Discussion of this kind of broadening involves the formalism for dealing with the scattering mechanisms. As the mechanisms are discussed in Chapter.7 in connection with the transport properties, the broadening of photoluminescence spectrum caused by scattering is also discussed in the same chapter.

5.9. Photoluminescence Spectrum

The line width of photoluminescence is related to the interface quality and alloy inhomogeneities as discussed in Section 5.8. It is often used to judge the quality of a grown structure. The luminescence spectrum may, however, be used to obtain more detailed information about the interface defects and impurities. The spectrum is required to be resolved into its fine structures for this purpose. Such resolution may be done by using high resolution spectroscopy to record the spectrum. More commonly, this is done by applying the technique of photoluminescence excitation spectroscopy[5.89]. In this technique the luminescence is detected at a fixed wavelength, conveniently located within the photoluminescence spectrum. Electrons are excited by using a source of variable wavelength. As the photon energy of the incident radiation resonates with a particular transition, the radiation is strongly absorbed and a large number of electrons are excited to the conduction band. The excited electrons are subsequently scattered to the lowest conduction level and produce luminescence near the band edge, which is detected by the detector. The detected signal therefore exhibits peaks for resonance wavelengths. Recently, the detection technique of photoluminescence has been so improved by using microscopic excitation that the output from small areas of the sample may be recorded instead of recording the output from the whole sample[5.90]. It is therefore possible to probe the luminescence outputs as affected by localized areas of defects and inhomogeneities.

All the three techniques, high-resolution photo- luminescence (PL) spectroscopy, photoluminescence excitation (PLE) spectroscopy and microprobe photoluminescence recording, have been used extensively to study defects and impurities in quantum well structures. A detailed discussion of these studies is beyond the scope of this book. We shall briefly mention only some important findings from such studies which have important bearing on the performance of the devices, using these structures.

Excitons, bound to neutral beryllium atoms were detected[5.91] in GaAs wells by resolving the photoluminescence spectrum. Four peaks were identified in the spectrum which correspond to the $e1 - HH1$ first electron-first heavy hole level free excitons, $e1 - HH1$ exciton bound to neutral beryllium (Be^0) atoms and $e1$ electrons recombining with Be^0 impurities in the interfaces and the centre of the well.

The PLE-spectroscopy gives spectrum showing transitions between higher quantized levels[5.92], nth electron level to the nth heavy hole or light hole levels. Some of the spectra exhibit transitions for $\Delta n \neq 0$. More importantly, the band-edge spectrum reveals details, which throw light on the interface roughness. It is observed that the photoluminescence peak is red-shifted from the PLE peak. The shift is due to the participation of phonons in the recombination process and is referred as the Stokes shift. It has been shown[5.93] that the line width of the PL spectrum is linearly related to the Stokes shift. This linear relation has been explained theoretically[5.92] by assuming a Gaussian distribution of the variation of well widths or alloy potentials, in which the excitons are localized. The line width of the PL spectrum may, therefore, be used as an indicator of the interface roughness. Further evidence of the localization has been obtained from the microscopic PL spectra[5.90] in thin (35 Å thick) quantum wells. Sharp lines are seen in the spectra which vary in energy and intensity as different areas of the sample are explored. It has been possible to map the variation of the well widths by utilizing the photoemission from the localized excitons.

Evidence of the interface defects and the bound excitons due to the well width fluctuations have been presented[5.90-97] also for GaAs/GaAlAs, (GaIn)As/Ga(PAs) and Si/Si_xGe_{1-x} systems. It may be concluded from these and other similar studies that the defect-related and the localized excitons play a significant role in the photoluminescence output. Such emission reduces the desired output at chosen wavelength and causes deterioration in the performance characteristics of the devices. PL, PLE or microscopic PLE provide convenient techniques for the characterization of such defects.

Photoluminescence provides also a useful tool for the study of the stresses in device structures. Uniaxial stress enhances the splitting of the heavy and the light hole energy levels and changes thereby the polarization of the emitted light. Microscopic examination of the polarization of the emitted light gives information about the stress in the particular position of the devices. This technique has been successfully used[5.98] to image stresses in GaAs diode lasers.

5.10. Conclusion

The various electron-photon interaction phenomena, discussed in this chapter, are made use of in different opto-electronic and photonic devices. Optical detectors and mixers use absorption, light-emitting devices and lasers use luminescence, optical modulators use the Stark effect and the nonlinear absorption is used in the optical bistable devices. Intense work is in progress for the development of the quantum well devices for all these applications as superior performance characteristics may be realized by using quantum well structures. Reports appear frequently about order-of-magnitude improvements. New materials and new structures are also being invented for these improvements.

The study of the physics of the phenomena, as observed in quantum wells, is also very rewarding since it is possible to investigate the electron-photon interaction under designed conditions. For example, excitonic interactions are hardly of interest at room temperature in bulk materials or in devices using bulk materials. On the other hand, excitons play a very important role in all the quantum well optoelectronic and photonic devices even at room temperature. Consequently, physics of the spectral content of excitons and of exciton dynamics are subjects of great current interest[5.99-101].

An extensive literature has been created on the various optical properties of quantum wells which deal with the intricate details of the physics of the various opto-electronic phenomena. References are given at the end which may be consulted by the interested reader. However, the physics of the phenomena, as explained in this chapter, may be considered adequate for the understanding of the devices, discussed in this book.

CHAPTER 6

TRANSPORT PROPERTIES

Quantum well structures attracted the attention of device designers because it is possible to segregate impurities from the carriers in these structures and thereby realize large carrier concentration without the associated reduction in the mobility. The well layer may be undoped, but carriers may be produced by doping the barrier layer. In contrast, carrier concentration may be increased in bulk materials only by increasing the concentration of impurities which causes a decrease in the carrier mobility. The larger conductivity resulting from the larger carrier concentration and larger mobility was expected to give higher switching speed and higher transconductance. This expectation has been fulfilled in High Electron Mobility Transistor(HEMT), also called Two-dimensional Electron Gas Field Effect Transistor(TEGFET) or Modulation Doped Field Effect Transistor(MODFET) or Selectively Doped Field Effect Transistor(SDFET). Emergence of these devices has made it important to study the characteristics of carrier transport in quantum wells.

The transport is controlled as in bulk materials by the various scattering[6.1,2] processes. All the scattering mechanisms of bulk materials are effective in quantum wells and in addition there are a few more mechanisms due to the multilayered structure. Further, electron gas in the quantum wells have only freedom of in-plane motion and may be considered for transport calculations as a two-dimensional gas (2DEG). It should, however, be remembered that the electron wave fubctions have finite extent in the direction of quantization. The quasi-two-dimensional character of the electrons modify the scattering rates and the expressions for the transport coefficients. The modification is further enhanced in narrow wells, as used in strained-layer systems, due to the spreading of the electron wave function into the barrier region[6.3]. The various scattering mechanisms are first discussed with special emphasis on the additional scattering processes and modifications due to the two-dimensional character. The methods used for the evaluation of the transport coefficients are then discussed.

The formalism used for the transport studies in quantum wells is, however, the same as used for bulk materials. Boltzmann transport equation is used at low fields and the same methods of solution are adopted. For high fields also, the analytic methods, e.g., the displaced-Maxwellian approximation method, or the Monte Carlo method is used. These methods are extensively discussed in the literature[6.2] and are, therefore, briefly considered here. Detailed discussion is, however, given for the application of the methods to quantum well systems and for the results obtained for different structures. Experimental results are also presented and compared with the theory.

6.1. Scattering Processes for 2DEG

Electrons are accelerated when an electric field is applied to a semiconductor sample, but for not-too-small samples the carrier velocity attains a steady value on the average, which is independent of the size. This limiting of the velocity is caused by the scattering processes. The accelerated electron loses the momentum gained from the field, when it is scattered. This loss may occur in one scattering event or in more than one event. Hence, it shows only an average gain in momentum which is due to its gain in between the scattering events. This situation prevails if the sample is long enough so that the electron is scattered a large number of times before it reaches an end of the sample. If, however, the sample is too short, the electron may move out of the sample before being significantly scattered. The average velocity is determined for such short samples not only by the scattering processes but also by the length and the terminal conditions of the sample. Samples used in the devices of interest in this book are, however, such that the former scattering-dominated transport prevails.

In a crystal with perfect periodicity, an electron is expected to move about freely and the transport would be determined by electron dynamics as in vacuum tubes. It is the imperfections which scatter the electrons and cause scattering-dominated transport. The scattering processes effective in the bulk materials are: lattice scattering of various kinds, i.e., deformation potential acoustic phonon, piezoelectric phonon, and polar and non-polar optic phonon scattering, ionized and neutral impurity scattering and alloy scattering in mixed compounds. All these scattering mechanisms are effective for the 2DEG in quantum well systems, but the interactions are modified. The momentum of the electron being quantized in the direction perpendicular to the well interfaces, when an electron interacts with the lattice or the impurity atom, the component of the lattice wave vector or of the Fourier wave number for the impurity potentials is not altered in the direction of quantization. The density of states is also different for the 2DEG. These two effects cause distinct changes in the effects of even the bulk scattering mechanisms.

In addition, the lattice vibrations are also modified by the quantum well structure. The physical constants of the constituent materials being different, discontinuities are introduced at the interfaces, which affect the modes and dispersion relations of lattice vibrations. The so-called surface and confined modes[6.4] are excited. These are particularly important in polar materials and changes the nature of interaction. Further, in narrow wells, the electron is not fully confined within the well, but penetrates into the barrier. Electron scattering in such wells is, therefore, a more complex phenomenon.

Impurity scattering is also different in nature in quantum wells, since the impurities are non-uniformly distributed. This gives rise to the so-called remote impurity scattering [6.5] in addition to the local impurity scattering.

There is in addition to the bulk scattering processes, another important scattering process in quantum wells. The interfaces of quantum wells have nonuniformities depending on the growth processes. Even in very well-grown quantum wells there are monolayer variations of different extents. This causes the so-called interface roughness scattering (IFRS) [6.6]. Its importance depends on the quality of the crystal and has been decreasing with the improvement in crystal growth technique; but even then it cannot be neglected, particularly in narrow wells at low temperatures.

It may also be noted that all these scattering processes may cause the final state to be in the same subband or in another subband. The former is referred as intra-subband scattering and the latter, intersubband scattering. These are similar to the intervalley and non-equivalent intervalley scattering in bulk materials and have similar effect.

6.2. Matrix Elements for 2DEG

The scattering interactions are dealt with in the transport calculations by the first order perturbation theory[6.1] which gives for the rate of scattering $S(\mathbf{k}, \mathbf{k}')$ of an electron from the k-state to the k'-state,

$$S(\mathbf{k}, \mathbf{k}') = (2\pi/\hbar)|M(\mathbf{k}, \mathbf{k}')|^2 \delta(E_{\mathbf{k}'} - E_{\mathbf{k}} \pm \Delta E), \qquad (6.1)$$

where $E_{\mathbf{k}}$ and $E_{\mathbf{k}'}$ are the energies corresponding to the wave vectors k and k', ΔE is the change in energy of the scatterer, $\delta(x)$ is the Dirac delta function. $M(\mathbf{k}, \mathbf{k}')$ is the element which relates k'-state with the k -state, in the matrix relating the perturbed state with the various equilibrium k-states and is usually referred as the matrix element.

The matrix element for electron- phonon interaction is given by[6.2],

$$M(\mathbf{k}, \mathbf{k}') = \int \psi_{\mathbf{k}'}^* \phi_{n'\mathbf{q}} \Delta V \psi_{\mathbf{k}} \phi_{n\mathbf{q}} d^3 r \qquad (6.2)$$

where $\psi_{\mathbf{k}}$ and $\psi_{\mathbf{k}'}$ are the electron wave functions for the initial k-state and the final k'-state respectively, while $\phi_{n\mathbf{q}}$ and $\phi_{n'\mathbf{q}}$ are the wave functions for the phonon system, $n_{\mathbf{q}}$ and $n'_{\mathbf{q}}$ being the phonon occupation probabilities for the wave vector q.

The scattering potential ΔV may, in general, be expressed for lattice vibrations as,

$$\Delta V = \sum_{\mathbf{q}} A(\mathbf{q})[a_{\mathbf{q}}\exp(i\mathbf{q}.\mathbf{r}) + a_{\mathbf{q}}^{\dagger}\exp(-i\mathbf{q}.\mathbf{r})], \qquad (6.3)$$

where $a_{\mathbf{q}}$ and $a_{\mathbf{q}}^{\dagger}$ are respectively the creation and the annihilation operator and $A(\mathbf{q})$ is the effective amplitude of the interaction potential corresponding to the phonon wave vector $\mathbf{q}$.

For three-dimensional (3D) systems the wave function $\psi_{\mathbf{k}}$ is the Bloch function $U_{\mathbf{k}}(\mathbf{r})\exp(i\mathbf{k}.\mathbf{r})$, $U_{\mathbf{k}}(\mathbf{r})$ being the cell-periodic part. On using this function, the matrix element reduces to,

$$M(\mathbf{k}, \mathbf{k}') = A(\mathbf{q})O_i(n_{\mathbf{q}} + 1/2 \pm 1/2)^{1/2}, \qquad (6.4)$$

where O_i is the overlap integral [6.2] defined as

$$O_i = \int U_{\mathbf{k}'}^*(\mathbf{r})U_{\mathbf{k}}(\mathbf{r})d^3r. \qquad (6.5)$$

It has a value of unity for parabolic energy bands for which $U_{\mathbf{k}}$ is taken to be independent of $\mathbf{k}$, but it may be significantly different from unity for nonparabolic bands[6.2]. The factor $(n_{\mathbf{q}} + 1/2 \pm 1/2)^{1/2}$ is due to the creation and annihilation operators, operating on the wave functions for the lattice vibrations. The plus sign corresponds to the creation or the emission of a phonon, while the minus sign corresponds to the annihilation or the absorption of a phonon. Also, as the integrands of the periodic functions $\exp i(\mathbf{k} - \mathbf{k}' \pm \mathbf{q}).\mathbf{r}$ is zero for sfficiently large crystals, unless the argument is zero, only one term of the summation has a non-zero value and $\mathbf{q}$ in Eq. (6.4) satisfies the relation,

$$\mathbf{k}' = \mathbf{k} \pm \mathbf{q}, \qquad (6.6)$$

which implies the conservation of total pseudomomentum of the electron-phonon system, and in effect defines the value of $\mathbf{q}$ which is involved in the scattering from the $\mathbf{k}$-state to $\mathbf{k}'$-state.

The interaction potential for defects and impurities may be expressed as

$$\Delta V = \sum_{\mathbf{q}} A(\mathbf{q})\exp(i\mathbf{q}.\mathbf{r}), \qquad (6.7)$$

where $A(\mathbf{q})$'s are the coefficients in the three-dimensional Fourier transformation coordinate $\mathbf{q}$, which plays the same role as the phonon wave vector.

The matrix element for these scattering mechanism is

$$M(\mathbf{k}, \mathbf{k}') = A(\mathbf{q})O_i, \qquad (6.8)$$

where $\mathbf{q}$ satisfies the relation,

$$\mathbf{k}' = \mathbf{k} + \mathbf{q}. \qquad (6.9)$$

Matrix elements for the scattering of 2D electrons are obtained by following the same procedure, but by replacing ψ_k and $\psi_{k'}$ by the 2D wave functions. As indicated in Section 4.1, the wave function for 2D electrons is expressed as,

$$\psi_{k_t} = U_0 F(z) \exp(i k_t.\rho), \tag{6.10}$$

where $F(z)$ is the envelope function, z-axis being chosen perpendicular to the hetero-interfaces; k_t and ρ are respectively the in-plane wave vector and the position coordinate; U_0 is the cell-periodic function for $k = 0$, corresponding to the particular energy band minimum to which the electrons belong. On substituting the wave function (6.10) in Eq. (6.2) we get for the matrix element for lattice scattering.

$$
\begin{aligned}
M(k, k') &= \int \{ U_0^* F^*(z) \exp(-i k'_t.\rho) \phi_{n'q} \\
&\quad \times [\sum_q (a_q \exp(i q.r) + a_q^\dagger \exp(-i q.r)] \\
&\quad \times U_0 F(z) \exp(i k_t.\rho) \phi_{n_q} \} d^2\rho \, dz.
\end{aligned} \tag{6.11}
$$

The overlap integral O_i has a value of unity since the cell- periodic part is assumed to be independent of k. Also, non-zero value of the periodic part is obtained only for

$$k'_t = k_t \pm q_t, \tag{6.12}$$

where q_t is the in-plane component of the phonon wave vector q. Thus, only the in-plane component of q is involved in the change of pseudomomentum of the electron due to the interaction. The longitudinal component of q, q_z is not directly involved in the transition and in effect transitions may occur from an energy E_{k_t} corresponding to the k_t-state by interactions with different modes of vibration characterized by different longitudinal components of q. It can be physically explained by visualizing that when the electron wave vector is quantized in the z-direction its momentum is diffuse, so that it can absorb or give up all possible momentum in that direction. This effect gives rise to a term in the matrix element, which is often called the form factor. It is given by[6.7]

$$G(q_z) = \int F'^*(z) \exp(i q_z z) F(z) dz, \tag{6.13}$$

Evidently, this is an additional term involved in the matrix element for 2D-electron scattering in comparison to that for 3D-electrons.

On taking account of this modification we find that the matrix element for the lattice scattering of the 2D-electrons for a particular mode is,

$$M(k_t, k'_t) = A(q_t, q_z) G(q_z)(n_q + 1/2 \pm 1/2)^{1/2}. \tag{6.14}$$

The total matrix element for all the modes having the same $\mathbf{q}$ is given by

$$|M(\mathbf{k}_t, \mathbf{k}'_t)|^2 = \sum_{q_z} |A(\mathbf{q}_t, q_z)|^2 |G(q_z)|^2 (n_q + 1/2 \pm 1/2), \qquad (6.15)$$

since each mode acts independently. The summation may, however, be written as an integral to obtain for lattice scattering,

$$|M(\mathbf{k}_t, \mathbf{k}'_t)|^2 = (L/2\pi) \int |A(\mathbf{q}_t, q_z)|^2 |G(q_z)|^2 (n_q + 1/2 \pm 1/2) dq_z, \qquad (6.16)$$

where L is the width of the infinite-barrier quantum well or the effective width of a single heterojunction , or a quantum well with finite barrier. We get similarly for the matrix element for defect scattering of 2D electrons,

$$
\begin{aligned}
M(\mathbf{k}_t, \mathbf{k}'_t) &= \int U_0^* F'^*(z) \exp(-i\mathbf{k}'_t.\rho) \sum_{\mathbf{q}} A(q) \exp[i\mathbf{q}.(\mathbf{r} - \mathbf{r}_i)] \\
&\times U_0 F(z) \exp(i\mathbf{k}_t.\rho) d^2\rho\,dz.
\end{aligned}
\qquad (6.17)
$$

It should be mentioned that the location of the defect or the impurity atom is not considered for bulk materials since electrons as well as these scattering centers are uniformly distributed in the sample, which is assumed to be infinite in extent. In 2D systems, on the other hand, electrons are confined in or near the well, and may not also be uniformly distributed. The relative position of the electron gas and the scattering centers is,therefore,important in 2D systems. We have assumed that the center is located at $\mathbf{r}_i$, the corresponding in-plane and z-coordinate being respectively ρ_i and z_i . We get, as for the lattice scattering,

$$\mathbf{k}'_t = \mathbf{k}_t + \mathbf{q}_t, \qquad (6.18)$$

for non-zero values of the matrix element. Expression for the matrix element may be simplified, by using this condition, to

$$
\begin{aligned}
|M(\mathbf{k}_t, \mathbf{k}'_t)|^2 &= |\sum_{q_z} A(q) G(q_z) \exp(-iq_z z_i)|^2 \\
&= |(L/2\pi) \int A(q) G(q_z) \exp(-iq_z z_i) dq_z|^2
\end{aligned}
\qquad (6.19)
$$

If $A(q)$ is independent of q, as in alloy scattering Eq. (6.19) may be simplified to

$$
\begin{aligned}
|M(\mathbf{k}_t, \mathbf{k}'_t)|^2 &= |(L/2\pi) A(q) \int F(z)^2 \exp[iq_z(z - z_i)] dq_z dz|^2 \\
&= |L \int A(q) F(z^2) \delta(z - z_i) dz|^2 = |L A(q) F(z_i)^2|^2.
\end{aligned}
\qquad (6.20)
$$

Expressions for the matrix element obtained by following the above procedure are given in Table 6.1 for the important scattering mechanisms.

6.3. Form Factor

The z-component of q , gives rise, as explained above, to the form factor,

$$G(q_z) = \int F'^*(z) \exp(iq_z z) F(z) \, dz. \qquad (6.21)$$

This integral may be evaluated analytically for wells of parabolic semiconductors with infinite barrier height. It is, however, required to be evaluated numeri-

Table 6.1 Matrix Elements for Scattering of Two-dimensional Electron

Scattering	Matrix Element Squared
1. Acoustic Phonon :	
Deformation	
Potential	$E_1^2 q^2 (\hbar/2V_c \rho \omega_q)(n_q + 1/2 \pm 1/2)|G(q_z)|^2$
Piezo electric	$(eh_{pz}/\epsilon_s)^2 (\hbar/2V_c \rho \omega_q)$
	$\times (n_q + 1/2 \pm 1/2)|G(q_z)|^2$
2. Optic Phonon :	
Non-polar	$D_0^2(\hbar/2V_c \rho \omega_0)(n_o + 1/2 \pm 1/2)|G(q_z)|^2$
Polar	$(e/q)^2(1/\epsilon_\infty - 1/\epsilon_s)(\hbar \omega_l/2V_c)$
	$\times (n_l + 1/2 \pm 1/2)|G(q_z)|^2$
3. Defect :	
Alloy	$(4\pi \Delta V_{all} r_0^3/3A_c)^2 \alpha(1-\alpha)|F(z_i)|^4$
Ionized Impurity	$(Ze^2/2A_c\epsilon_s)^2 I_{imp}^2$
	$I_{imp} = \int |F(z_i)|^2 q_t^{-1} \exp(-q_t|z - z_i|) \, dz$
Surface Roughness	$\pi(\Delta\Lambda/A_c)^2 \exp(-q_t^2 \Lambda^2/4)\Delta V_{sr}^2$
	$\Delta V_{sr} = (\partial E/\partial L)\Delta$

Meanings of symbols : E_1 - Acoustic phonon deformation potential constant, q - 2π times the phonon wave number, $\hbar = (1/2\pi)$ times Plank's constant, V_c - sample volume, ρ - Sample density, $\omega_q, \omega_o, \omega_l$ - Phonon frequencies for acoustic, non-polar optic and polar- optic phonons. n_q, n_o, n_l - Occupancy of acoustic, non-polar, polar optic phonons, $G(q_z) = \int_{-\infty}^{\infty} |F(z)|^2 \exp(iq_z zZ) \, dz$, $F(z)$ is the z-dependent part of the envelope function and q_z is the z-component of phonon wave vector q, e-electron charge, h_{pz} - Average piezoelectric constant, ϵ_∞ - High frequency and ϵ_s - Static, permittivity, D_o - Non-polar optic deformation potential, ΔV_{all} - Alloy potential, r_0 - Radius of alloy potential sphere, $N(z_i)$ - Number of scattering centers per unit volume at z_i , $\alpha(z_i)$ - Fraction of AC material in the compound $A_\alpha B_{1-\alpha}$ C at $z = z_i$, Z- Ionization number, q_t - In-plane component of q, λ_{2D} - 2D Debye length, Δ - Average displacement of the interface, Λ -The range of spatial variation of the interface. A_c - Interface area. V- Potential at z.$+, -$ signs in $\pm$ are for emission and absorption processes respectively.E_0 -Quantized energy level.L-Well width.

cally for wells of nonparabolic semiconductors or of finite barrier height. $G(q_z)$ is equal to unity for $q_z = 0$, but decreases with increasing q_z . The form factor was approximated as unity in some very early publications[6.8,9]. It is evident from the nature of the variation of $G(q_z)$ that this approximation will not introduce much error if scattering is predominated by low-q phonons or if $A(q)$ increases with decrease in q as in polar optic phonon, piezoelectric phonon or impurity scattering(see Table 6.1). On the other hand, for scattering mechanisms in which $A(q)$ is independent of q as in acoustic phonon or alloy scattering, the reduced value of form factor for large values of q_z will reduce the scattering rate and hence increase the mobility. This will be more so in narrow wells and wells with smaller barrier height.

Single heterojunctions are usually analyzed by using the variational envelope functions given in Section 4.1. This function gives for $G(q_z)$ [6.10]

$$G(q_z) = b^3/(b^2 + q_z^2)^{3/2}. \tag{6.22}$$

The effect of the deviation of $G(q_z)$ from unity will increases in single heterojunctions with increase in b or in the carrier concentration.

The effect of the wave-function penetration into the barrier is not considered in Eq. (6.22). The effect would be important for the material systems used in quantum well devices and such studies should be useful. It may, however, be commented here that the effect is expected to cause an increase in mobility for material systems with low barrier height.

6.4. Screening for 2DEG

Expressions given for $A(q)$ in Table 6.1 do not include the screening effect of the free electrons. Electron concentration in quantum well devices is usually very high and the screening effect cannot be neglected. The procedure used for the evaluation of the effect may be illustrated by considering a point charge e located at $z = z_i$ in an ideal two dimensional system, which has a sheet of charge at $z = 0$. We neglect for the sake of simplification the difference in the permittivity of the two materials constituting the well. The potential U of the point charge, as modified by the surrounding charges, is obtained by solving the Poison equation,

$$\nabla^2 U(z, \rho) = -\rho_c/\epsilon_s, \tag{6.23}$$

where ϵ_s is the permittivity and the charge density ρ_c includes the induced charge in addition to the point charge. The density of induced charge, ρ_{ind} may be expressed, by using a model analogous to the 3D Thomas-Fermi model, as

$$\rho_{ind} = e[N_s(U) - N_s(0)]\delta(z), \tag{6.24}$$

where $eN_s(U)$ is the sheet charge density for a potential U. It may be expressed in terms of U by assuming that the potential is weak and slowly-varying as

$$\rho_{ind} = -2(1/\lambda_{2D})U\delta(z), \tag{6.25}$$

where $1/\lambda_{2D} = (e^2/2\epsilon_s)(dN_s/dE_F)$, has the significance of screening length for 2D electrons. N_s and E_F are respectively the electron concentration and the Fermi energy.

The resulting potential, satisfies the equation,

$$\nabla^2 U(z,\rho) + 2(1/\lambda_{2D})U\delta(z) = e\delta(z - z_i)/\epsilon_s. \tag{6.26}$$

It may be shown that the Fourier coefficient for the expansion of the potential U satisfying the above equation in the plane of the point charge is

$$A(q_t) = (e/4\epsilon_s)\exp(q_t z)/(q_t + 1/\lambda_{2D}). \tag{6.27}$$

On comparing this expression with that in the absence of screening charges it is found that, the 2D electrons modify the potential by introducing the factor

$$S_c = q_t/(q_t + 1/\lambda_{2D}) \tag{6.28}$$

In practical 2D systems, however, electrons are not exactly two- dimensional but are quasi-two-dimensional, as they are distributed over a finite width, although their motion is two-dimensional. The finite extent changes the screening length by introducing a form factor, $F(q)$, given by,

$$F(q) = \int_{-\infty}^{\infty} dz \int_{-\infty}^{\infty} dz' F(z)F(z')\exp(-q_t|z - z'|). \tag{6.29}$$

In addition, the assumptions in the Thomas-Fermi model of screening that the potentials being screened are also slowly varying i.e., $q = 0$ and static are not generally valid. These limiting conditions are removed in the random phase approximation (RPA), in which the response of an electron to the external field and all other electrons is worked out to define an effective dielectric constant[6.11],

$$K(q,w) = K_s + (1/2)(q^2 - K_s\omega^2/c^2)^{1/2}\chi(q,\omega), \tag{6.30}$$

where K_s is the static lattice dielectric constant, ω is the frequency of the external field, $\mathbf{q}$ is the wave vector of the excitation, c is the velocity of light in free space, $\chi(q,\omega)$ is the polarizability determined from the response function. The full expression for $\chi(q,\omega)$ is rarely applied and mobility is usually analyzed by using the static approximation, for which[6.12]

$$K(q,0) = K_s[1 + (e^2/2\epsilon_s q)F(q_t)\pi(q_t)], \tag{6.31}$$

where $F(q)$ is the form factor given by Eq. (6.29) and $\pi(q)$ is the static polarization. In the RPA [6.12]

$$\pi(q) = (m^*/\pi\hbar^2)\{1 - H(q_t - 2k_F)[1 - (2k_F/q_t)^2]^{1/2}\}, \qquad (6.32)$$

where m^* is the electron effective mass, H is the Heaviside unit function and k_F is the 2D-wave vector corresponding to the Fermi- energy.

Expression (6.31) may be simplified when only the ground state is occupied and when the electron concentration is degenerate, $\pi(q) \approx m^*/\pi\hbar^2$. The simplified expression is

$$K(q,0) = K_s[1 + F(q_t)/q_t\lambda_{2D}]. \qquad (6.33)$$

Consequently, the screening factor is modified to

$$S_c = q_t[q_t + F(q_t)/\lambda_{2D}]^{-1}, \qquad (6.34)$$

since for degenerate materials,

$$1/\lambda_{2D} = e^2m^*/2\pi\epsilon_s\hbar^2. \qquad (6.35)$$

The dependence of S_c on temperature and the carrier concentration is mostly determined by the effective q_t as λ_{2D} is independent of these two parameters. In degenerate materials the effective q_t is the same as k_F,which is proportional to $\sqrt{N_s}$. The screening constant S_c therefore decreases with increase in the carrier concentration. In other words, the potential is much more screened with decrease in the carrier concentration. This result is opposite to the physical expectation. The anomaly has not yet been satisfactorily resolved although Eq. (6.34) has been widely used to asses the effect of screening.

6.5. Collision Integral for 2DEG

Low-field transport coefficients are evaluated by using the Boltzmann transport equation, which may be written for wells with uniformly distributed carriers of isotropic effective mass as follows[6.2]

$$(e\mathcal{E}/\hbar).\nabla_{k_t}f = -(A_cD') \int \{S(\mathbf{k}_t, \mathbf{k'}_t)f(\mathbf{k}_t)[1 - f(\mathbf{k'}_t)]$$
$$-S(\mathbf{k'}_t, \mathbf{k}_t)f(\mathbf{k'}_t)[1 - f(\mathbf{k}_t)]\} d^2k'_t \qquad (6.36)$$

where $\mathcal{E}$ is the externally applied field and f is the distribution function, D' is the density of states for the 2DEG and d^2k_t is the two-dimensional element in the wave vector space. The left-hand side gives the change in f, caused by the field $\mathcal{E}$, while the right-hand-side gives the change due to scattering and in the steady-state the two terms are balanced.The right-hand-side term is often referred as the collision integral. The first term in the integral gives the number of electrons scattered out from the k_t-state to different k'_t-states and is calld the out-scattrig term. The second term in the integral gives the number of electrons scattered from different k'_t-states into the k_t-state and is called the in-scattering term.

The equation is solved by expanding f with the spherical harmonics with the direction of $\mathcal{E}$ as the polar axis. The series is terminated at the second term at low fields and is written as,

$$f(k_t) = f_0(k_t) - (e\hbar\mathcal{E}/m^*)k_t \cos\theta (\partial f_0/\partial E)\phi(k_t) \tag{6.37}$$

where E is the energy corresponding to k_t, θ is the angle between $\mathcal{E}$ and k_t, f_0 and ϕ are the two energy-dependent functions, m^* is the band-edge effective mass. Equation (6.35) may be reduced for low fields, by using this expansion to,

$$\int \{[S(k_t, k'_t)f_0(k_t)[1 - f_0(k'_t)] - S(k'_t, k_t)f_0(k'_t)[1 - f_0(k_t)]\}\, d^2k'_t = 0, \tag{6.38}$$

and

$$\int [1 - f_0(k'_t)][1 - f_0(k_t)]^{-1}S(k_t, k'_t)[\phi(k_t) - (k'_t \cos\beta/k_t)\phi(k'_t)]d^2k'_t$$
$$= (m*/m_v) \tag{6.39}$$

It is assumed for arriving at these equations that the principle of detailed balance holds and the integrand is balanced for each value of k'_t . Further, the angle θ' between k_t and $\mathcal{E}$ has been written in terms of θ and β, the angle between k_t and k'_t . The term containing $\sin\beta$ is zero, because of symmetry, while the $\cos\beta$ term may become non-zero for some scattering mechanisms. The velocity effective mass m_v has also been substituted for $[(1/\hbar^2 k_t)\nabla_{k_t} E]^{-1}$.

The symmetric component of the collision integral gives the steady-state unperturbed distribution function f_0 , which for low fields, is the Fermi-Dirac function for degenerate electron concentration or the Maxwellian function for non-degenerate electron concentration. The asymmetric part of the distribution function ϕ is obtained by solving Eq. (6.39). The integrated scattering rate is required to be evaluated for this purpose. These are discussed in the following section.

6.6. Scattering Rate for 2DEG

Expressions for the two components of scattering, out-scattering and in-scattering are further simplified by using the property of the delta function. The out-scattering component of lattice scattering is considered first. Replacing $S(k_t, k'_t)$ by Eq. (6.1) ad the matrix element by the expression of Eq. (6.15) we get for the out-scattering rate,

$$S_{out}(\mathbf{k}_t) = \sum_{+,-}(V_c/8\pi^3)(2\pi/\hbar)\int\int |A(|\mathbf{k}'_t - \mathbf{k}_t|, q_z)|^2 |G(q_z)|^2$$
$$\times \delta(E_{\mathbf{k}'_t} - E_{\mathbf{k}_t} \pm \hbar\omega)[1 - f_0(\mathbf{k}'_t)][1 - f_0(\mathbf{k}_t)]^{-1}$$
$$\times \{n[(|\mathbf{k}'_t - \mathbf{k}_t|^2 + q_z^2)^{1/2}] + 1/2 \pm 1/2\} d^2 k'_t dq_z \qquad (6.40)$$

The density of states D' has been replaced by $(1/2\pi)^2$ as the electron gas is two-dimensional and spin-flip scattering is assumed to be absent[6.13,14]. The plus and minus signs correspond respectively to the emission and the absorption of a phonon. The in-plane component of $\mathbf{q}, \mathbf{q}_t$, has been replaced by $\mathbf{k}'_t - \mathbf{k}_t$. The phonon occupation probability has been written as $n[(|\mathbf{k}'_t - \mathbf{k}_t|^2 + q_z^2)^{1/2}]$ as it is a function of the magnitude of the phonon wave vector. ΔE has been replaced by the energy of a phonon $\hbar\omega$.

The integration over k'_t is carried out putting

$$d^2 k'_t = (m_v/\hbar^2) dE_{\mathbf{k}'_t} d\beta \qquad (6.41)$$

and using the property of the delta function. We obtain for the out-scattering rate,

$$S_{out}(\mathbf{k}_t) = \sum_{+,-}(V_c/8\pi^3)(2\pi/\hbar)\int\int |A(|\mathbf{k}'_t - \mathbf{k}_t|, q_z)|^2 |G(q_z)|^2$$
$$\times [1 - f_0(E_{\mathbf{k}'_t})][1 - f_0(E_{\mathbf{k}_t})]^{-1}$$
$$\{n[(|\mathbf{k}'_t - \mathbf{k}_t|^2 + q_z^2)^{1/2}] + 1/2 \pm 1/2\}(m_v/\hbar^2) d\beta dq_z \qquad (6.42)$$

where now,

$$|\mathbf{k}'_t - \mathbf{k}_t| = (k_t'^2 - 2k'_t k_t \cos\beta + k_t^2)^{1/2} \text{ and } E_{\mathbf{k}'_t} = E_{\mathbf{k}_t} \pm \hbar\omega. \qquad (6.43)$$

The magnitudes of $\mathbf{k}'_t$ correspond to $E_{\mathbf{k}_t} \pm \hbar\omega$ and is determined by the $E - \mathbf{k}$ relation for the material of the well.

Expressions for the out-scattering rates for defects and impurities are derived by using the same procedure. Using the matrix element of Eq. (6.19), we get for the scattering rate due to a center at ρ_i, z_i ,

$$S_{out}(\mathbf{k}_t) = (A_c/(4\pi^2))(2\pi/\hbar)\int |\int (L/2\pi)A(|\mathbf{k}'_t - \mathbf{k}_t|, q_z)G(q_z)\exp(-iq_z z_i) dq_z|^2$$
$$\delta(E_{\mathbf{k}'_t} - E_{\mathbf{k}_t}) d^2 k'_t \qquad (6.44)$$

The term $[1 - f_0(E_{\mathbf{k}'})][1 - f_0(E_{\mathbf{k}})]^{-1}$ and ΔE have been omitted, since defect scattering is mostly elastic as the electron does not gain or loose any significant amount of energy in such collisions.

On integration over $d^2k'_t$ followig the same procedure as discussed for lattice scattering and summing the contribution of all the scattering centers in the whole structure we get ,

$$S_{out}(\mathbf{k}_t) = \int N_i(z_i)A_c \int |\int (L/2)A(|\mathbf{k}'_t - \mathbf{k}_t|, q_z)G(q_z)\exp(-iq_z z_i)dq_z|^2$$
$$\times (A_c/4\pi^2)(2\pi/\hbar)(m_v/\hbar^2)d\beta dz_i \qquad (6.45)$$

Also,

$$E_{\mathbf{k}'_t} = E_{\mathbf{k}_t}, |\mathbf{k}'_t - \mathbf{k}_t| = [2k_t^2(1 - \cos\beta)]^{1/2}. \qquad (6.46)$$

The integrals of Eq. (6.42) and (6.45) may be evaluated only numerically when the finite barrier height of the wells and the nonparabolicity of the well material are taken into account. Analytic expressions may, however, be obtained for wells with infinite barrier height and for materials with parabolic bands. These expressions are collected in Table 6.2.

Table 6.2 Scattering Rates for Two-dimensional Electrons

Scattering Mechanism	Scattering Rate
1. Acoustic Phonon : Deformation Potential	$(E_1^2 k_B T m_v/2\pi\hbar^3 \rho s^2)\int_{-\infty}^{\infty}|G(q_z)|^2\, dq_z$
Piezo electric	$(e^2 P^2 k_B T m_v/2\pi^2 \epsilon_s \hbar^3)\int_0^{2\pi} d\beta$ $\times \int_{-\infty}^{\infty}|G(q_z)|^2(q_t^2 + q_z^2)^{-1}\, dq$
2. Optic Phonon : Non-polar	$(D_0^2 m_v/4\pi\hbar^2 \rho\omega_o)[n_o + (n_o + 1)$ $\times H(E - \hbar\omega_o)]\int_{-\infty}^{\infty}|G(q_z)|^2\, dq_z$
Polar	$[e^2(1/\epsilon_\infty - 1/\epsilon_s)\omega_l m_v/8\pi\hbar^2][n_l + (n_l + 1)$ $\times H(E - \hbar\omega_l)]\int_0^{2\pi}\int_{-\infty}^{\infty}|G(q_z)|^2(q_t^2 + q_z^2)^{-1}\, dq_z d\beta$
3. Defect : Ionized Impurity	$(Z^2 e^4 m_v/8\pi\hbar^3 \epsilon_s^2)\int_{-\infty}^{\infty}N_I(z_i)$ $\times\int_0^{2\pi}[\int_{-\infty}^{\infty}|F(z)|^2 q_t^{-1}\exp(-q_t|z - z_i|)\, dz]^2 d\beta\, dz_i$
Alloy	$(m_v/\hbar^3)\int[4\pi\Delta V_{all}(z_i)r(z_i)^3/3]^2[N(z_i)/2]$ $\times\alpha(z_i)\{1 - \alpha(z_i)\}|F(z_i)|^4\, dz_i$
Surface Roughness	$(\pi\Delta^2 2m_v^*/\hbar^3)\Delta V_{sr}^2$

Meanings of symbols : k_B - Boltzmann constant, T-Lattice temperature, s-Longitudinal sound velocity, P-Piezoelectric constant, $(h_{pz}^2/\rho s^2 \epsilon_s)$, $q_t = |\mathbf{k}_t - \mathbf{k}'_t| = 2k_t^2(1 - \cos\beta)$, $\mathbf{k}_t$, $(\mathbf{k}'_t)$ being the electron wave vector before (after) collision, $\cos\beta = \mathbf{k}_t.\mathbf{k}'_t/k_t k'_t$, E-Electron energy, $H(x)$-Heaviside Unit function, m_v - Velocity effective mass, $N_I(z_i)$-Concentration of impurity atoms at $z = z_i$, $N(z_i)$- Number of atoms per unit volume at $z = z_i$, All other symbols have the same meaning as given in Table 7.1.

The energy dependence of the scattering rates for 2D electrons are very different from that for bulk materials. In fact, the scattering rate does not vary much with energy in 2D systems. This characteristic arises mainly from the step density of states and makes the temperature dependence of mobility very different in quantum wells.

In the discussion of out-scattering rates, so far, carriers have been assumed to populate only the lowest subband and only intra-subband scattering has been considered. Expressions may be generalized to include inter-subband scattering.

The form factors and the relations between $E_{\mathbf{k}'_t}$ and $E_{\mathbf{k}_t}$ are altered to the following forms

$$G(q_z) = \int F'^{*}_l(z) \exp(iq_z z) F_l(z) dz \qquad (6.47)$$

$$E_{\mathbf{k}'_t} = E_{\mathbf{k}_t} + E_l - E_{l'} \pm \hbar\omega. \qquad (6.48)$$

The initial and final subbands are indicated respectively by the indices l and l' ; E_l and $E_{l'}$ are the respective subband energies.

The integrals for the in-scattering rates may be evaluated by following the same procedure. Only the additional factor $\cos\beta$ has to be included for elastic scattering and the factor $(k'_t/k_t)\cos\beta$ for inelastic scattering. This term becomes zero in randomizing collisions e.g., in non-polar optic phonon scattering, in idealized deformation-potential scattering and in alloy scattering for which the matrix element is independent of q.

6.7. Solution of the Transport Equation for 2DEG

The solution of the transport equation is first discussed with the assumption that the wells are so narrow that the separation between the lowest and the next-to-lowest subband is a few times the energy corresponding to $k_B T$(k_B is the Boltzmann constant and T is the lattice temperature). Carriers may then be assumed to populate only the lowest subband, a condition which is often referred as the extreme quantum limit(EQL). Extension of the analysis to conditions in which more than one subband may be populated will be done after the discussion of the EQL solution.

The equation has been resolved into two equations in Section (6.5), the first [Eq. (6.38)] of which gives the symmetric part f_0 as the distribution function in the absence of the external field. The perturbation due to the external field is obtained as the solution of the second equation, [Eq. (6.39)]. It may be written by using the results of Section 6.6 as follows:

$$\sum_i S_{\text{out}i}\phi(E_{\mathbf{k}_t}) - \sum_i S_{\text{in}i}\phi(E_{\mathbf{k}'_t}) = m^*/m_v, \qquad (6.49)$$

where $\sum S_{\text{out}i}$ is the sum of the out-scattering rates for all the different scattering mechanisms, and $\sum S_{\text{in}i}$ is the in-scattering rate for the ith scattering mechanism for which the initial energy is $E_{\mathbf{k}'_t}$.

Solution of this equation is trivial for elastic (acoustic phonon and defect) and randomizing (non-polar optic phonon) scattering events. In elastic scattering, since $E_{\mathbf{k}'_t} = E_{\mathbf{k}_t}$ the solution of the equation is simply,

$$\phi(E_{\mathbf{k}_t}) = (m^*/m_v)[\sum_i (S_{\text{out}i} - S_{\text{in}i})]^{-1} \tag{6.50}$$

For randomizing scattering, on the other hand, $S_{\text{in}i} = 0$ and the solution is

$$\phi(E_{\mathbf{k}_t}) = (m^*/m_v)[\sum_i S_{\text{out}i}]^{-1} \tag{6.51}$$

It is only in the case of non-randomizing inelastic scattering (polar optic phonon), that the solution is more involved. The equation may be written for polar-optic phonon scattering as

$$\sum_i S_{\text{out}i}\phi(E_{\mathbf{k}_t}) - \sum_{+,-} S_{\text{in}i}(E_{\mathbf{k}_t} \pm \hbar\omega_l) = m^*/m_v \tag{6.52}$$

where the first term is the sum of all the out-scattering rates including that due to the polar-optic phonon scattering. The second term is the sum of the two in-scattering terms contributed by polar-optic phonons, the plus and minus sign corresponding respectively to the emission and the absorption of a polar-optic phonon, the frequency of which is ω_l. The equation is a difference equation and cannot be solved straightaway, and different techniques, e.g.,variational method, matrix method and iteration methods[6.2] have been used in the past. The iteration method, however, appears to be the most convenient as all the complexities of scattering terms can be easily included and the numerical evaluation does not involve propagation of computational errors. The iteration may be done either by the Ritz method or the Gauss-Siedel method. In the Ritz method, values of $\phi(E_{\mathbf{k}_t} + n\hbar\omega_l)$ are obtained for different values of n in the ith step of iteration by using the iteration formula,

$$\phi^i(E_{\mathbf{k}_t} + n\hbar\omega_l) = (\sum_i S_{\text{out}i})^{-1}[(m^*/m_v) + \sum_{+,-} S_{\text{in}i}\phi^{i-1}(E_{\mathbf{k}_t} + (n \pm 1)\hbar\omega_l). \tag{6.53}$$

The starting value for $i = 0$ is taken to be zero. In the Gauss-Siedel method the iteration formula is

$$\phi^i(E_{\mathbf{k}_t} + n\hbar\omega_l) = (\sum_i S_{\text{out}i})^{-1}\{(m^*/m_v) + S_{\text{in}i}\phi^i[E_{\mathbf{k}_t} + (n-1)\hbar\omega_l]$$
$$+ S_{\text{in}i}\phi^{i-1}[E_{\mathbf{k}_t} + (n+1)\hbar\omega_l]\} \tag{6.54}$$

The Ritz method or the Gauss-Siedel method may be used to obtain results of similar accuracy, but the latter is a little faster. The methods are usually convergent, but the convergence is assured and fast when the diagonal terms, i.e S_{outi} is larger than the other two terms, which is usually the case.

Solutions for ϕ are illustrated in Figure 6.1 for the important mechanisms in a 10 nm wide double-junction quantum well of $Ga_{0.47} In_{0.53} As/InP$ systems. The perturbation ϕ is relatively independent of energy for the elastic scattering mechanisms. It varies strongly with energy for the polar-optic phonons scattering, and drops drastically down when the carrier energy equals the optic phonon energy and phonon emission starts, but it does not vary much till that energy is reached. It may, however, be recalled that the carriers populate mostly the energies below $\hbar\omega_l$ because of the step density of states, unless it is so degenerate that the Fermi energy is higher than $\hbar\omega_l$. The relative independence of energy is the characteristic of 2-D scattering, which makes Hall ratio close to unity and causes a small value of magnetoresistance.

The above analysis may be easily extended when the carriers are distributed over different subbands, if the strength of intersubband scattering is negligible. Carriers in each subband may then be considered to constitute isolated systems and characterized by a different distribution function, for which the kinetic energy of the carriers may be measured from the subband energy. Specifically, the symmetric part

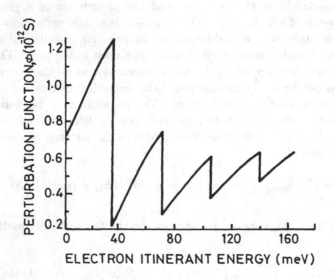

Figure 6.1. The perturbation function , ϕ, for a 10 nm double-junction quantum well of $InP/Ga_{0.47}In_{0.53}As$. [After B. R. Nag and S. Mukhopadhyay , *Jpn. J. Appl. Phys.* **31**, 3287 (1992)].

of the distribution function for the lth subband would be given by

$$f_{0l} = [\exp(E + E_l - E_F)/k_B T + 1]^{-1} \tag{6.55}$$

The asymmetric part ϕ_l would be given by the same equation as for the first subband, i.e., Eq. (6.52). The scattering rates would, however, be different due to differences in the form factor, $G(q_z)$, and the effect of nonparabolicity. The perturbation ϕ may however, be obtained by solving Eq. (6.49) following the same procedure as outlined for the first subband.

The solution gets involved, when the effect of inter- subband scattering is not negligible. Equation(6.49) is then modified to

$$\sum_i S_{\text{outi}} \phi_l(E_{k_t}) - \sum_i S_{\text{ini}} \phi_l(E_{k'_t}) - \sum_{l',i} S_{\text{ini}} \phi_{l'}(E_{k_{t'i}}) = m^*/m_v, \tag{6.56}$$

where now the out-scattering term includes in addition to the intra-subband contribution, the inter-subband contribution. The intra-subband in-scattering term is unaffected, but an additional term arises from inter-subband scattering. These equations may be solved in principle by using an iteration scheme similar to that used for a single subband. The iteration formula is then as given below :

$$\phi_l^i(E_{k_t}) = (\sum_i S_{\text{outi}})^{-1}[m^*/m_v + \sum_i S_{\text{ini}} \phi_l^{i-1}(E_{k'_{ti}}) + \sum_{l',i} S_{\text{ini}} \phi_{l'}^{i-1}(E_{k'_{t',i}})] \tag{6.57}$$

Such detailed solution has not, however, been reported by including all the scattering mechanisms. Results are reported only for the polar optic phonon scattering[6.15].

6.8. Mobility

The current density is given by

$$\mathbf{J} = (2/4\pi^2)(e/\hbar) \int \nabla_{k_t} E f d^2 k_t. \tag{6.58}$$

Replacing f by [see Section 6.5]

$$f = f_0 - (e\hbar\mathcal{E}/m^*)(\partial f_0/\partial E)\phi k_t \cos\theta, \tag{6.59}$$

we get for the mobility, $\mu = |J/ne\mathcal{E}|$, (n is the areal electron density),

$$\mu = -(|e|/m^*) \int \phi m_e(\partial f_0/\partial E)E dE[\int f_0 m_v dE]^{-1}, \tag{6.60}$$

where $m_E = (\hbar k_t)^2/2E$ and $m_v = \hbar^2 k_t/(dE/dk_t)$. The formula is applicable to parabolic as well as to nonparabolic bands. In the latter case m_E and m_v are required to be obtained by using thev proper $E - k$ relation[6.3].

This expression may be generalized when more than one band is occupied as follows:

$$\mu = -(|e|/m^*) \sum_l \int \phi_l m_{El}(\partial f_{0l}/\partial E) E dE / \sum_l \int f_{0l} m_{vl} dE. \qquad (6.61)$$

The index l indicates the lth subband. The distribution function f_{0l} is given by Eq. (6.55). The perturbation functions are required to be obtained by solving Eq. (6.56) as discussed in Section 6.7. It should be noted that m_{vl} and m_{El} are to be evaluated by considering the total energy $E_l + E$.

Electron mobility is evaluated by using Eq. (6.60) when only one subband is occupied or by using Eq. (6.61) when more than one subband is occupied. The perturbation function, ϕ has the dimension of time, and has the significance of the commonly used relaxation time, which applies to randomizing and elastic collisions. A relaxation time cannot be defined for more complex scattering mechanisms, but ϕ gives for such mechanisms a measure of the time in which the momentum gained from the external field is randomized.

The integral in Eq. (6.60) may be evaluated only numerically in the general case when all the complexities of nonparabolicity, finite barrier height and screening are taken into account. The Gaussian quadrature method is often used for this purpose. The total energy range is often broken into groups spaced by $\hbar\omega_l$ (ω_l is the optic phonon frequency) to facilitate the iterative solution of Eq. (6.49) as discussed in Reference 6.2. Such solutions are discussed at the end of this section along with the experimental results.

Analytic solutions may be obtained, however, for some special cases. The function $-(\partial f_0/\partial E)$ behaves as a delta function when the material is extremely degenerate, which is often true for the electron concentrations used in quantum well devices. Under such conditions, Eq. (6.61) may be simplified to,

$$\mu = -(|e|/2\pi m^* n \hbar^2) \sum_l m_{El}(E_F)\phi_l(E_F)E_F. \qquad (6.62)$$

For large carrier concentration, E_F has a large positive value, and it does not vary much with the temperature. The temperature dependence of mobility is then determined only by the temperature dependence of the coupling constant and is not significantly affected by the energy dependence of ϕ. It is hence expected that alloy scattering will be independent of temperature, deformation potential scattering will vary as T^{-1} and polar-optic phonon scattering will vary as $[\exp(\hbar\omega_l/k_B T) - 1]^{-1}$.

The integral may be evaluated analytically also for wells with non-degenerate carrier concentration, infinite barrier height and parabolic bands, including the effect of screening. The integrated expressions are given in Reference 6.10, for single hetrojunctions. Alloy scattering-limited mobility is independent of temperature and deformation-potential scattering-limited mobility varies as T^{-1} as for degenerate electron concentration. The temperature-dependence for other scattering mechanisms can not be simply expressed by a power law.

Mobility of electrons and holes in single heterojunctions and in quantum wells has been extensively studied [6.16-86] since the publication of the first paper by Dingle et al[6.16], showing the improvement in mobility achieved by modulation doping. A major part of the studies are related to the mobility at very low temperatures, as the improvement due to the reduction of the effect of impurities is expected to be prominent at such temperatures. Such studies are useful for the assessment of the quality of the sample, but, mobility values near room temperature are more relevant to the performance of the QW- devices.

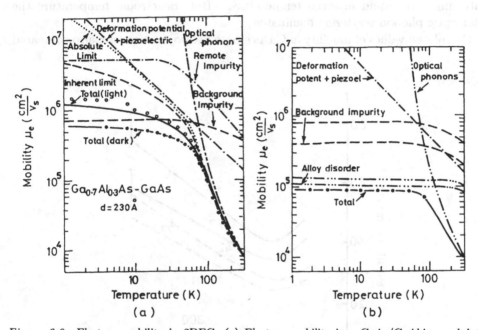

Figure 6.2. Electron mobility in 2DEG. (a) Electron mobility in a GaAs/GaAlAs modulated single-junction well at different temperatures for a spacer width of 230 Å between the junction and doped GaAlAs. Full and open circles give the experimental points respectively for the electron density of 2.2×10^{11} cm^{-2} and 3.8×10^{11} cm^{-2}. (b) Electron mobility in In$_{0.53}$Ga$_{0.47}$/ Al$_{0.52}$In$_{0.48}$As modulated single-junction heterostructure with spacer width of 80 Å. Dots are experimental points. Component mobilities are calculated values for electron density of 4.5×10^{11} cm^{-2}. The upper and lower curves for impurity scattering are respectively for impurity concentration of 0.5×10^{16} cm^{-3} and 1×10^{16} cm^{-3}. Alloy scattering potentials are 0.55 eV and 0.63 eV respectively for the upper and the lower curve.[After W. Walukiewicz, H. E. Ruda, J. Lagowski and H. C. Gatos, *Phys. Rev. B* **30**, 4571 (1984); Copyright (1984) by the American Physical Society].

6.8.1. ELECTRON MOBILITY IN AlGaAs/GaAs AND AlGaAs/InGaAs SINGLE HETEROJUNCTION, AND InP/GaInAs QW's

In Fig. 6.2 are illustrated the variation of electron mobility with temperature in GaAs/GaAlAs and AlInAs/GaInAs single heterojunctions[6.10] and in InP/GaInAs[6.3] quantum wells. Figure 6.2 shows that ionized impurity scattering contributes significantly only at very low temperatures. Contribution of piezo-electric scattering is also not very significant.

Mobility is mostly determined by the acoustic phonon scattering and alloy scattering near liquid nitrogen temperature. But, near room temperature the polar-optic phonon scattering dominates.

Calculated values of mobility in GaAs single heterojunctions are also compared

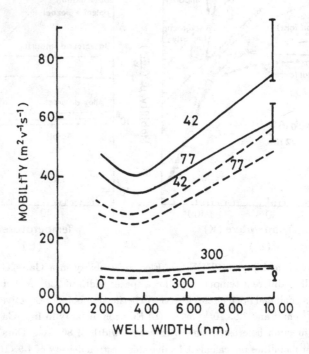

Figure 6.3. Calculated values of electron mobility in InP/Ga$_{0.47}$In$_{0.53}$As double-junction wells of different widths. The dashed lines are for very small carrier density of 0.97×10^{12} m^{-2}. Numbers on the curves give the lattice temperature. The bars give the range of experimental values. Values for Al$_{0.48}$In$_{0.52}$As wells are shown by open circles. Deformation potential acoustic and optic phonon scattering and alloy scattering were included in the calculations. [After S. Mukhopadhyay and B. R. Nag, *Phys. Rev. B* **48**, 17960 (1993); Copyright (1993) by the American Physical Society].

with the experimental results in Fig. 6.2(a,b) and 6.3.

The theory as discussed here is found to agree with the experimental results fairly well, provided some of the coupling constants are adjusted. The main difficulty in explaining the transport results by theory arises from our insufficient knowledge about the coupling constants. The dielectric constants, acoustic phonon deformation potential and alloy potential are not known with sufficient accuracy. On the other hand, experimental transport coefficients may be fitted by adjusting their values and this procedure has been used to estimate their values, which has resulted in controversies raging over decades. The only justified conclusion is that the theory gives the right kind of temperature and energy dependence, and helps us to estimate the relative importance of different scattering mechanisms.

In-plane electron mobility have been studied in recent years for quantum wells using a few other systems. Mention may be made of the $In_{0.5}Ga_{0.5}P/GaAs$ [6.79-81] and AlGaN/GaN[6.82-86] quantum wells.

6.8.2 ELECTRON MOBILITY IN InGaP/GaAs QW's

Replacement of AlGaAs by InGaP with lattice matching composition $In_{0.5}Ga_{0.5}P$ is expected to remove many of the problems involved with aluminum, e.g., high growth temperature, low doping efficiency and persistent photocurrent effects due to the presence of DX centers. It was also expected that the quality of the interface may also be better in this system in comparison to that in AlGaAs/GaAs structures, particularly in the inverted structure. Studies have been made to examine this possibility by studying the electron mobility at low temperatures, e.g.,20 K. The mobility is dominantly controlled for such temperatures by the IFRS and therefore gives an indication of the surface roughness. We find from the expression for scattering rate given in Table 6.2 for IFRS the mobility is proportional to the inverse of the sqare of the differential coefficient of energy eigenvalues with respect to the well width, e. g., $(\partial E/\partial L)^{-2}$. Since the eigenvalue of energy e varies as $L^{-}2$ in infinite-barrier wells , the IFRS-limited mobility is expected to vary as L^6. The so-called L^6 law is approximately true for the AlAs/GaAs wells [6.79]. On the other hand, the electron mobility in InGaP/GaAs wells were found [6.80] to be much larger for wells of small width than that obtained from the L^6 law. These large values of mobility were accepted as evidence of large values of correlation lengths of roughnesses in these wells.It has, however, been found [6.81] that the deviation from the l^6 law may also be due to the low value of the barrier height in this system. The barrier height has not yet been conclusively established for this system and hence the problem remains open at present. It may be mentioned that experimental or theoretical results are not available for InGaP/GaAs single heterojunctions or quantum wells for higher temperatures although a number of studies have been reported for low temperatures.

6.8.3. ELECTRON MOBILITY IN AlGaN/GaN QW's

Results on the electron mobility in AlGaN/GaN QW's have been reported by Redwing et al [6.82]. The mobilities were found to be 1700 cm^2/V.s at 300 K and 7500 cm^2/V.s at 77 K. It was also found rthat the mobility varies with the aluminum content of Al$_x$Ga$_{1-x}$N and peaks at composition close to $x = 0.2$ for MOCVD-grown layers , and for $x = 0.4$ for MBE-grown layers. Theoretical values of mobility have also been studied for this system [6.85]. The theoretical values of low-temperature mobility are higher by about two orders of magnitude. The descrepancy has been attributed to residual scattering centers. The variation of mobility with the composition of AlGaN is also not yet well understood [6.86]. It may be concluded that the from the available data that the material properties in this system are not yet properly known for comparison with the theory.

6.8.4 GENERALIZED EXPRESSION FOR MOBILITY

The generalized expression for the mobility when more than one subband is populated is given by Eq. (6.61). This equation is required to be used for large well widths.

The structures used in quantum well devices, however, are of small effective widths. Population of higher subbands may, therefore, be considered unimportant for the estimation of mobility in quantum well devices.

The wells for the devices, however, may be so narrow that the effects of wave function penetration becomes important. Wave function penetration increases mobility by reducing effectively the form factor, and at the same time decreases the mobility by increasing the effective value of m_v since the barrier mass is usually larger than the well mass. The net effect may be assessed only by detailed calculations. Some calculations [6.3] reported in the literature indicate that the effect may be important and will then be required to be incorporated into the theory. This may be particularly important for wells of strained- layer structures in which the well width is required to be narrower than 4 nm for preventing the appearance of crystal dislocations.

6.9. High-field Velocity of 2DEG

Electron mobility is independent of field for low fields and Ohm's law applies. The velocity increases linearly with the field. It is, however, found in quantum wells as in bulk materials, that the relation becomes nonlinear for high values of the field. The velocity saturates and in some cases decreases with increase in the field at high fields. Since the operating condition of the devices usually extends to this region, the switching speed and the transconductance are very much dependent on this saturation velocity or the high-field velocity characteristics. Velocity-field characteristics become nonlinear for high fields due to the so-called hot-electron effect. A very extensive literature exists on this effect and the interested reader may consult Reference 6.2. One way to explain the effect is as follows. Equation (6.38) was obtained from the transport equation, Eq. (6.36), by assuming that the field is small and consequently the product of the field and the perturbation term, which is proportional to $\mathcal{E}^2$ may be neglected. This assumption evidently fails at high fields. Equation (7.38) is modified to the following equation when this term is retained.

$$-(e\mathcal{E})^2(1/m^*)(m_E/m_v)\partial_E(\phi\partial_E f_0) = \int \{S(\mathbf{k}_t, \mathbf{k}'_t)f_0(\mathbf{k}_t)[1 - f_0(\mathbf{k}'_t)]$$
$$-S(\mathbf{k}'_t, \mathbf{k}_t)f_0(\mathbf{k}'_t)[1 - f_0(\mathbf{k}_t)]\}d^2k'_t. \tag{6.63}$$

Evidently for high fields the symmetric part of the distribution function f_0 , is a function of the electric field and is different from the zero-field function. As a result, electron mobility, being proportional to $(\partial f_0/\partial E)$ varies with the field. The nature of variation is determined by the characteristics of the scattering mechanisms. Since the average energy of the electrons usually increases with the field, the mobility is expected to increase with the field for those mechanisms for which the scattering rate decreases with increase in the electron energy as in impurity scattering or polar-optic phonon scattering in parabolic-band materials. The mobility may decrease with increase in the field and lead to velocity-saturation if the scattering rate increases with increase in the electron energy as in deformation-potential acoustic phonon or alloy scattering or for other scattering mechanisms in bulk nonparabolic materials. However, at high fields since the electron energy is much enhanced, some scattering mechanisms, normally ineffective for low fields, become important and cause velocity saturation and in some cases a decrease of velocity with increasing field. The most effective such mechanism is the non-equivalent intervalley scattering in bulk materials.

In quantum wells, on the other hand, conditions are more complex. The average energy of the electrons may exceed the barrier potential and then the electrons do not remain confined within the wells, but may move about in the barrier layers. This so-called electron transfer effect[6.52,53] causes a decrease of velocity with increasing field since the mobility in the barrier material is lower than that in the well. Even for those electrons which remain confined, the wave functions become very different and the scattering is of different nature. Also, as in bulk materials non-equivalent intervalley scattering becomes effective, but in this case a 2D electron may be scattered into a 3D state or the vice-versa. Inter-subband scattering also plays a more and more dominant role as the average energy increases. Because of all these factors, it is very difficult to work out or to predict the high-field velocity characteristics of 2DEG in quantum well structures. The straightforward expectation, that since for most of the scattering mechanisms the scattering rate is independent of energy or decreases slowly with energy in quantum wells the velocity will increase linearly or runaway at high fields, is not seen in experiments. Only a few experiments have, however, been performed. Some of them show saturation of velocity but a few show decreasing velocity with increasing field. Considering that the knowledge of the characteristics is very important for understanding the devices, the available experimental and theoretical results are presented, although these are far from being complete for a thorough discussion of the subject.

6.9.1. THEORY

The theory developed for the evaluation of hot-electron characteristics of bulk materials is applicable also to the 2D EG in quantum well structures. The 2D character only changes the density of states and the formulae for the scattering rates and the other integrals are required to be modified accordingly. Analyses reported in the literature have been done by using the displaced Maxwellian approximation method and the Monte Carlo method.

Displaced-Maxwellian Approximation
The functional form of f_0 is assumed in this approximation to be displaced Maxwellian, as written below:

$$f_0 = I_n \exp[-\hbar^2/2m^*)(|\mathbf{k}_t - \mathbf{k}_{t0}|^2)/k_B T_e]. \tag{6.64}$$

where $I_n = [\int f_0 d^2 k_t]^{-1}$ is the normalization integral, $\mathbf{k}_{t0}$ is the average value of the wave vector in the direction of the field and T_e is the effective temperature. This function is chosen from the considerations that as the electrons are energized by the field, their average velocity and energy increase. However, for large electron concentration due to frequent electron-electron collisions their velocity is shared. Hence, electrons have identical average velocity although a particular electron may attain a different velocity due to the difference in actual collision time. Also, when electron-electron collisions are frequent the distribution function must be Maxwellian (for non-degenerate concentration), but as the average energy is increased from the no-field value, i.e.,$k_B T_L$ (T_L is the lattice temperature) the electron temperature has a different value. It is because the value of the temperature is higher, that the high-field transport is often referred as hot-electron transport.

The average value of the wave vector $\mathbf{k}_{t0}$ and the electron temperature T_e , are obtained by balancing the energy and momentum gained from the field to those lost through collisions. The necessary equations are obtained by multiplying the two sides of Eq. (6.36) by E and $\mathbf{k}_t$ and integrating over the in-plane phase space. The common wave vector $\mathbf{k}_{t0}$ and the drift velocity are obtained by solving the resulting equations.

Monte Carlo Method

The application of the displaced-Maxwellian approximation method, although justified for very high electron concentration as exists in quantum well devices, is difficult when all the different scattering mechanisms, energy band nonparabolicities and partial confinement effects are to be taken into account. The Monte Carlo method is considered to be more versatile and has been used extensively to derive the hot- electron characteristics of quantum wells as for bulk materials. The method is discussed in details in Reference 6.2. However, a brief outline is given below.

The basic idea is to simulate on the computer the trajectories of a single or an ensemble of electrons. In the first step of simulation the electron is given an energy E_i and wave vector $\mathbf{k}_{ti}$ which may be chosen arbitrarily if the interest is in the calculation of the steady-state velocity. If, however, the interest is in the calculation of the time and position- dependence of the average velocity after the field is applied, the initial energy and wave vector are chosen according to the zero-field room temperature distribution.

In the second step, the time of the subsequent scattering is computed by using a uniformly distributed random number r_1 , and the following formula.

$$t_c = -\Gamma \log(1 - r_1), \tag{6.65}$$

where Γ is a suitably chosen constant, larger than the total scattering rate due to all the scattering mechanisms.

After determining the collision time, the energy and the wave vector of the electron before collision are computed by using the equations :

$$\mathbf{k}_{tf} = \mathbf{k}_{ti} + (e\mathcal{E}/\hbar)t_c \tag{6.66}$$

and

$$(\hbar k_{tf})^2/2m^* = F(E_f), \tag{6.67}$$

where $\mathbf{k}_{ti}$ is the initially chosen wave vector, k_{tf} is the wave vector at the collision time t_c , immediately before collision; $\mathcal{E}$ is the applied field, m^* is the band edge mass and $F(E)$, is a function relating energy E to the wave vector. For parabolic bands $F(E) = E$, whereas for nonparabolic bands obeying the simplified Kane relation $F(E) = E(1 + \alpha E)(\alpha$ is the nonparabolicity parameter). It may be chosen to be more complex as required by the material properties. The scattering probabilities for the different scattering mechanisms are next computed for the electron energy at time t_c. These may be computed directly at this stage or may be obtained by interpolation from look-up tables prepared before the start of the simulations. A second random number is then generated to determine which kind of scattering occurs by using the inequality.

$$\sum_{i=1}^{n-1} S_i(\mathbf{k}_{tf}) < r_2 \leq \sum_{i=1}^{n} S_i(\mathbf{k}_{tf}), \tag{6.68}$$

where the left-hand side sum is the total scattering probability for $(n-1)$ scattering mechanisms and the right-hand side sum is the value obtained by including another scattering mechanism. When this inequality applies, the nth mechanism is taken to be effective. If, however, r_2 is larger than the sum, including all the scattering mechanisms, the scattering is taken to be the so- called self scattering, which is really no-scattering.

After knowing the kind of scattering in the second step, explained above, the energy and wave vector of the electron after the scattering are determined by generating a third random number r_3 and by using the equations:

$$k_{ti} = (2m^*/\hbar^2)[F(E_f \pm \Delta E)]^{1/2}, \tag{6.69}$$

$$r_3 = \int_0^\beta S(\mathbf{k}_t, \mathbf{k'}_t)d\beta / \int_0^{2\pi} S(\mathbf{k}_t, \mathbf{k'}_t)d\beta, \tag{6.70}$$

$$k_{ti} = \hat{\mathbf{i}}\, k_{ti} \cos(\theta + \beta) + \hat{\mathbf{j}}\, k_{ti} \sin(\theta + \beta), \tag{6.71}$$

where θ is the angle between $\mathbf{k}_{tf}$ and $\mathcal{E}$ and β that between $\mathbf{k}_{tf}$ and $\mathbf{k}_{ti}$ after collision and $\pm \Delta E$ is the change in energy of the electron in the particular scattering.

This process is repeated several thousand times to generate data about $\mathbf{k}_{tf}$ and $\mathbf{k}_{ti}$ and the free time t_c in between the scattering events. These data are then utilized to evaluate the different transport coefficients by using the relevant descriptors. The drift velocity is given by

$$v_d = \sum(E_f - E_i)/\hbar \sum |\mathbf{k}_{tf} - \mathbf{k}_{ti}|. \tag{6.72}$$

The diffusion constant is given, on the other hand, by,

$$D_t = (1/2)(< v_t^2 > - < v_t >^2)t_s, \tag{6.73}$$

where t_s is the sampling time, and $< v_t >$ is the average velocity during this time.

In materials, for which the electron concentration is such that the nondegenerate statistics applies and electron- electron scattering is negligible, the simulation may be done by considering one electron and recording its trajectory over a long time. If, however, electron-electron scattering be important, the ensemble simulation is required since inclusion of this scattering requires knowledge of the energy distribution function.

The method has been used extensively to work out the drift velocity in both single heterojunctions and quantum wells. In some of these simulations, degeneracy, electron-electron scattering, electron transfer effect and intersubband and nonequivalent scattering have been included. The effect of electron-electron scattering is not very significant on the drift velocity , as may be expected. Electron-electron scattering really changes the distribution function and causes sharing of the momentum between the electrons without causing a change in the total

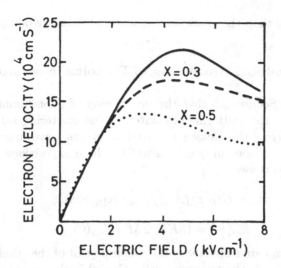

Figure 6.4. Hot-electron velocity-field Characteristics at 300 K. The dashed curves are for 2DEG in $Al_xGa_{1-x}As/GaAs$ samples with x = 0.5. The full curve is for bulk GaAs with electron density of 1×10^{15} cm^{-3}. [After W. T. Masselink, *Semicond. Sci. Technol.* 4, 503 (1989); Copyright IOP Publishing Ltd.].

momentum of the system and the drift velocity is not much sensitive to the exact form of the distribution function. The simulations indicate that at high fields, the electron velocity decreases in quantum wells with increase in field as in bulk materials.

6.9.2 EXPERIMENTAL RESULTS

Experimental results on high-field characteristics are given in Fig. 6.4. The determination of the characteristics in the negative differential mobility region is difficult for quantum wells both experimentally and theoretically. The time-of-flight techniques, both dc and microwave, which have been used for bulk materials to obtain reliable characteristics, cannot be applied conveniently to obtain the in-plane drift velocity. Data have been obtained by either measuring the current-voltage characteristic of dumbbell shaped samples[6.60-62] or by shining with microwaves and measuring the low-field dc conductivity[6.63].

On the theory side, simulations for the energies corresponding to the peak velocities involve many approximations. The quantum confinement is likely to be significantly altered at such energies and this alteration occurs dynamically. Exact formulation of the intersubband and intervalley scattering is also difficult as the quantized picture of the higher-valleys is not very clear. Because of these reasons, the high-field drift velocity obtained from these studies should be considered as guiding values rather than the exact values, when used in the design of devices.

6.10. Scattering-induced Broadening of Photoluminescence Spectrum

It was mentioned in Section 5.5 that the broadening of photoluminescence spectrum is partly due to the scattering of excitons. The scattering rate for excitons is computed by applying the Golden rule and using the wave functions and dispersion relations for excitons in quantum wells. These expressions, given in Section 5.3, are repeated below:

$$\psi_{ex} = G(\rho, z) f_e(z_e) f_h(z_h) \exp(i\mathbf{K}.\mathbf{R}), \tag{6.74}$$

$$E_{ex}(K) = (\hbar K)^2 / 2M + E_{ex}(0), \tag{6.75}$$

where $f_e(z_e)$ and $f_h(z_h)$ are respectively the z-component of the envelope functions for electrons and holes in the quantum well. $G(\rho, z)$ is the envelope function for the exciton, ρ and z being respectively the in-plane and the z component of the relative coordinates of electrons and holes. $\mathbf{K}, \mathbf{R}, M$ are the wave vector, the in-plane coordinate of the center of mass and the effective mass of the exciton. These are related to the corresponding quantities for electrons and holes, indicated respectively by the subscripts e and h, by the following relation:

$$\mathbf{K} = \mathbf{k}_e + \mathbf{k}_h; M = m_e^* + m_h^*; \mathbf{R} = (m_e^* \rho_e + m_h^* \rho_h)/M. \tag{6.76}$$

E_{c0} and E_{v0} are the magnitudes of the energies of the lowest electron and hole subbands measured from the band edges. The scattering rate for lattice scattering for an exciton with the initial wave vector K_i is given, according to the Golden rule by,

$$
\begin{aligned}
S(\mathbf{K}_i) = & (2\pi/\hbar) \sum_{\mathbf{q}, \mathbf{K}_f} |M(\mathbf{K}_i, \mathbf{K}_f)|^2 [N_f(E_{exf}) + 1] N_i(E_{exi}) \\
& \times [n_{\mathbf{q}} \delta(E_{exf} - E_{exi} - \hbar\omega_{\mathbf{q}}) + (n_{\mathbf{q}} + 1) \delta(E_{exf} - E_{exi} + \hbar\omega_{\mathbf{q}})],
\end{aligned} \tag{6.77}
$$

where $\mathbf{K}_f, \mathbf{K}_i$ and E_{exi}, E_{exf} are respectively the final and initial wave vectors and energies of the exciton, $N_f(E_{exf})$ and $N_i(E_{exi})$ are the final and the initial exciton populations, $n_{\mathbf{q}}$ and $\mathbf{q}$ are the phonon occupation factor and wave vector. $M(\mathbf{K}_i, \mathbf{K}_f)$ is the matrix element for scattering from the $\mathbf{K}_i$ -state to the $\mathbf{K}_f$ - state. It is also assumed that the excitons are created by a strong source of illumination so that $[N_f(E_{exf}) + 1] \approx N_i(E_{exi}) \approx 1$.

The central task in evaluating the scattering rate is the formulation of the matrix element. The matrix element is formulated by suitably modifying the electron-phonon interaction matrix elements discussed in Section 6.2. It should, however, be noted that the excitons are strongly coupled electron-hole pairs, and as such are not expected to be affected by the electronic potential originating from the different kinds of imperfections which scatter free electrons or free holes if the electrons and holes are equally affected. However, as the electrons and holes belong to different bands, the imperfections affect them differently. For example, the acoustic phonon deformation potentials are different for the electrons and the holes. As a consequence of this difference, electrons and holes are affected by different amounts and a net effect is observed. Further, in the dynamics of excitons, although the electron-hole pair moves as one particle, the motion being indicated by the center of mass, the individual motions of the electron and the hole are different. This difference also causes a scattering for the scattering mechanisms for which electrons and holes see the same potential.

Evaluation of the matrix element by including all the complexities of the excitonic wave function, particularly for quantum well systems, is very difficult and has not been attempted so far. Some illustrative results have been obtained by introducing simplifying approximations. But, even with these approximations, the analysis is too complex and not much will be gained by discussing them. The interested reader may consult Reference 6.64, for acquaintance with these results. We discuss instead the basic approach used for the analysis and the final expressions which may be used to explain the experimental results or for designing the opto-electronic devices.

We consider deformation-potential acoustic-phonon scattering for explaining the basic approach. In principle, the matrix elements is given by,

$$M(\mathbf{K}_i, \mathbf{K}_f) = \int \psi_{exf} \Delta V \psi_{exi} d^3 r. \qquad (6.78)$$

It may be easily appreciated that evaluation of the above expression is almost impossible by substituting the expressions for ΔV for the different scattering mechanisms, discussed in Section 6.6 and for ψ_{ex} the expression discussed in Section 5.1.3. It may however, be shown that the integral may be separated into two components: one involving the interaction with the electron and the other with the hole. As the electron and the hole have opposite but equal charge, and have, therefore, equal but opposite potential energy in a common potential, the net matrix element is the difference of the two components.

The component for the interaction with the electron may be written as

$$M(\mathbf{K}_i, \mathbf{K}_f) = (\hbar/2V\rho\omega_q)^{1/2} E_e q \int \exp(-i\mathbf{K}_f.\mathbf{R})[G(\rho', z)f_e(z_e')f_h(z_h')]$$
$$\times \exp(\pm i\mathbf{q}.\mathbf{r}) \exp(i\mathbf{K}_i.\mathbf{R})$$
$$\times [G(\rho, z)f_e(z_e)f_h(z_h)]d^2\rho d^2\rho' dz_e dz_h dz_e' dz_h'. \qquad (6.79)$$

where $\rho', z_e', z_h' (\rho_e, z_e, z_h)$ are the coordinates of the exciton after (before) scattering. $G(\rho, z)f_e(z_e)f_h(z_h)$ is the excitonic envelope function. E_e is the deformation potential for the electron. These coordinates are not affected by the scattering; only the energy and wave vector of the exciton and the phonon change.

The integral may hence be written as,

$$M_e(\mathbf{K}_i, \mathbf{K}_f) = (\hbar/2V\rho\omega_q)^{1/2} E_e q \int [G(\rho, z) f_e(z_e) f_h(z_h)]^2$$
$$\times \exp i(q_z z_e + \alpha_h \mathbf{q}_t.\rho) d^2\rho dz_e dz_h, \quad \alpha_h = m_h/M. \quad (6.80)$$

The exponential term within the integral is obtained by considering that a non-zero value of the integral is obtained only if,

$$(\mathbf{K}_i + \mathbf{q}_t).\mathbf{R} - \mathbf{K}_f.\mathbf{R} = 0. \quad (6.81)$$

The integral is evaluated by introducing the assumptions that the barrier potential is infinite, the well is very narrow and the excitonic envelope function is given by (see Section 5.1.3 d)

$$G(\rho, z) = I_n \exp(-\beta\rho/2), \quad (6.82)$$

where β is a variational parameter to be determined for the minimization of the energy.

We may write,

$$\int [f_h(z_h)]^2 dz_h = \int [f_e(z_e)]^2 dz = 1. \quad (6.83)$$

$$\int \exp(iq_z z_e)[f_e(z_e)]^2 dz_e = \int \exp(iq_z z_h)[f_h(z_h)]^2 dz_h = G(q_z). \quad (6.84)$$

The component of the matrix element due to the electron may now be simplified to

$$M_e(\mathbf{K}_i, \mathbf{K}_f) = (\hbar/2V_c\rho\omega_\mathbf{q})^{1/2} E_e q \gamma_e G(q_z)(n_\mathbf{q} + 1/2 \pm 1/2)^{1/2}, \quad (6.85)$$

where

$$\gamma_e = [1 + (\alpha_h|\mathbf{K}_i - \mathbf{K}_f|/\beta)^2)]^{-3/2}. \quad (6.86)$$

The component due to the hole may be evaluated by following the same procedure and the total matrix element written as,

$$M(\mathbf{K}_i, \mathbf{K}_f) = (\hbar/2V_c\rho\omega_\mathbf{q})^{1/2} q(\gamma_e E_e - \gamma_h E_h) G(q_z)(n_\mathbf{q} + 1/2 \pm 1/2)^{1/2} \quad (6.87)$$

The scattering rate is now obtained by substituting the above expression for $M(\mathbf{K}_i, \mathbf{K}_f)$ into Eq. (6.77), converting the summation to integration and carrying out the integrations. We obtain,

$$S(\mathbf{K}_i) = (2\pi/\hbar)(V_c/8\pi^3)(\hbar/2V_c\rho s^2)(2k_B T/\hbar)(M/\hbar)^2$$
$$\times \int (\gamma_e E_e - \gamma_h E_h)^2 d\theta \int |G(q_z)|^2 dq_z. \quad (6.88)$$

Equipartition approximation has been used, i.e. we have assumed that $n_\mathbf{q} \approx n_\mathbf{q} + 1 \approx k_B T/\hbar\omega_\mathbf{q}$. The phonon dispersion relation has been assumed to be linear, i.e. $\omega_\mathbf{q} = sq$, where s is the sound velocity. Integration over $\mathbf{K}_f$ is carried out by using the property of the delta function. The scattering is assumed to be elastic i.e., $\hbar\omega_q$ is considered negligible in comparison to the kinetic energy, of the exciton. Expression (6.88) may be further simplified by considering that β is about $1/2a_B$ (a_B is the Bohr radius) and its numerical value is about 10 m^{-1} . The value of K_i is about an order lower. We may hence either neglect $(\alpha_h K_i/\beta)^2$ or $(\alpha_e K_i/\beta)^2$ in comparison to unity, or replace $\gamma_e(\gamma_h)$ by suitable average values $\bar{\gamma}_e(\bar{\gamma}_h)$.

Equation (6.88) then simplifies to

$$S(K_i) = (3k_B TM/2\rho s^2 \hbar^3 L)(\bar{\gamma}_e E_e - \bar{\gamma}_h E_h)^2, \qquad (6.89)$$

where we have put

$$(1/2\pi) \int |G(q_z)|^2 dq_z = 3/2L. \qquad (6.90)$$

Expression (6.90) is the same as that for the scattering rate for electrons multiplied by $(\bar{\gamma}_e - \bar{\gamma}_h E_h/E_e)^2$. The scattering rate for excitons may, therefore, be estimated from the values for free electrons, by multiplying by a factor. It is, however, difficult to accurately estimate the value of this factor due to incomplete knowledge about E_e and E_h . There being no established direct method for the determination of their values, indirect methods have been used and the reported values differ by large factors. This difference is further enhanced in the case of excitons as the scattering rate is proportional to the square of the difference of the two deformation potentials.

Similar results are obtained for other elastic scattering mechanisms: piezoelectric scattering, alloy scattering and interface roughness scattering. The factors are respectively, $(\bar{\gamma}_e - \bar{\gamma}_h h_e/h_h)^2$, $(\bar{\gamma}_e - \bar{\gamma}_h \Delta E_h/\Delta E_e)^2$ and $(\bar{\gamma}_e - \bar{\gamma}_h m_e^*/m_h^*)$, where $h_e(h_h)$, $\Delta E_e(\Delta E_h)$ are the piezoelectric constants and alloy potentials for electrons(holes). As for deformation potential scattering exact estimation of these factors is rendered difficult by our incomplete knowledge of the different coupling constants. The scattering rates for the impurities and the polar optic phonons are still more difficult to formulate. Location of the impurities play an important role. Although no detailed theory has yet been attempted, it may be expected that the scattering rate will be given by the free electron scattering rate multiplied by a factor as for the other elastic scattering processes. The factor would be $(\bar{\gamma}_e - \bar{\gamma}_h)^2$. However, impurities play a relatively unimportant role in quantum wells as these are usually segregated from the active layer. Effect of impurities on the broadening may, therefore, be considered insignificant. Polar optic phonons, on the other hand, would be very important particularly near room temperature, as observed from the analysis of the mobility of free carriers in quantum wells. Analysis of the effect of this scattering on excitons is, however, extremely complicated. The energy of the polar optic phonons (30-40 meV) is much larger than the binding energy of excitons ($\sim$10 meV). Hence, the scattering may result in the ionization of the exciton, transition to an excited state or to just an increase in the kinetic energy associated with the motion of the center of mass. Although, some expressions have been given in the literature[6.65] for the estimation of polar optic phonon scattering rate, these should be considered incomplete as all the above-mentioned possibilities are not considered. The only acceptable conclusion is that the rate will be proportional to the phonon occupation probability i.e., to $[\exp(\hbar\omega_l/k_BT) - 1]^{-1}$.

It should be evident from the above discussion that estimation of scattering for excitons is less accurate than that for the free carriers. This is partly due to the complicated nature of the exciton wave functions, small binding energies and closely spaced excited levels, and partly because, the inaccurate knowledge of the interaction constants affect strongly the computed values. We may, however, conclude that the rate will vary with the well width and the temperature like the scattering rates for the free electrons. Accordingly, it is expected that the variation of the line width with temperature will be given by the following relation.

$$\Gamma(T) = \Gamma_0 + \Gamma_1 T + \Gamma_2[\exp(\hbar\omega_l/k_BT) - 1]^{-1}, \qquad (6.91)$$

where Γ_0 represents the contribution from alloy scattering, interface roughness and other structural defects, discussed in Section (6.6), $\Gamma_1 T$, the contribution from the acoustic phonons and the third term from the polar optic phonons. An experimental[6.66] curve for $\Gamma(T)$ is given in Fig. 6.5 for a 102 Å wide well. The curve may be fitted by taking $\Gamma = 2$ meV, $\Gamma_1 = 0$ and $\Gamma_2 = 5.5$ meV. The line-width broadening may, therefore, be ascribed for temperatures below 100 K to structural imperfections and for temperatures above 100 K, mainly to polar optic phonon scattering.

The broadening may, therefore, be expected to vary with the well width like

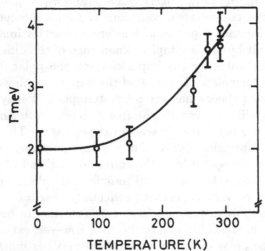

Figure 6.5. Temperature dependence of the half-width of heavy-hole exciton peak. Experimental points are shown by circles and error bars. The solid lines give the calculated values with adjusted values of Γ_0 and Γ_2 of Eq. (6.91).[After D. S. Chemla, D. A. B. Miller, P. W. Smith, A. C. Gossard and W. Wiegmann, *IEEE J. Quantum Electron.* **QE-20**, 265 (1984); Copyright: (=A9 1984=IEEE)].

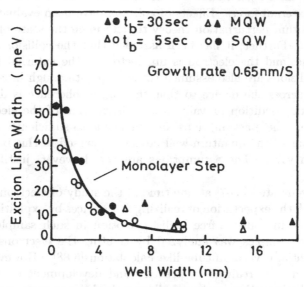

Figure 6.6. Photoluminescence full width at half maximum (FWHM) for different well widths. Solid line is the calculated FWHM for an average well width variation of one monolayer. t_b - Interruption time between the growth of $Ga_{0.47}In_{0.53}As$ and InP.[After D. Grützmacher, K. Wolter, H. Jürgensen, P. Balk and C. W. T. Bulle Liewma, *Appl. Phys. Lett.* **52**, 872 (1986); Copyright American Institute of Physics].

the inverse of optic-phonon scattering- limited mobility near room temperature and as the inverse of defect scattering-limited mobility at low temperatures. Experimental data on the dependence of room temperature line width on well width are not, however, available.

Data are available for near helium temperatures[7.78] and these are presented in Fig. 6.6. The line-width is almost independent of the well width down to about 7 nm, but increases with decrease in the well width for smaller widths. Attempts have been made to explain the variation by invoking the various broadening mechanisms, particularly the interface roughness scattering. However, exact agreement can hardly be expected, since the parameters of the scattering mechanisms which cause broadening at low temperature (see Section 6.6) may not remain the same for samples of different widths as these depend on growth conditions.

6.11. Ballistic Transport

The theory of transport, discussed in the preceding sections, has been based on the Boltzmann transport equation and a semiclassical description of the electron motion. Electrons have been assumed to move as particles in between collisions in response to the external forces. Effects of collision have been evaluated, however, by using the quantum perturbation theory to determine the state of the electron after the collision. Further, it has been assumed that the collisions are isolated in space and time, and the electron is unaffected by the external forces during the collision. Collisions are also assumed to be frequent enough in comparison to the transit time across the device so that the phase coherence is destroyed and a quasi-equilibrium condition prevails in the distribution of the electrons in the momentum space. These assumptions are applicable to sufficiently long samples and the dimensions of the quantum well devices discussed in this book, are such that they remain valid. The assumptions are not, however, justified for short samples.

Interest was generated [6.87] at one time in the study of electron transport in short samples with the expectation of realizing fast devices by exploiting the higher velocity obtainable in collision-free ballistic motion in such samples. Electron transport in short samples was analyzed by treating the electrons as particles and making essentially vacuum-tube-like calculations[6.88]. However, the ideal conditions could not be realized in practice and development of devices using collision- free ballistic motion in short bulk samples has not made much progress.

A different kind of ballistic transport is, receiving attention currently, which may occur in short quantum-wire-like structures. The transport is analyzed by a formalism, which is often referred as the Buttiker-Landauer formalism[6.89-94]. In this formalism, the analysis of the interaction of the electron with a scattering center is not based on the perturbation theory but on the scattering matrix for the electron wave. The method was applied to calculate the resistance of a one-dimensional medium with a series of independent scatterers[6.90]. Recent experiments on electron transport in ballistic one-dimensional channels have added further importance to this method. This aspect of electron transport is not, however, discussed here as the phenomenon is not relevant to the devices of interest in this monograph.

CHAPTER 7

HIGH ELECTRON MOBILITY TRANSISTOR (HEMT)

The superior opto-electronic and transport properties of quantum well structures have been utilized to realize better performance characteristics of transistors, lasers and nonlinear optic devices. The basic principles of operation of these devices are the same as those of the corresponding devices using bulk materials. But, the operating characteristics are required to be worked out by taking into account the confinement and the 2D motion of the carriers.

In addition, new devices have also been invented by utilizing the peculiar tunneling properties in quantum well structures. These properties may be used to realize oscillators in the tera-hertz range, for realizing multi-functional devices or for high speed switching.

The principles of operation and the functional characteristics of these devices are discussed in this and in the following chapters. The basis of this discussion are the physical properties explained in the preceding chapters.

Field-effect trnasistors were developed using single- junction heterostructures as soon as the high mobility property of electrons in these structures were known. These transistors were named high electron mobility transistor (HEMT) by one group of workers[7.1]. Several other names, TEGFET[7.2], MODFET[7.3,4], and SDFET[7.5] are used for these devices[7.6]. The name, HEMT, being most popular it is used in this book. is used instead The structure,the operating principle and the experimental results of this device are discussed in the following sections.

7.1. Structure and Principle of Operation

The structure used for HEMT's is shown in Fig. 7.1. Basically, the structure consists of a semi-insulating GaAs substrate, on which is first grown a thin GaAs layer; then an undoped AlGaAs layer, then a doped AlGaAs layer and finally another thin GaAs layer. Typical widths of these layers are shown in Fig. 7.1. The transistor is finally realized by depositing aluminum to form a Schottky barrier and serve as the gate, and by providing two ohmic contacts to serve as the source

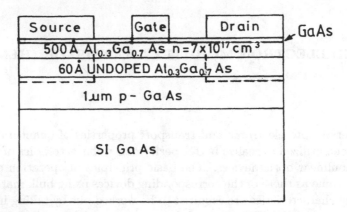

Figure 7.1. Schematic diagram showing the structure of a HEMT. Quantum well is formed in the GaAs layer near the interface with the $Al_{0.3}Ga_{0.7}As$. Carriers are supplied by the doped $Al_{0.3}Ga_{0.7}As$ layer.

and the drain.

The transistors may be operated in two modes, normally-on mode and normally-off mode. These are determined by the thickness of the AlGaAs layers. When these layers are thick enough, charge is supplied by the layer to fill up the surface states at the interface between it and the gate metal and also to the GaAs layer for the alignment of the Fermi levels. The transistor is normally-on. If, on the other hand, the barrier layer is made thin, then the charge available in the layer is not enough to cause alignment of the Fermi levels and the GaAs layer is required to supply additional charge. The layer being depleted, the transistor is normally-off. Both these kinds of transistors are required for different functional circuits. These may be constructed on the same chip and such realizability is considered to be a great advantage in using HEMT's.

The current between the source and the drain contacts is controlled, as in other FET's, by applying a voltage between the gate and the source. Application of the volatage causes a redistribution of the charges and in effect alters the carrier density and hence the current through the channel formed by the quantum well in the GaAs layer near the heterointerface. An expression for the current is obtained by first working out the potential distribution in the structure and determining the carrier density, and then using the mobility-field relations to evaluate the current as explained below.

7.2. Potential Distribution and Accumulated Charge Density

The potential distribution in the presence of a negative voltage V_g between the gate metal and the GaAs layer is as shown in Fig. 7.2. The Fermi level is aligned in the AlGaAs and GaAs layer, whereas that in the metal is lower by V_g. Two depletion regions are produced, one near the gate metal (region I) and the other near the heterointerface (region II). In region I, electrons collect on the metal surface, the Fermi level in which is initially lower than in AlGaAs, to produce a negative electric field $\mathcal{E}_s$. The positively charged ionized donors in the AlGaAs layer produce a positive field gradient which counteracts $\mathcal{E}_s$ and at some distance d_1, completely annuls it to produce equilibrium. The consequential band bending causes the alignment of the Fermi levels as shown. The potential energy of the electron in this region, measured from the Fermi level, is given by,

$$|e|\, V_I(z) =|e|\,(V_s - V_g) + |e|\,\mathcal{E}_s z + (|e|^2/\epsilon_2)\int_0^z dz' \int_0^{z'} [N_d(z'') - n(z'')]dz'' \quad (7.1)$$

The direction perpendicular to the interfaces has been chosen as the z direction, the x direction being chosen along the length of the device. The doping and the electron concentration are taken respectively as $N_d(z)$ and $n(z)$. The first term is

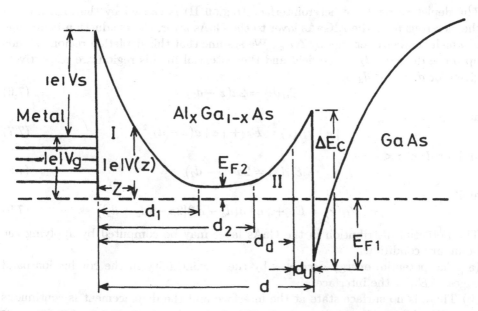

Figure 7.2. Potential distribution in a HEMT. ΔE_c - Conduction-band offset between GaAs and $Al_x Ga_{1-x}As$. E_{F1} and E_{F2} are the Fermi levels in $Al_x Ga_{1-x}As$ and GaAs layers respectively. A voltage V_g is assumed to be applied to the gate.

the potential at $z = 0$, due to the surface barrier potential V_s and the applied gate voltage V_g . The second term gives the potential drop at some distance z due to the surface field $\mathcal{E}_s$. The third term is the increase caused by the charges in the layer, e and ϵ_2 being respectively the charge of the electron and the permittivity of AlGaAs. We assume that the AlGaAs layer is doped uniformly upto the distance d_d and is completely undoped for another length d_u , adjacent to the heterointerface. The expression for $V_I(z)$ may be simplified by neglecting $n(z)$. The simplified expression is as follows:

$$V_I(z) = V_s - V_g + \mathcal{E}_s z + az^2, \tag{7.2}$$

where

$$a = | e | N_d / 2\epsilon_2. \tag{7.3}$$

The field in this region is given by

$$\mathcal{E}_I = \mathcal{E}_s + 2az. \tag{7.4}$$

At the equilibrium distance d_1 , the field is zero and $V_I(d_1)$ is equal to the conduction-band edge energy of AlGaAs, E_{F2}. Putting these conditions we get

$$\mathcal{E}_s = -2ad_1. \tag{7.5}$$

The depletion near the heterointerface (region II) is caused by the migration of the electrons from the AlGaAs layer to the GaAs layer, the conduction-band edge of which is lower in energy by ΔE_c . We assume that this depletion region extends upto the distance d_2 . The field and the potential in this region are respectively given for $d_2 < z < d_d$ by,

$$\mathcal{E}_{II}(z) = 2a(z - d_2) \tag{7.6}$$

and

$$| e | V_{II}(z) = E_{F2} + | e | a(z - d_2)^2, \tag{7.7}$$

and for $d_d < z < d$ by,

$$\mathcal{E}_{II}(z) = 2a(d_d - d_2), \tag{7.8}$$

and

$$| e | V_{II}(z) = E_{F2} + | e | a(d_d - d_2)[2z - (d_d + d_2)]. \tag{7.9}$$

The potential distribution in the GaAs layer may be computed by applying the boundary conditions:

(a) The potential energy decreases by the discontinuity in the conduction-band edges, ΔE_c, at the interface.

(b) There is no surface state at the interface and the displacement is continuous across the interface.

(c) The GaAs layer is completely undoped.

The field in the GaAs layer is then given by,

$$\mathcal{E}(z) = (\epsilon_2/\epsilon_1)2a(d_d - d_2) - (|\,e\,|\,/\epsilon_1)\int_d^z n(z)dz, \tag{7.10}$$

where $n(z)$ and ϵ_1 are respectively the electron concentration and the permitivity in GaAs and d is the total width of the AlGaAs layer.

The field decreases with the distance, and the corresponding potential energy increases from its value at the interface to the bulk value at some distance d_∞, where the field becomes zero, i.e.,

$$\int_d^{d_\infty} n(z)dz = N_d(d_d - d_2), \tag{7.11}$$

or when the total accumulated charge is equal to the depletion charge.

The potential distribution may be altered by changing the gate volatage. However, the gate voltage changes only the depletion layer width in the metal side so long as d_1 remains less than d_2. At some threshold voltage, d_1 becomes equal to d_2 and the whole AlGaAs layer is depleted. Further change in the gate voltage controls the accumulated surface charge density and the device then works as a FET.

The threshold voltage for charge control, V_{th1}, is obtained by eliminating $d_1 = d_2$ from Eq. (7.2) and Eq. (7.7) and using the relation (see Fig. 7.2),

$$|\,e\,|\,V_{II}(d) - \Delta E_c + E_{F1} = 0, \tag{7.12}$$

where E_{F1} is the Fermi energy in the GaAs layer measured from the conduction-band edge at the interface as shown in Fig. 7.2. The threshold voltage is,

$$V_{th1} = V_s - E_{F2}/\,|\,e\,|\,-a(d - d_0)^2, \tag{7.13}$$

where

$$ad_0^2 = ad_u^2 + (\Delta E_c - E_{F2} - E_{F1})/\,|\,e\,| \tag{7.14}$$

Charge accumulated in the GaAs layer for operating conditions may be obtained from the field produced at the interface under the complete depletion regime of the AlGaAs layer. The field and the potential are respectively given by

$$\mathcal{E}(d) = \mathcal{E}_s + 2ad_d. \tag{7.15}$$

$$V(d) = V_s - V_g + \mathcal{E}_s d - ad_d^2 + 2add_d. \tag{7.16}$$

The surface field $\mathcal{E}_s$ is evaluated by using Eq. (7.16) and relation (7.12). It is given by,

$$\mathcal{E}_s = (1/d)(V_g - V_s + \Delta E_c/|\,e\,| - E_{F1}/\,|\,e\,| + ad_d^2) - 2ad_d. \tag{7.17}$$

Putting in Eq. (7.11) we get,

$$\mathcal{E}(d) = (1/d)(V_g - V_s + \Delta E_c/\,|\,e\,| - E_{F1}/\,|\,e\,| + ad_d^2). \tag{7.18}$$

The accumulated areal charge density is, hence

$$Q_s = (\epsilon_2/d)(V_g - V_s + \Delta E_c/\mid e \mid -E_{F1}/\mid e \mid +ad_d{}^2).\qquad(7.19)$$

The Fermi level E_{F1}, however, depends on the surface electron density, n_s and for degenerate conditions, as is usually the case, it may be expressed as[7.7,8],

$$E_{F1} = E_{F0} + \alpha n_s,\qquad(7.20)$$

where $\alpha = 1.25\times10^{-17}$ eV.m^2 for the GaAs/AlGaAs system. The accumulated areal charge density is, therefore,

$$\mid Q_s \mid = (\epsilon_2/d_{eff})(V_g - V_{th2}),\qquad(7.21)$$

where

$$V_{th2} = V_s - \Delta E_c/\mid e \mid -E_{F0}/\mid e \mid -ad_d{}^2, d_{eff} = d + \epsilon_2\alpha/e^2.\qquad(7.22)$$

Operation of the device requires that charge should be accumulated at the interface and at the same time it should be controlled by the gate voltage.For this purpose, the gate voltage is required to be between the two threshold gate voltages, V_{th1} and V_{th2}. For normally-off HEMT's, the thickness of the AlGaAs layer, d, is such that V_{th2} is positive and $V_{th1} > V_{th2}$. Charge, therefore, accumulates when V_g is larger than V_{th2} and remains controllable upto $V_g < V_{th1}$. On the other hand, for normally-on HEMT's, the AlGaAs layer thickness is such that V_{th2} is negative and $\mid V_{th2} \mid > \mid V_{th1} \mid$. The limits of the operating negative gate voltage are therefore given by $\mid V_{th2} \mid > \mid V_g \mid > \mid V_{th1} \mid$.

7.3. Current-Voltage Characteristic

We assume that a voltage V_d is applied between the drain and the source and a gate voltage V_g is applied between the gate and the source. The gate has a length L and the zero of the x-coordinate is taken where the gate starts. The voltage varies from the source to the drain and let it be indicated by $V(x)$ at a distance x from the gate end.

The effective gate voltage at x is then

$$V_{eff} = V_g - V(x)\qquad(7.23)$$

and the surface charge density at x is

$$\mid e \mid n(x) = (\epsilon_2/d_{eff})(V_{eff} - V_{th2})\qquad(7.24)$$

Current through the sample L is continuous and its magnitude is given for any value of x by

$$I = \mu(\mathcal{E})\mid e \mid n(x)[dV(x)/dx].w,\qquad(7.25)$$

where w is the width of the gate, $\mu(E)$ is the mobility of the electrons for a field

$$\mathcal{E} = -dV(x)/dx, \tag{7.26}$$

and $n(x)$ is given by Eq. (7.19).

Relation between V and I is obtained by integrating Eq. (8.3), with the condition that the current I is continuous. This requires knowledge of $\mu(E)$. However, even though there has been extensive interest in HEMT's, reliable data for $\mu(E)$ in single heterojunctions or quantum wells are not available, particularly for high fields as discussed in Section 6.9. Also, the concentration of electrons being very large, the mobility should depend on this concentration and hence on the transverse field as observed in silicon inversion layers[7.9]. No data are available for taking this dependence into account. The $V - I$ chracteristics are obtained instead by using a simple model[8.10] for the velocity-field characteristic illustrated in Fig. 7.3.

The mobility is assumed to have a field-independent value μ_0 upto a threshold field $\mathcal{E}_{th}$, beyond which the velocity is assumed to have a constant value v_s.

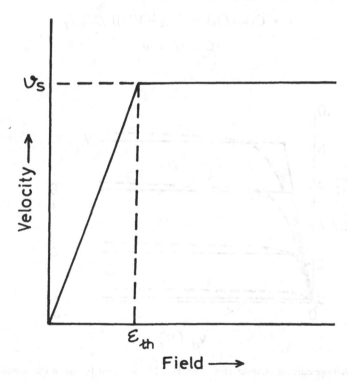

Figure 7.3. Simplified velocity-field characteristic of GaAs. v_g - saturation velocity. $\mathcal{E}_{th}$ - threshold field for saturation.

Equation (7.3) may be integrated to obtain,

$$V(x) = V_g - V_{th2} - \{[V_g - V_{th2} - V(0)]^2 - (2Id/\epsilon_2\mu_0 w)x\}^{1/2}, \qquad (7.27)$$

where $V(0) = R_s I$, R_s being the resistance of the channel between the source contact and the front end of the gate. The voltage across the length of the gate is

$$V(L) = V - (R_d + R_s)I = V_g - V_{th2} - [(V_g - V_{th2} - R_s I)^2 - (2/\epsilon_2\mu_0 w)ILd]^{1/2}, \quad (7.28)$$

where R_d is the resistance of the channel between the drain contact and the rear end of the gate.

The current-voltage characteristic obtained from this model is illustrated in Fig. 7.4. The current initially increases nearly linearly with the drain voltage V and then saturates as the voltage exceeds a threshold value. The saturation current I_s is obtained by evaluating the field, $(-dV(x)/dx)$ from Eq. (7.26) and putting it equal to $\mathcal{E}_{th}$. We get accordingly,

$$I_s = (\epsilon_2 w v_s/d)\{[(V_g - V_{th2} - V(0))^2 + \mathcal{E}_{th}^2 L^2]^{1/2} - \mathcal{E}_{th}L\}. \qquad (7.29)$$

I_s may be simplified for large values of L to

$$I_s = C(v_s/L)(V_g - V_{th2} - V(0))^2/2\mathcal{E}_{th}L, \qquad (7.30)$$

$$C = \epsilon_2 w L/d \qquad (7.31)$$

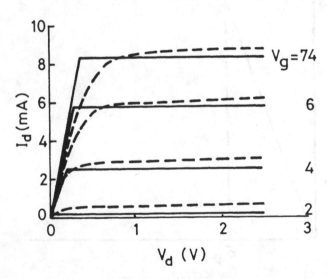

Figure 7.4. Voltage-current characteristic of a HEMT, obtained by using the simplified velocity-field characeric of Fig. 7.3.[After T. J. Drummond, H. Morkoc, K.Lee and M. Shur, *IEEE Electron Dev. Lett.* **EDL-3**, 338 (1982). Copyright: (=A9 1982=IEEE)].

and the saturation current is proportional to the square of the effective voltage.

On the other hand, for small values of L, the terms containing L may be neglected and

$$I_s = C(v_s/L)(V_g - V_{th2})(1 + R_sCv_s/L)^{-1}. \tag{7.32}$$

The saturation current, thus varies linearly with the gate voltage.

The intrinsic transconductance, g_m, is given by

$$g_m = C(v_s/L)(1 + R_sCv_s/L)^{-1}. \tag{7.33}$$

The transconductance increases with $(L/v_s)^{-1}$, or inversely as the transit time of the electrons across the gate under saturation conditions. For large values of the transconductance or for fast switching, the gate length is therefore required to be small and the saturation velocity should be large. The source resistance should also be small. The constant C is the capacitance between the gate metal and the quantum well layer. The current gain is therefore given by

$$I_s/ \mid I \mid = g_m/wC = f_T/f, \tag{7.34}$$

where

$$f_T = (1/2\pi)(g_m/C). \tag{7.35}$$

The cut-off frequency for the current gain increases like the transconductance with decrease in the transit time (L/v_s). The basic parameters limiting the performance of the HEMT's are thus obtained from the simplified model. However, in practice, the characteristics are significantly affected by the parasitic circuit components and the deviations from the ideal conditions on which the analysis is based.

Attempts have been made to improve the accuracy of the computed characteristics by using a more realistic velocity-field, $(v - \mathcal{E})$, characteristic than that assumed above. The relations used in the literature[7.11,12] are as follows,

$$v = \mu_0 E(1 + \mu_0 \mathcal{E}/v_s)^{-1}; \tag{7.36}$$

$$v = v_s[1 - \exp{(\mu_0 \mathcal{E}/v_s)}], \tag{7.37}$$

where μ_0 is the low-field mobility and v_s is the saturation velocity.

It should, however, be noted that the basic characteristics are not significantly altered when the above more realistic velocity-field relations are used. Accurate evaluation of the characteristics requires knowledge of the source resistance R_s , drain resistance R_d , the concentration of donors, N_d, modification of the $v - \mathcal{E}$ relation by the transverse field due to alteration in the well width and screening by the accumulated electrons, the velocity over-shoot effect and various other physical constants. All these information being not available, more accurate models are not really relevant at this stage. It is also of interest to note that the two-dimensional

character of the electron gas does not play any essential role in the operation of
the HEMT's. The 2D character is a consequence of the large values of electron
density in these devices and the device is required to be analysed by taking into
account this character, which affects electron mobility and the Fermi level. In all
other aspects the device may be considered to be a metal-insulator-semiconductor
FET, the depleted AlGaAs layer playing the role of the insulator.

7.4. Experimental Results

The first experimental HEMT was reported by Mimura et al[7.1] in 1980. They
used a MBE grown structure consisting of a Cr-doped semi-insulating GaAs substrate, on which was first grown an undoped GaAs layer, then a doped
$(N_d = 6.6 \times 10^{17}/\text{cm}^3)$ $Al_{0.32}Ga_{0.68}As$ layer and finally an undoped GaAs layer.
The devices showed an increase in the instrinsic transconductance by a factor of
3 at 77 K in comparison with the conventional bulk devices.

Improvements in the device characteristics were made by using essentially the
same structure, but with an undoped AlGaAs layer[7.3] between the undoped
GaAs and doped AlGaAs layer to reduce the remote ionized impurity scattering.
Physical dimensions of the device structure and the growth of materials were,
however, optimised to realize higher and higher values of power-delay product and
cut-off frequency. Values have been reported[7.13-16] for the transconductance,
for the power-delay product (pdp) and for the delay time τ_D for 1 μm gate lengths
lying respectively between 170 and 280 mS/mm, between 16 and 22 fJ/stage and
between 8.5 and 20 ps. Improvement has been obtained by using the 0.35 μm gate
technology, for which the values of pdp and τ_D are respectively 10.5 fJ and 10.2
ps at 300 K and 10.2 fJ and 5.8 ps at 77 K [7.13-16].

Further improvement has been realized by utilizing sophisticated fabrication
techniques with short gate lengths. Cut-off frequencies of 350 GHz [7.17], noise
figure of 2.1 dB and 6.3 dB gain at 94 GHz have been achieved[7.18].

Devices have been used to realize frequency dividers and microwave amplifiers[7.19]. Frequency dividers have been operated upto about 13 GHz, while
microwave amplifiers have been operated upto 35 GHz with a noise figure of 2.7
dB at 300 K. Substantially lower noise figure of 0.8 dB has also been reported at
18 GHz with a cut-off frequency f_T of 80 GHz.

All the above performance characteristics were realized with the AlGaAs/GaAs
system. But this system exhibits some anomalous behaviour : the so-called collapse[7.20] of the $I - V$ characteristic at cryogenic temperatures, a reduction in
drain current for large voltages and a shift in the threshold voltage. This deterioration in the characteristics is thought to be due to the injection of charges into the
AlGaAs layer and subsequent trapping. The $Al_xGa_{1-x}As$ layer contains a defect
centre (DX) which has quite a large barrier to electron capture and emission at

cryogenic temperature. This centre gives rise also to persistent photoconductivity. The centres capture the electrons injected into the barrier layer for a long enough time to produce the observed phenomenon.

Alternative material systems have been used to construct HEMT's, in an attempt to avoid the current-collapse problem and also to realize higher transconductance. Devices have been developed[7.21] by using the lattice matched $Al_{0.48}In_{0.52}As$ /$Ga_{0.47}In_{0.53}As$ system. The higher mobility and the higher saturation velocity in $Ga_{0.47}In_{0.53}As$ make it possible to realize higher transconductance and higher cut-off frequency. A noise figure as low as 1.2 db has been demonstrated[7.22] at 94 GHz. Current gain cut-off frequencies exceeding 200 GHz have been reported[7.23-32].Values of $f_{max} = 455$ GHz [7.29], a noise figure of 1.7 dB with 7.7 dB gain at 94 GHz [7.30] and $f_T > 340$ GHz [7.31] have been reported for the lattice-matched $In_{0.53}Ga_{0.47}As/Al_{0.48}In_{0.52}As$ system. These devices like the AlGaAs/GaAs

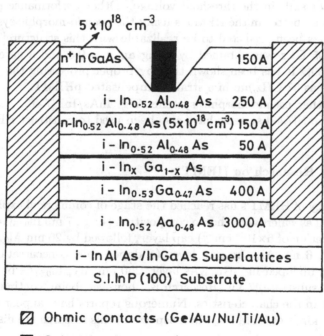

◪ Ohmic Contacts (Ge/Au/Nu/Ti/Au)

■ Schottky Contact (Ti/Au)

☰ Strained Channel

Figure 7.5. Schematic diagram showing the structure of a pseudomorphic HEMT. The 2DEG is formed in the $In_xGa_{1-x}As$ layer at the interface with the $In_{0.52}Al_{0.48}As$ layer. [After R. Lai, P. K. Bhattacharya, D. Yang, I. Brock, S. A. Alterovitz and A. N. Downey, *IEEE Trans. Electron Dev.* **39**, 2206 (1992); Copyright:(=A9 1992 IEEE)].

and Silicon-on-insulator metal-oxide FET (SOI-MOSFET) suffer from an anomalous increase in the drain current at a certain drain-to-source voltage. This phenomenon , known as kink phenomenon , has been studied for AlInAs/GaInAs system by Suematsu et al [7.33] and it has been concluded that the kink is produced due to a modification of the parasitic source resistance induced by the hole accumulation.

Attention has also been given[7.34-35] to the pseudo-morphic $Al_yGa_{1-y}As/In_x Ga_{1-x}As$ and $In_xGa_{1-x}As/In_{0.52}Al_{0.48}As$ systems, in which the two layers are not lattice matched.This system is expected to enhance the performance due to the larger value of the conduction band discontinuity[7.36], a higher low-field electron mobility [7.37] and a higher electron peak velocity The effects of mismatching are minimized by keeping the well material thin enough for absorption of the strain without creating dislocations. One such structure[7.35] is shown in Fig. 7.5. A transconductance of 290 mS/mm has been reported for 1 μm gate length. The devices, often referred as pHEMT show no current collapse or persistent effect of illumination or shift in the threshold voltage. Other performance parametrs are also found to be better in the HEMT's using the pseudo-morphic system. A value of 500 GHz has been predicted to be realizable with this structure[7.38]. A recor high f_T of 305 GHz was obtained by using an $x = 0.8$ and a gate length L_g 0f 0.065 μm [7.39]. It has been shown that by proper processing the $f_T.L_g$ may be made as high as 57 GHz.μm in a strain-compensated pHEMT.

Various other pseudo-morphic systems, e.g, InAs/InP[7.40], AlSb/InAs[7.41] and AlGaN/GaN[7.42] systems have also been used to construct HEMT's.

7.5 Current Research on HEMT's

The technology of HEMT's has reached the stage of commercial exploitation. The devices as well as wafers with chosen composition [e. g., 7 nm $In_{0.53}Ga_{0.47}As$ (doped to a concentration of 6×10^{18} cm^{-3}) cap layer, followed by 20 nm $Al_{0.48}In_{0.52}As$ (undoped) layer, 6 nm $Al_{0.48}In_{0.52}As$ layer (δ doped to a concentration of 5×10^{12} cm^{-2}), 20 nm undoped $In_{0.53}Ga_{0.47}As$ layer, 250 nm $Al_{0.48}In_{0.52}As$ buffer layer on a 450 μ m InP substrate)[7.43]. Research is , however, being continued for further improvement in the characteristics. Numerous reports have appeared in the literature on new kinds of HEMT's. Main aspects of this research are discussed in this section.

7.5.1 TEMPERATURE DEPENDENCE OF THE CHARACTERISTICS

The temperature dependence of the characteristics of AlGaAs/GaAs HEMT's arise mainly from the presence of trap states and DX centers. It was studied theoretically by Valois et al [7.44] and Subramanian[7.45], and experimentally by Gobert and Salmer[7.46]. On the other hand, variation of electron mobility and the gating function ($n_s - V_g$ relation) are the main causes of tempersature dependence of the p-HEMT characteristucs.Mizutani and Maezawa7.47] studied the temperature dependence of the high frequency small signal characterics. The large signal characteristics of AlGaAs/InGaAs pHEMT's have been studied theoretically as well as experimentally by Zurek et al[7.48] over the temperature range 300-405 K. The analysis is on the same lines as explined in Section 7.3. The current is given by the same expression as (7.25) and integrated to obtain the $I - V_g$ characteristic. However, the mobility is obtained by solving the momentum and the energy balance equations as explained in Section 6.9.1. The electron concentration is determined by first computing the conduction band edge and electron densaity profile with a program named C-BAND[7.49]. The electron density, so obtained, is integrated over the cross-section of the device. It was found that the InGaAs channel works as HEMT while the AlGaAs layer also contributes to the current by acting as a MESFET. The zero temperature coefficient (ZTC) of the characteristics was found to be close to the threshold voltage. Consequently the ZTC for these devices may be realized at the working voltages only by using suitably chosen external circuit elements.

7.5.2. THEORETICAL MODELS AND SIMULATORS

HEMT simulators have been developed [7.50], which allow prediction of DC and small signal microwave performance of HEMT's taking into account the hot-electron effects , parasitic MESFET conduction, quantum effects , substrate injection phenomena and effects of DX ceners and other traps.A numerical sanalysis of the performance characterics of a $In_y Al_{1-y}As/In_x Ga_{1-x}As$ HEMT has been published[7.51] A capacitance-voltage model has been used by Sen et al [7.52] to analyze the performance of AlGaAs/GaAs HEMT. A unified model for δ-doped and uniformly doped HEMT's has been developed by Karmalkar and Ramesh[7.53].

7.5.3. HIGH-POWER MICROWAVE HEMT's

An important aspect of the evolution of HEMT's has been devising structures to control the gate leakage current and the gate-drain or the gate-source breakdown for large gate voltages in microwave and millimeter wave power HEMT's[7.54-56]. Typical performance characteristics of a power microwave p-HEMT using a gate length of 0.25 μm are: power output-1 W, associated gain 11 dB, power added efficiency 60 % at 10 GHz for a gate width of 1200 μm, extrinsic transconductance of 400 mS/mm and maximum current density 550   mA/mm. The device features an asymmetric double recess gate, delta doping of the barrier and a narrow channel with low indium content on a $Al_{0.24}Ga_{0.76}As$ 1000 Å buffer on intrinsic GaAs buffer [7.57].

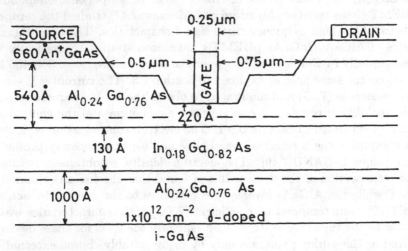

Figure 7.6. Schematic diagram of the cross section of an AlGaAs/InGaAs double-heterostructure pHEMT. [After J. J. Brown, J. A. Pusl, M. Hu, A. E. Schmitz, D. P. Docter, J. B. Shealy, M. G. Case, M. A. Thompson and L. D. Nguyen, *IEEE Microwave Guided Wave Lett.* **6**, 91 (1996); Copyright (=A9 1996=IEEE)].

The breakdown voltage has been controlled in different devices by using symmetric or asymmetric recessed gate or a symmetric or asymmetric double-recessed gate. One such structure is illustrated in Fig. 7.6.

The off-state breakdown in power pHEMT's has been studied by Somerville et al [7.58]. It is commonly believed that the off-state breakdown is determined by the field between the gate and the drain. But, Somerville et al showed by measurements and simulations that the electrostatic interaction of the source seriously degrades the gate-drain breakdown and that the key aspect ratio is the gate length divided by the depletion region length on the drain.

The effect of gate recess dimension on the breakdown voltage and the high frequency characteristics have been studied by Higuchi et al [7.59]. It has been shown by analysis and confirmed by experiments that a breakdown voltage of 14 V and f_{max} of 127 GHz may be realized by using a gate length of 0.66μm and optimal gate recess.

An alternative technique has been proposed to enhance the breakdown voltages by Enoki et al [7.60]. A composite channel combining a thin layer of $Ga_{0.47}In_{0.53}As$ and InP was shown to be effective in enhancing the breakdown voltage. Carriers are confined in the $Ga_{0.47}In0.53As$ channel at low voltages , while for high voltages the electrons are transferred to the InP layer and the high breakdown voltage and the saturation velocity are utilized. Further improvement in the characteristics has been realized by making the $Ga_{0.47}In_{0.53}As$ layer thin enough so that quantization increases the effective band gap [7.61]. A triple channel HEMT has been studied [7.57], in which a third InGaAs channel and a quaternary carrier supplier have been introduced to improve the performance characteristics [7.62]. It has been shown that by using a gate length of $0.8\mu m$, a g_m of 275-325 mS/mm, a static voltage gain of 20 and a cut-off frequency of 40 GHz may be realized with this structure.

7.5.4. CONTROL OF HEMT CHARACTERISTICS BY LIGHT

The characteristics of HEMT's may be altered by shining light on the device. This effect has been utilized to cotrol gain of amplifiers, for oscillator tuning,locking and switching[7.63,64]. Optical control of monolithic microwave integrated circuits(MMIC's), especially oscillators have been demonstrated in AlGaAs/GaAs HEMT[7.65]. Such control has been demonstrated also in InAlAs/GaInAs HEMT's [7.66,67]. Recently, the current-voltage relation of InAlAs/InGaAs HEMT has been studied by using both CW 1.3 μm laser light and also by vmodulated light by Takanashi et al[7.68,69]

7.6 Conclusion

The basic structure and the principle of operation of HEMT's have been presented in this Chapter. HEMT is, perhaps, the quantum well device, which has found maximum applications as a low-signal high-gain and low-noise device, as well as a high power device upto microwave and millimeter wave frequencies.

The device,however, is of great interest to researchers. Novel structures, e.g., Si channel SiGe n-HEMT [7.70] and HEMT incorporating a resonant tunneling diode [7.71] have been proposed. Work is also being continued for evolving test procedures for the performance characteristics[7.43,72].

CHAPTER 8

RESONANT TUNNELING DIODE

The tunneling properties of double-barrier quantum well heterostructures may be conveniently utilized to produce differential negative resistance. Such negative resistance devices make it possible to realize oscillators or amplifiers like Gunn diodes or impatts. The cut-off frequency is, however, much higher in these devices, and it has been possible to construct oscillators working in the THz range with these devices and also to realize various kinds of functional circuits by building-in the tunnel device in the emitter or the base of a transistor.

The device has, received great attention and exhaustive literature has been created about the diodes. Interested reader may consult Reference 8.1 for detailed discussions. We shall concentrate in this book only on the physics aspects of the devices.

8.1. Introduction

Possibility of realizing a negative resistance device by using a double barrier tunneling structure was first demonstrated by Chang et al[8.2] in 1974 . The structure consisted of two undoped barrier layers which sandwiched an undoped well layer and was provided with doped contact layers on the two outer surfaces as shown

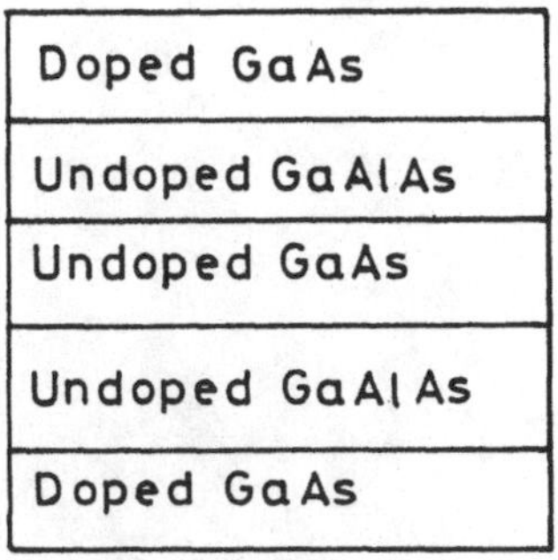

| Doped GaAs |
| Undoped GaAlAs |
| Undoped GaAs |
| Undoped GaAlAs |
| Doped GaAs |

Figure 8.1. Structure of a resonant tunneling diode. Composition of the layers is indicated in the figure.

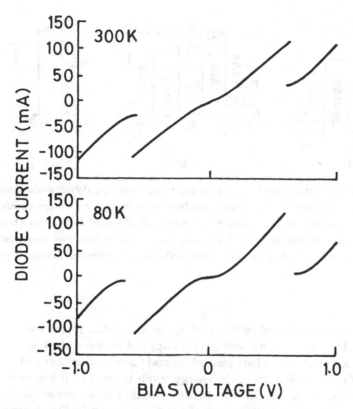

Figure 8.2. Typical voltage-current characteristic of a resonant tunnel diode. One resonant peak is shown. The device being symmetric the characteristic is identical for both the polarity of the voltage.

in Fig. 8.1. The current through the structure initially increased with the voltage across the two end layers, reached a peak and then decreased with further increase in the voltage. At still higher voltages, the current showed a second peak. Results of a later experiment are illustrated in Fig. 8.2. The characteristics may be explained qualitatively as follows.

The potential distribution in the structure is illustrated in Fig. 8.3. The double-barrier quantum well structure acts as a Fabry-Perot resonator, the transmission coefficient of which is a maximum at the resonant frequencies, which are in this case the quasi-quantized energy levels of the quantum well. An electron from one contact may tunnel through the structure to the other contact if its energy is close to a quantized level. No current therefore flows through the device so long

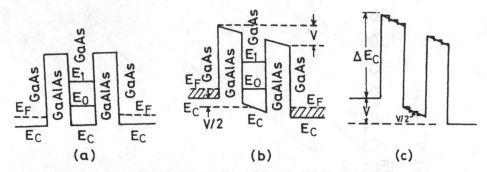

Figure 8.3. Potential distribution in a resonant tunneling diode. (a) Potential distribution in the absence of the external voltage. E_c - Conduction-band edge energy. E_F - Fermi energy. E_0, E_1 - Quantized lowest and the next higher energies in the well. (b) Potential distribution when a voltage V is applied between the end contacts. The barrier layers and the well layer being undoped and of equal width the applied voltage is distributed as shown. (c) Potential variation simulated by steps.

as the lowest quantized level in the well E_0 is above the Fermi level E_F in the contact layer. However, when a voltage V is applied between the two end contact layers, energy levels at one end are pushed up and those in the other end are pushed down by $|e|V/2$, with reference to the energy levels in the well if it is assumed that the barriers are identical and the voltage is distributed uniformly across the device. Consequently, when $E_F + |e|V/2$ is equal to E_0 as shown in Fig. 8.3(b), current starts flowing and continues to flow till E_0 is less than $E_c + |e|V/2$. As E_0 falls below the conduction band edge of the contact layer for larger voltages, the current drops to zero. A current peak is therefore expected near $V = 2(E_0 - E_c)/|e|$, Similar peaks are also expected at higher voltages near the higher lying quantized levels. The details of the characteristics involve other factors which are discussed later. It is. however, clear that the current in the device is produced basically by the tunneling of the electrons through the barriers.

Current in the device being produced by the tunneling of the electrons, the frequency and the speed limitation of the device is controlled by the tunneling time and the effective device capacitance. Since these two parameters are very small in this structure, the operating frequency and the switching speeds of the device were expected to be much higher than that of other available devices. Work on the device, however, did not progress till the middle of eighties as the technology of growing the structures was not easily available. Rapid progress has been made only during the last two decades and the device has been perfected and used in practical circuits.

We present in this section the detailed theory of operation and the current status of the device.

8.2. Tunneling Characteristic

It is assumed that the conduction band offset is ΔE_c and the end layers are degenerate as shown in Fig. 8.3(a). The barrier layers and the sandwiched well layer are assumed to be undoped. The transmission probability for the structure is evaluated by using the envelope function formalism (discussed in Chapter 4), in which the wave function of the electron is taken as,

$$\psi_{\mathbf{k}} = U_0(\mathbf{r}) \exp[i(k_z z + \mathbf{k}_t.\rho)], \tag{8.1}$$

where $U_0(\mathbf{r})$ is the cell-periodic function corresponding to the conduction-band edge, k_z and $\mathbf{k}_t$ are respectively the z-component and the in-plane component of the wave vector, corresponding position coordinates being z and ρ. The energy corresponding to the wave vector is,

$$E = E_{ci} + (\hbar k)^2/2m^*, \tag{8.2}$$

where m^* is the band-edge mass and

$$k^2 = k_z^2 + k_t^2. \tag{8.3}$$

The nonparabolicity of the band may be taken into account by using the Kane nonparabolic dispersion relation,

$$(\hbar k)^2/2m^* = (E - E_{ci})[1 + \alpha(E - E_{ci})], \tag{8.4}$$

where α is the nonparabolicity parameter. The relation is assumed to be valid in the conduction band as well as in the forbidden band.

The transmission coefficient is evaluated by assuming that an electron wave of unit amplitude is incident on the structure from the left. It is partially reflected with amplitude R and partially transmitted with amplitude T_u. The transmission coefficient is then given by $T_u^* T_u [(\partial E/\partial k_{lt})(\partial E/\partial k_{li})^{-1}]$, where the term within the parenthesis is the ratio of the outgoing and the incident group velocities, k_{li} and k_{lt} being respectively the longitudinal wave number of the incident and the transmitted electron; $T_u^* T_u$ gives the probability density of the outgoing electron. The tunneling coefficient T_u may be obtained by solving analytically the envelope function equation for the potential distribution in the device. But, such solutions involve the Airy functions, which make the computations involved. A more convenient method is the so-called matrix method[8.3,4] in which the potential is assumed to vary in steps as shown in Fig. 8.3(c), instead of varying linearly. Solutions for the envelope functions are then obtained for each segment, which behave as rectangular barriers or wells. These solutions are matched at the interfaces between the sections by applying the conditions of the continuity of the probability density, the current probability density and the energy. This procedure yields the following matrix equation relating R, T_u and the incident wave.

$$\begin{bmatrix} 1 \\ R \end{bmatrix} = \mathbf{M}_0 \cdot \mathbf{M}_1 \cdot \mathbf{M}_2 \ldots \mathbf{M}_n \cdot \begin{bmatrix} T_u \\ 0 \end{bmatrix} = \mathbf{M} \cdot \begin{bmatrix} T_u \\ 0 \end{bmatrix} \qquad (8.5)$$

where the matrix for the pth segment is

$$\mathbf{M}_p = (1/2) \begin{bmatrix} \exp(-ik_p d_p) & 0 \\ 0 & \exp(ik_p d_p) \end{bmatrix} \cdot \begin{bmatrix} 1 + r_p k_{p+1}/k_p & 1 - r_p k_{p+1}/k_p \\ 1 - r_p k_{p+1}/k_p & 1 + r_p k_{p+1}/k_p \end{bmatrix} \qquad (8.6)$$

k_p, d_p are the wave vector and width of the pth section, the width of the zeroth section being taken to be zero. The wave vector k_p is given in general by

$$k_p^2 = (2m_p^*/\hbar^2)(E - V_p)[1 + \alpha(E - V_p)] - k_t^2, \qquad (8.7)$$

where m_p^* and V_p are the band-edge effective mass and the band-edge energy (including the contribution of the applied voltage) of the pth section, α_p being the corresponding nonparabolicity parameter, $\mathbf{k}_t$ is the in-plane component of wave vector which is identical in all the sections because of the continuity of the probability density, r_p is the ratio of the velocity effective mass of the pth and the $(p+1)$th section and is given by,

$$r_p = (m_p^*/m_{p+1}^*)[1 + 2\alpha_p(E - V_p)][1 + 2\alpha_{p+1}(E - V_{p+1})]^{-1}. \qquad (8.8)$$

The value of T_u is given by

$$T_u = 1/m_{11}, \qquad (8.9)$$

where m_{11} is the first element of the equivalent matrix $\mathbf{M}$.

The method can be implemented on a computer straightaway when complex numbers are admissible. It should be noted that k_p is imaginary in the barrier layers and real in the well layers. The method involves only the evaluation of V_p for the selected number of segments n, evaluation of r_p and k_p and straightforward matrix multiplication.

Some computed results are presented in Fig. 8.4. A symmetric AlGaAs/GaInAs structure with $d_1 = d_2 = d_3 = 10$ nm was considered and computations were done with the physical constants corresponding to the lattice temperature of $300\ K$ as given below:

m^* (AlInAs)$= 0.08\ m_0$, m^* (GaInAs)$= 0.042\ m_0$

α(AlInAs) $= 0.571$ (eV), α (GaInAs) $= 1.167$ (eV)$^{-1}$,

$\Delta E_c = 0.5$ eV, $m_0 = 9.1 \times 10^{-31}$ kg

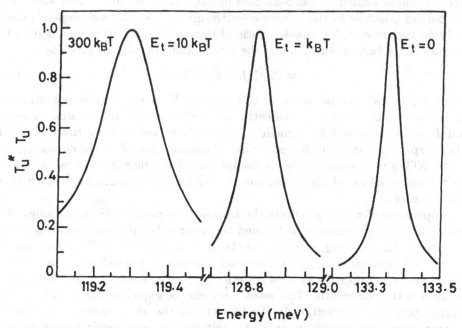

Figure 8.4. Energy dependence of the tunneling probability in a double-barrier heterostructure, calculated by the matrix method. The curves are for 300 K for different transverse energies as indicated on the curves, k_B and T being respectively the Boltzmann constant and the lattice temperature.

We first consider the transmission probability curve in the absence of the voltage. The curve has a peak of finite width at an energy of about 133.35 meV, when the in-plane component of the wave vector i.e., k_t is taken to be zero. Barriers being of finite height, act as leaky reflectors. The transmission coefficient is unity when the electron energy resonates with the quasi-eigenvalue of energy in the well; but the transmission curve has a finite width because of leak in the barriers.

The effect of a finite value of the in-plane component of the wave vector is to increase the bandwidth at resonance since it effectively reduces the barrier height approximately by $(\hbar k_t)^2/2)(1/m_w^* - 1/m_b^*)$, m_w^* and m_b^* being respectively the effective mass of the well and the barrier layer. The effect of the energy band nonparabolicity is not separately displayed, but it is significant and alters the energy for peak transmission as well as the bandwidth.

Some conclusions may be made about the variations in the tunneling characteristics with changes in the dimension of the diode structure. For example, the bandwidth of transmission should increase as the well becomes more leaky i.e., as the electron life time, τ_l , is reduced since the bandwidth, $\Delta E = \hbar/\tau_l$. Such increase will be caused by the reduction in the barrier and in the well width or in the barrier height or in the barrier effective mass. The leakiness depends essentially on the transmission characteristic of the barriers, which is governed by the product of the barrier width L and the attenuation constant α, given by

$$\alpha = (2m_b^*/\hbar^2)(E_b - E_r)^{1/2}, \qquad (8.10)$$

where m_b^* is the effective mass in the barrier, E_b is the barrier potential and E_r is the resonant energy. Evidently, total attenuation will be smaller as m_b^* and L are reduced and E_r is made to approach E_b by reducing the well width. These expected changes in the bandwidth of transmission affect the characteristics of the RTD's very significantly. A major part of the development work [8.5-10] for the optimization of the diode parameters have been carried out with these considerations.

Application of a voltage affects the transmission probability in two ways. The energy for peak transmission is lowered and approaches the left hand conduction band edge. The lowering is about half the applied potential, $|e|V/2$. This may be understood by considering that the average potential in the well region is $-|e|V/2$. The second effect appears as a decrease in the peak value of transmission and an increase in the bandwidth. This result may also be explained by considering the average potential distribution in the barrier and the well regions. The barriers become asymmetric in the presence of a voltage. An asymmetric system behaves as a Fabry-Perot resonator with reflectors of different characteristics and hence the effect of reflections are not compensated and the wave inside the resonator cannot build to the level required for producing unit transmission. The resonator is also more leaky as the barrier heights are reduced on the average. When the applied voltage is more than $2E_0/|e|(E_0$ is the energy of the peak transmission in the absence of a voltage) only a few electrons will tunnel through and the number will die down as the voltage is made larger.

Experimental current-voltage characteristics may be explained by considering the tunneling characteristic discussed above, by using the expression for the current density.

8.3. Current-voltage Characteristic

The current density through the device is given by the product of the incident flux of electrons and the transmission coefficient. It may, hence be expressed as

$$\mathbf{J} = \int (2/8\pi^3)(e/\hbar)(\nabla_{k_l}E)T_u^*T_u[f(E) - f(E + |e|V)]d^3k, \qquad (8.11)$$

where k_l is the component of electron wave vector in the direction perpendicular to the junction interfaces; E is the electron energy as measured from the bottom of the conduction band on the incidence side, V is the applied voltage, d^3k is a volume element in the wave vector space.

The distribution function for the electrons is the Fermi-function $f(E)$, given by

$$f(E) = \{\exp[(E - E_F)/k_B T] + 1\}^{-1} \tag{8.12}$$

where E_F is the Fermi energy, k_B is the Boltzmann constant and T is the lattice temperature. It may be noted that the device is bipolar and hence electrons may flow either way. The energy of the electrons being conserved in tunneling, an electron having an energy E with respect to the band edge on the left hand side has an energy $(E + |e|V)$ with respect to the band edge on the right hand side, as the edge is lowered by the applied voltage by the amount $|e|V$. The distribution function for the transmission side is therefore given by $f(E + |e|V)$.

Expression for **J** may be transformed by noting that the transverse component of **k**, is conserved i.e., $k_{ti} = k_{lt}$, and the energy may be split into two components E_l and E_t , corresponding to the wave-vector components k_l and k_t . It is then obtained that,

$$J = (e/\hbar)(2/8\pi^3)2\pi(m^*/\hbar)\int_0^\infty (dE_l/dk_l)T_u^*T_u[f(E) - f(E + |e|V)]dE_t dk_l \tag{8.13}$$

On substituting (9.12) into (9.13), writing $E = E_l + E_t$ and carrying out the integration over E_l , we obtain

$$J = (em^* k_B T/2\pi^2\hbar^3)\int T_u^* T_u \ln\{1 + \exp[(E_F - E_l)k_B T]\}$$
$$\{1 + \exp[(E_F - |e|V - E_l)/k_B T\}^{-1}dE_l. \tag{8.14}$$

The resonance feature of the experimental characteristics, which were explained qualitatively in Section 8.2 are also exhibited by the curves computed by using Eq. (9.14),as illustrated in Fig. 8.5. It is of interest to note that the bandwidth of transmission, ΔE, is about 1-2 meV, while the Fermi energy is about 20-30 meV. Hence, electrons within the bandwidth ΔE take part in producing the peak current. The peak current, therefore, increases as ΔE is increased by reducing the barrier voltage or the electron effective mass in the barrier or by reducing the well width or the effective mass in the well layer. Such variations have been observed in the experimental characteristics[8.5-8].

The model based on the assumption that the electron travels through the structure without any impedance in its motion thus explains the basic features of the current-voltage characteristics of RTD's. In a real diode, however, there are many other important mechanisms which play a significant role in determining the shape of the characteristic. Electrons may be scattered as they travel in the barrier or in the well region. The coherence of the electron wave is destroyed by scattering and the tunneling model discussed above becomes inapplicable. However, even when such scattering occurs, the essential feature of the characteristic, the negative

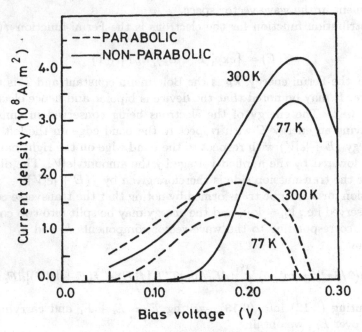

Figure 8.5. Voltage-current characteristic of a resonant tunneling diode for the lattice temperatures of 77 K and 300 K . Curves obtained by assuming a parabolic and a nonparabolic energy band are shown respectively by the solid and the dashed lines.

differential conductivity, may be explained on the basis of the tunneling of the 3D electrons from the end contact into the well, where the motion is restricted to be two-dimensional[8.11]. Since such tunneling is controlled by the condition of conservation of transverse momentum and energy we have

$$E = E_c + \hbar k_l^2/2m^* + \hbar k_t^2/2m^* = E_r + \hbar k_t^2/2m_w^*, \qquad (8.15)$$

where E_c and E_r are respectively the conduction-band edge energy and the quantized energy level in the well. The longitudinal and the transverse components of the wave vector are given by k_l and k_t. It should be noted that k_t and E remain the same as the electron tunnel from the contact layer into the well. We may also assume that the electron levels are occupied up to the Fermi energy, E_F in the contact layers. The longitudinal component of the wave vector k_l is related to k_t as,

$$k_l^2 = (2m^*/\hbar^2)(E_F - E_c) - k_t^2 = k_F^2 - k_t^2, \qquad (8.16)$$

where k_F is the wave vector corresponding to the Fermi energy E_F . Evidently, electrons cannot tunnel through so long as $E_r > E_F$, as no electron level is occupied above E_F . However, as E_r is pushed down by the applied voltage V by $|e|V/2$, flow of current becomes possible for a threshold voltage V_t for which,

$$E_r - |e|V_t/2 = E_F. \tag{8.17}$$

The value of k_l for this condition is

$$k_l = (2m^*/\hbar^2)(E_r - |e|V_t/2 - E_e)^{1/2} = (2m^*/\hbar^2)^{1/2}(E_F - E_c)^{-1/2}. \tag{8.18}$$

The corresponding value of k_t is zero and the number of electrons with a particular value of k_l being proportional to k_t^2 , the current remains zero for V_t . However, as the applied voltage is further increased k_l decreases and k_t increases. Increasing current flows with increasing k_t . When the applied voltage is larger than the cut-off value V_c, given by,

$$E_r - |e|V_c/2 = E_c, \tag{8.19}$$

current stops flowing as there is no occupied state in the contact layer below E_c .

Hence, the NDC in the characteristics of RTD's may be explained also by using the so-called sequential model[8.12]. It is assumed in this model that the electrons tunnel through the structure in two steps. In the first step, it tunnels through the first barrier coherently, it is then thermalized in the well and finally tunnels through the second barrier in the second step.

Which of the two models will apply to a particular experiment should be determined by the relative values of the electron life time and the scattering time. The coherent model would apply when the life time τ_l is much smaller than the scattering time τ_s and the sequential model would apply when $\tau_s < \tau_l$. Experiments appear to confirm this expectation[8.13].

A further complication in the calculation of the characteristics arises from the distortion in the potential profiles due to the accumulation and depletion of the charges. Charges collect at the emitter end to produce single-junction quantum wells and also in the well to bend the band edge. At the same time, charges are depleted on the collector side to produce a parabolic potential profile. These distortions explain the deviation of the experimental values of the voltage for peak current from $2E_r/|e|$. Such charge accumulation also plays a role in determining the actual form of the current-voltage characteristic, which may show step changes during the downfall or hysteresis.

It may be noted further that although the basic features of the current-voltage characteristic are explained by the resonant tunneling current, certain features require consideration of other modes of current transport. For example, the current does not drop to zero as the resonant energy goes below the emitter-band edge. In fact, the so-called valley current may be fairly large. No exact calculation of the valley current is available. It is, however, suggested that it is partly due to tunneling through the non-resonant energies and partly due to thermionic emission

over the barriers. The scattering of the carriers in the barrier and the well region
may also play an important role.

8.4. Experiments

The basic phenomenon of resonant tunneling, as mentioned earlier[8.2], was demonstrated in 1974. However, intensive experiments were carried out since about
1983[9.14], after the epitaxial techniques for growing the structures were perfected.
Experiments were first carried out with the GaAs/GaAlAs system.

Structures were grown on n^+ (Si-doped to 10^{18} cm^{-3}) GaAs substrates with
composition as shown in Fig. 8.6. The double- barrier structure is isolated from
the end-contacts formed by 1 μm thick GaAs(Si-doped to about 0.8×10^{18} cm^{-3}) by
15 Å thick undoped GaAs layers to prevent diffusion of the impurities to
the subsequent layers. Contacts to the diode are finally formed at the two
ends by AuGe/Au on well-defined geometries. The area of the diodes is about
2×10^{-5} cm^2 .

The figures of merit for the diodes are the peak current density and the peak-to-
valley current (PVC) ratio. Best values of these parameters for the GaAs/GaAlAs
system are 1- 2×10^5 A/cm^2, and 3.6 at 300 K and 21 at 80 K. The peak current does
not vary much with the temperature. Its value increases as the barrier thickness
is reduced. But such increase is accompanied with a deterioration in the value of
the PVC ratio as the valley current increases faster than the peak current.

Higher figures of merit may be conveniently realized[8.15-18] for the GaInAs/
AlInAs system. Constructional features for these diodes are the same as for the

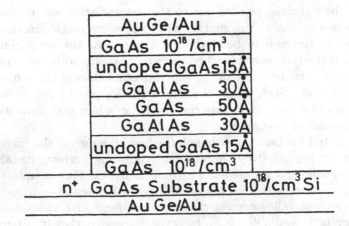

Figure 8.6. Schematic diagram showing the structure of an experimental resonant tunnel diode.

GaAs/GaAlAs systems. Only GaInAs replaces GaAs and AlInAs replaces GaAlAs. The peak current density is reported to be 5.5×10^4 A/cm^2 and the PVC ratio is 21.6 at 77 K and 6 at 300 K. Even a PVC ratio of 14 has been reported[9.19] at 300 K by using the GaInAs/AlAs system. The improvement for this system is essentially due to a larger value of the barrier potential and lower value of the effective mass in the well.

Some typical results obtained with the structures of different systems are collected in Table 8.1.

Suggestive calculations have been made for the III-V/II-VI systems[8.20] to predict higher speed of operation. However, detailed experiments are not reported for such systems. Experiments have been reported on the InAs/AlSb system[8.21,22] which gave a peak current density of 10^5 A/cm^2 and PVC ratios of 11 at 300 K and 28 at 77 K.

8.5. Applications

The diodes have been used to realize oscillators, mixers, frequency multipliers, switches and some special circuits[8.23-33]. The attractive feature of these diodes for such applications is the high value of the cut-off frequency, f_{max}. Value of f_{max} is controlled by the intrinsic time scale of charge relaxation and tunneling time. It is also determined by the contact resistance, terminal capacitance and the time delay in the anode contact. In the steady state of transmission certain amount of charge collects in the well for any applied voltage. As the voltage is changed, this collected charge has to change by the leakage through the barrier. The relaxation time is estimated to the about 0.6 ps per meV band width. The tunneling time is a small fraction of this time.

The contact resistance of the diodes is about 10-15 Ω, while the terminal capacitance may be about 1.6 fF/μ m^2 and the time delay is 175 fs. The value of f_{max} for these values is about 270 GHz. But, these parameters may be changed by heavy doping of the cathode layer, and lengthening of the anode depletion region by low doping. It is possible to push f_{max} controlled by these parameters to near 1000 GHz by proper adjustment of their values.

The value of f_{max} is, therefore, controlled ultimately by the charge relaxation and tunneling time. It is estimated that by suitable design, the value of f_{max} may be pushed beyond 300 GHz.

The power output of the diodes is, however, very low. The estimated value is 225 μW for a 4 μm diameter diode and a typical value is 0.2 μW at 200 GHz. The diodes are, therefore, likely to be useful for very special applications in the high frequency end of the spectrum.

The use of the diodes as self-oscillating mixers and more recently in avalanche

Table 8.1 Structural parameters for the best Figure of merit
of Resonant Tunneling Diodes.L_{b1},L_w,L_{b2} and I_{peak} are respe-
-tively the first barrier width,the well width , the second barrier
width and the peak current

Semiconductor System	L_{b1} (nm)	L_w (nm)	L_{b2} (nm)	$I_{peak} \times 10^{-5}$ (A/ nm^2)	PVC 300 K	Ratio 77 K
AlGaAs/GaAs[a]	23	50	23	0.15	3	11
In$^{0.52}$Al$^{0.48}$As/						
In$^{0.53}$Ga$^{0.47}$As[b]	45	61	45	0.1	6.1	21.6
	50	50	50		4	15
AlSb/InAs[c]	28	65	28	0.04	11	28
	19	65	19	1.6	1.8	5

a. Reference 9.6
b. References 9.18 and 9.15
c. Reference 9.22

photo diodes[8.25] has been demonstrated successfully. Frequency multipliers and
some very sophisticated function generators have also been demonstrated by integrating the diodes vertically and horizontally[8.24,34-38]. Resonant tunneling
structures have been incorporated in the emitter or in the base region of conventional transistors to realize resonant tunneling transistors [8.38-42]. Recently, a
pseudomorphic InGaAs/AlAs/InAs RTD was incorporated into the source of an
InGaAs/InAlAs HEMT [8.43]. The RTD had high current density and a large
peak-to-valley ratio, and the HEMT achieved high gain and high speed of operation. It was possible to generate by using this device low order harmonics of a
considerably higher power levels than those produced by conventional transistors,
and extremely high order harmonics with significant power.

Efforts are being made to exploit the picosecond switching capabilities of RTD's
by combining them with high-performance heterostructure bipolar transistor based
logic cicuits [8.44]. A 10 GHz monolithic latching comparator has been demonstrated by using InAs/AlAs/GaSb RTD's. An exclusive-or circuit withe capability
of running at 25 GHz has realized by using RTD's with pHEMT's [8.45]. A very
low-power RAM cell has been produced, which combines FET's for reading and
writing and RTD's for charge state storage (replacing the storage capacitor).
Silicon-based RTD's are also being studied for memory applications [8.46]. A
novel memory has been proposed by using a double tunneling transistor, in which
current flows in the in-plane direction and is switched from one quantum well to
the other by a surface gate. The source terminal contacts with the top quantum
well, while the drain terminal is contacted to the lower well.

In summary, The current researches on RTD's are aimed at applications to
high-speed signal processing , e.g., three dimensional image processing. Efforts are
also directed to the development a Silicon-based RTD which may be used to replace

the usual storage capacitors in dynamic random access memories (DRAM's), as the RTD causes storage by changing the voltage level, which switches a transistor in the output read line and the transistor provides the current for the output sense amplifier. Very little charge being required to set the voltage level, it is expected that with a silicon-based RTD it will be possible to realize a static RAM cell with the speed and packing density of DRAM's [8.47]. Switching of a RTD by using long-wavelength intersubband transitions has also been demonstrated recently [8.48]

It may be concluded that the future applications of RTD's lie mostly in switching circuits and memories rather than in high-frequency oscillators.

CHAPTER 9

QUANTUM WELL LASER

The basic principle of operation of quantum well lasers (QWL's) is the same as that of bulk lasers. But, the change in dimension of the carrier motion causes improvement in the characteristics, e.g., ultra-low threshold current, narrow gain spectrum and high characteristic temperature. The density of states (DOS) in bulk materials increases with energy from zero at the band edge. On the other hand, in quantum wells, as discussed in Chapter 4, the DOS has the same value at the band edge as for higher energies due to the two-dimensional character of carrier motion. The energy-independent DOS causes greater net emission and reduced temperature-dependence of the Fermi level and thereby leads to better performance characteristics.

Extensive studies have been made of QWL's during the last decade. Many kinds of structures have been developed for improving the performance and detailed theories have been worked out to explain the characteristics for different structures. A complete discussion of these studies is beyond the scope of this book. The discussion is confined only to the physics of the devices with special emphasis on the factors which cause improvement in the performance of QWL's. Up-to-date experimental results are, however, briefly described.

9.1. Operating Principle

Structure of a quantum well laser is illustrated schematically in Fig. 9.1. A 100-200 Å thick lower-band gap material, e.g., GaAs, is sandwiched between two cladding layers of a lattice-matched higher band gap material, e.g., $Ga_xAl_{1-x}As$, which are doped p-type and n-type. The sandwiched layers are provided with contacts for passing current through the device. Current may be passed through a stripe contact for lateral confinement of the injected charge carriers as shown in Fig. 1(a). Alternately, the quantum well may be sandwiched between two higher band gap material also in the lateral direction for lateral confinement as shown in Fig. 1(b). The former is referred as gain-guided and the latter as index-guided laser.

The band diagram of the device is shown in Fig. 9.2 in the presence of a forward biasing voltage. Electrons and holes coexist throughout the well region and the well acts as the active layer. It is in this region that electrons and holes

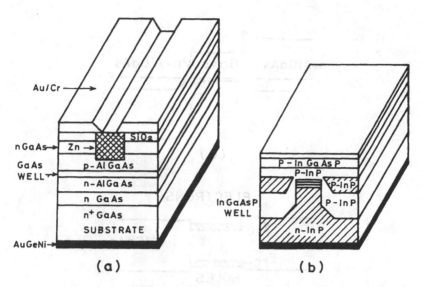

Figure 9.1. Schematic diagram of a quantum well laser (QWL). (a) Gain-guided QWL> Current flows through the Zn-diffused region of GaAs and the injected carriers are thereby confined to the central region of the well. The p-type AlGaAs and n-type AlGaAs layer together with the sandwiched GaAs well forms the laser diode. (b) Index-guided QWL. The InGaAsP/InP multi-quantum- structure is flanked by p-type InP region which has a separate refractive index and confines the optical signal in the MQW. The top p-type InP and the bottom n-type InP together with the sandwiched InGaAsP/InP MQW forms the laser diode.

recombine to generate photons of frequencies corresponding to the difference of their energies. The associated waves travel back and forth inside the Fabry-Perot cavity formed by the two partially reflecting end surfaces in the longitudinal direction. These waves are confined in the vertical as well as in the lateral direction by the refractive index differential between the well and the surrounding material. It should be mentioned that the refractive index and the band gap are inversely related[9.1,2], so that a material with a higher band gap has a lower refractive index. The difference in the refractive indices is about 10%, which is, however, sufficient to cause confinement of the light energy.

The waves are partly absorbed as they travel due to interband absorption, free-carrier absorption[9.3] and other loss processes. At the same time, they are also enhanced by coherent stimulated emission from the recombining electrons and holes. Under conditions, in which the enhancement dominates over the loss, the

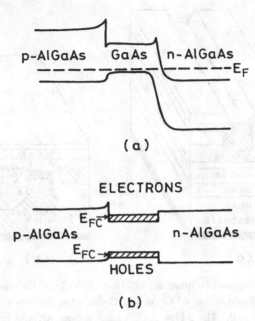

(a)

(b)

Figure 9.2. Band diagram of a QWL. (a) The band diagram in the absence of an external voltage.Discontnuities at the interfaces are the band offsets. (b) The band diagram for an applied positive bias voltage. Inlected carriers are confined in the well region.

diode acts as a laser, producing coherent light at a definite frequency within a small bandwidth.

9.2. Laser Equation

The basic laser equation is obtained by considering that under the equilibrium lasing condition, the electromagnetic wave should remain unaltered after a round trip. We obtain from this condition,

$$(R_1R_2)^{1/2}\exp(2ikL) = 1, \tag{9.1}$$

where R_1 and R_2 are the reflectivities of the end faces in the longitudinal direction, k is the complex propagation constant and L is the length of the active region. We assume, for the sake of simplicity, that the wave is a plane wave. The actual wave will be more complex and the wave front may be curved[9.4], because the wave propagates in the wave guide formed by the cladding layers and the lateral confining layers.

Equation (9.1) gives for the free-space wavelength λ and the power absorption coefficient α,

$$\lambda = 2n_r L/n, \tag{9.2}$$

$$\alpha = -(1/2L)\ln(1/R_1R_2), \tag{9.3}$$

where n_r is the refractive index of the laser material. The first condition gives the possible wavelengths of oscillation corresponding to different values of the integer n. It should be noted that the laser operates at wavelengths in the range of 1 μm and the length L is a few hundred μm's. The value of the integer n for operation at 1.4 μm is about 1250, assuming $n_r=3.5$, $L=250$ μm. The laser therefore operates in a very high order mode of the Fabry-Perot cavity. Evidently, the phase condition given by Eq. (9.2) will be satisfied by a very large number of values of λ around the above value. The separation between these values of λ is

$$\Delta\lambda \approx (\lambda/n)(n_r/n_g), \tag{9.4}$$

where n_g is the group refractive index. The value of $\Delta\lambda$ is 1 nm for $n_g=4$. Evidently, optical signals will be generated at different wavelengths separated by 1 nm around 1.4 μm, for which the attenuation constant which is required to be negative, satisfies Eq. (9.3). In other words, the medium should amplify rather than attenuate. But, as the gain varies with the frequency, radiation over a small band of wavelengths among all the possible wavelengths given by Eq. (9.2) for different values of n, are excited. Further, with increase in the amplitude of the signals, optic signals of wavelengths for which the gain is maximum develops, while signals of other modes are suppressed as in electronic oscillators with multi-frequency resonant circuits through nonlinear interaction[9.5]. The laser then operates mostly at a single wavelength, for which the gain is maximum.

9.3. Operating Characteristics

Optical signal is generated in a diode laser due to the recombination of the carriers injected by applying a forward bias voltage. The resulting photon population is controlled by four processes : (a) the generation by stimulated emission, (b) the interband absorption[9.6], (c) absorption through various other processes, e.g., free-carrier absorption, interface scattering, inter-valence band absorption and (d) the spontaneous emission. The rate equation for the photons may be written as,

$$dN_{\rm ph}/dt = GN_{\rm ph} - \gamma N_{\rm ph} + R_{\rm sp}, \tag{9.5}$$

where N_{ph} is the total photon number in the cavity, GN_{ph} is the net rate of generation of photons resulting from generation through stimulated emission and loss through interband absorption, γN_{ph} is the net rate of photon decay through intrinsic absorption and end-surface loss, R_{sp} is the rate at which photon number is enhanced by spontaneous emission.

The steady-state photon number is given by

$$N_{ph} = R_{sp}(\gamma - G)^{-1} \tag{9.6}$$

The photon number is, therefore very small so long as G is less than γ, since R_{sp} is very small. It increases rapidly as G approaches γ and becomes very large when $G = \gamma$. The equation indicates that N_{ph} should go to infinity for this condition. In practice, however, other processes, not considered here, limit the photon number. It should, however, be noted that G cannot be larger than γ, since then N_{ph} is required to be negative which is unrealistic. The gain and hence the carrier density remain nearly pinned to the threshold value.

The current density corresponding to the threshold condition for stimulated emission is

$$J_{th} = |e| dn_{th}/\tau_e(n_{th}), \tag{9.7}$$

where d is the active layer width, n_{th} is the carrier density under the threshold condition and $\tau_e(n_{th})$ is the carrier- recombination time corresponding to the threshold carrier density. The stimulated recombination being negligible, $\tau_e(n_{th})$ may be written as[9.7]

$$\tau_e(n_{th}) = (A_{nr} + Bn_{th} + Cn_{th}^2)^{-1}, \tag{9.8}$$

where the first term accounts for non-radiative recombination through impurity states, surface roughness etc., the second term is the radiative recombination term and the third term is the Auger recombination term.

For current densities larger than the threshold value, the additional injected carriers cause rapid increase of photon density n_{ph} which is then given by

$$n_{ph} = \eta_i \tau_p (J - J_{th})/|e|d, \tag{9.9}$$

where τ_p is the photon lifetime and η_i is the internal quantum efficiency, given by the ratio of the radiative and the total recombination rate including the nonradiative recombination.

The light output from the device from one facet, assuming the two facets to be identical with reflectivity R is

$$P_{out} = (1/2)\hbar\omega n_{ph} V \alpha_m v_g, \tag{9.10}$$

where V is the total volume of the active layer, v_g is the group velocity at the lasing frequency ω and α_m is defined as

$$\alpha_m = -(1/L)\ln R \tag{9.11}$$

Expressing $n_{\rm ph}$ in terms of the total current I we get for the light output,

$$P_{\rm out} = (\hbar\omega/2|e|)\eta_i\alpha_m(\alpha_m + \alpha_i)^{-1}(I - I_{\rm th} - \Delta I_L), \qquad (9.12)$$

where τ_p has been replaced by $(\alpha_m + \alpha_i)^{-1}$, expressing the facet losses as a distributed loss factor α_m , α_i being the internal loss due to the various phenomena mentioned earlier.I is the total current, which may have a component, the so-called leakage current, which does not pass through the active layer.ΔI_L is the change in the leakage current, as the current increases from $I_{\rm th}$ to I.

The light output increases almost linearly with the diode current, when the diode starts lasing. However, for very large currents the light output saturates through various mechanisms, not discussed here. A characteristic parameter of the light output-current curve is the differential (external) quantum efficiency defined as the ratio of the photon escape rate and the generation rate , multiplied by the internal quantum efficiency. It is given by[9.8],

$$\eta_d = \eta_i\alpha_m(\alpha_m + \alpha_i)^{-1} = (2|e|/\hbar\omega)(dP_{\rm out}/dI) \qquad (9.13)$$

The characteristics of the laser discussed above applies equally well to the quantum well lasers. The distinctive deviation occurs only in the magnitude and the temperature dependence of the threshold current. This aspect is discussed in the following sections.

9.4. Threshold Current

The component of the threshold current due to the recombining carriers is given by,

$$J_{\rm th} = n_{\rm th}|e|d/\eta_i\tau_r, \qquad (9.14)$$

where $n_{\rm th}$ is the threshold carrier density, τ_r is the recombination time and η_i is the internal quantum efficiency defined in Eq. (9.9). The threshold current may be computed by determining the value of $n_{\rm th}$ for which the gain is equal to the loss. The value of τ_r may be evaluated by using the expressions for the electron-photon interaction discussed in Chapter 5. The quantum efficiency, however, depends on the perfections of the diode structure and is treated in most cases as an experimentally determined parameter. Some of the non-radiative processes, e.g., Auger recombination[9.9-11], which becomes important in low band gap materials and hence in long wavelength lasers, have been theoretically studied. We shall not discuss the theories, but we shall note only that the recombination rate due to the Auger processes is proportional to n^3 , while non-radiative recombination due to the other processes is proportional to n and the radiative recombination rate is proportional to n^2 (n is the carrier density). As the Auger processes are more effective in low band gap materials the magnitude of n_{th} is much higher in these materials and this is one of the major difficulties in designing a room-temperature InGaAsP laser.

9.4.1. CONFINEMENT FACTOR

Electromagnetic energy although waveguided in the active layer is not fully confined in it. Consequently the active medium affects the optic signal only partially. The effective gain may be written as Γg, where g is the gain in the active layer and Γ is the confinement factor, defined as the ratio of the electromagnetic energy within the well and the total energy.

Evaluation of the threshold carrier density n_{th} involves the computation of the confinement factor Γ and the gain g as a function of the carrier density. Extensive analysis[9.12,13] has been carried out for double-heterostructure lasers to calculate Γ. It is found that Γ may be considered to be the product of a transverse confinement factor Γ_T and a lateral confinement factor Γ_L. These factors give the electromagnetic energy in the active region as a ratio of the total energy, as determined by the variation of the refractive index in the two directions. The factors may be evaluated theoretically in principle. However, the analysis is complicated by the complex geometry, non-homogeneity, variation of the electron density, attenuation and gain in different regions. A complete analysis is not available, but some simplified relations have been deduced, which may be used to estimate the value of Γ. The simplified expressions for Γ_L and Γ_T are as given below[9.8].

$$\Gamma_L = W^2/(2 + W^2), \tag{9.15}$$

where $W = (2\pi/\lambda)[(n_r^{\text{in}})^2 - (n_r^{\text{out}})^2]^{1/2}w$, n_r^{in} and n_r^{out} are respectively the refractive indices inside and outside the active layer, w is the width and λ is the operating wavelength. The width w is about 2 μm and considering that $\lambda \approx 1\mu$m, $n_r \approx 4$ and $\Delta n_r \approx 0.1 n_r$, the value of W works out to about 10. The value of Γ_L is therefore close to unity and Γ may be taken to be the same as Γ_T. Γ_T is given by,

$$\Gamma_T \approx D^2/(2 + D^2), \tag{9.16}$$

where $D = (2\pi/\lambda)(n_{ra}^2 - n_{rc}^2)^{1/2}d$, n_{ra} and n_{rc} are respectively the refractive indices of the active layer and the cladding layers; d is the thickness of the active layer. For QWL's $d \approx 150$ Å and putting $\lambda = 1\mu$m, $n_{ra} = 3.5$, $(n_{ra}^2 - n_{rc}^2)/n_{rc}^2 = 0.1$, we obtain $D \approx 0.1$. Γ may, therefore, be approximated as

$$\Gamma_T = 4\pi^2 n_{ra}\Delta n_{ra}(d/\lambda)^2 \tag{9.17}$$

The value of d being typically 1.5 nm we find that the confinement factor lies between 0.003 and 0.004 for wavelengths of 1-1.5 μm. Such low values of Γ partially annuls the increased value of g for quantum wells. Attempts have been made to increase the value of Γ, retaining the two-dimensional character of the electron gas by devising essentially two kinds of structures : the multi-quantum well (MQW) or the modified multi-quantum well (MMQW) and the graded-index separately confined heterostructure (GRIN-SCH) wells shown in Fig. 9.3.

The confinement factor for multi-quantum wells may be approximated as[9.8],

$$\Gamma = (2\pi^2/\lambda^2)(d_w/d)(Nd)^2(\bar{n}_{ra}^2 - n_{rc}^2), \tag{9.18}$$

where $\bar{n}_r = [n_{ra}d_w + (d - d_w)n_{rc}]/d$, d is the total thickness of the individual quantum well structure, d_w is the thickness of each well and N is the total number of wells. Expression (9.18) is obtained by considering that the refractive index of the active region has an average value determined by the width of the active layer and the barrier layer and that the active region is a fraction d_w/d of the total active layer. For wells of equal thickness, it is found that Γ is effectively the value for the individual well, multiplied by the square of the number of wells. The confinement factor may, therefore, be increased by large factors by using multiple quantum wells, keeping at the same time the advantage of the two- dimensional motion of carriers.

The confinement factor for the GRIN-SCH systems is also given by the simple expression[9.8]

$$\Gamma = (2\pi^2/\lambda^2)(d/d_w)(\bar{n}_{ra}^2 - n_{rc}^2), \tag{9.19}$$

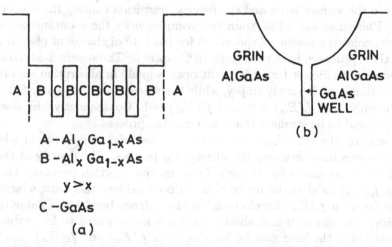

(b)

(a)

Figure 9.3. Quantum well structure for the enhancement of optical cdonfinement. (a) Modified multi-quantum well (MQW). Carriers are confined in the GaAs wells, while the optical field is confined in the region between the two other AlGaAs layers. (b) Graded-index separate confinement heterostructure (GRIN-SCH). The carriers are confined in the well, While the optical field is confined in the GRIN region, due to variation of the refractive index.

where d_w is the width of the central well and d is the total width of the active layer, $\bar{n}_{ra}$ is the average refractive index of the active layer. The confinement factor is thus multiplied in the GRIN-SCH structure by approximately (d/d_w). If, however, d is made large, then Γ_T is given by the simple expression[9.14]

$$\Gamma_T = (2/\pi)^{1/2}(d_w/d) \tag{9.20}$$

The MQW or GRIN-SCH structures increase the value of the confinement factor by factors of about 10-30 and a typical value of Γ for these structures may be taken to be 0.1.

These structures improve the performance of QWL's through an additional process. Carriers are injected in the active layer in the high energy end. These are then scattered downwards to the band edge before they may recombine and generate photons. This settling time is not available for narrow wells and capture of the injected carriers take long time in SQL's. This affects the dynamic performance of the lasers[9.15] In a MQW or GRIN-SCH structure, on the other hand, this problem is mitigated and most of the injected carriers are available for recombination in the active layer immediately.

9.4.2. GAIN

The phenomenon of electron-photon interaction has two components : the emission of photons due to the recombination of electrons from the conduction band with the holes in the valence band and the reverse transition causing the absorption of photons. Photon emission has again two components : the spontaneous emission and the stimulated emission. Expressions for the rate of change of photon density through these processes have been given in Chapter 5. These were used to calculate the absorption coefficient for an incident optic signal. In absorption experiments, the conduction band is mostly empty, while the valence band is mostly full, so that we may assume that $f_c(E_{ck}) \approx 0$, and $f_v(E_{vk}) \approx 1$. Consequently, the absorption process is found to be stronger than the emission process.

In lasers, on the other hand, we are concerned with conditions in which the emission process dominate over the absorption process. The density of the carriers, both electrons and holes, is made large by the injection process. The Fermi functions $f_c(E_{ck})$ for electrons in the conduction band has, therefore, a large value. Also, the function $f_v(E_{vk})$ for electrons in the valence band has a value different from unity. The emission and absorption rates are required to be evaluated for the calculation of the laser gain by keeping both $f_c(E_{ck})$ and $f_v(E_{vk})$.

The gain coefficient g is the difference of the emission and the absorption rate. It is obtained by changing the sign of the absorption coefficient given by Eq. (5.143) .

The gain coefficient is therefore given by

$$g = V_c(2/8\pi^3)(2\pi/\hbar)(e^2 A_0^2/4m_0^2)(n_r/c)$$
$$\int |\hat{a} \cdot p_{cv}|^2 [f_c(E_{ck}) - f_v(E_{vk})]\delta(E_{ck} - E_{vk} - \hbar\omega)d^3k. \qquad (9.21)$$

The k-selection rule has been assumed to apply and the integral is therefore required to be evaluated by using the joint density of states.

Evaluating the integral as for absorption we get

$$g = (e^2 m_r/Ln_r\epsilon_0 cm_0^2\hbar^2\omega)|\hat{a} \cdot p_{cv}|^2$$

$$\sum_n \{\exp[(\hbar\omega - E_{gn})(m_r/m_c) + E_{cn} - E_{cF}] + 1\}^{-1}$$
$$-\{\exp[(E_{gn} - \hbar\omega)(m_r/m_h) + E_{vn} - E_{vF}] + 1\}^{-1}H(\hbar\omega - E_{gn}), \qquad (9.22)$$

where E_{gn} is the difference between the nth level energies of electrons and holes. It is assumed that the injected electrons and holes reach a quasi-equilibrium condition so that their distribution is also given by Fermi functions with the quasi- Fermi levels E_{cF} and E_{vF} .

Computed values[9.16] of g are illustrated in Fig. 9.4 for a GaAs/AlGaAs system. The gain, g decreases faster than $(1/\hbar\omega)$, since f_c decreases while f_v increases with increase in $\hbar\omega$. The gain, however, shows step increases as $\hbar\omega$ becomes equal to the subsequent values of E_{gn}.

It may also be noted that the quantized levels are broadened by intraband scattering. The effect of such scattering is taken into account by multiplying the gain function by a Lorentzian distribution function and then integrating over the energy level to obtain the convolved gain function, $G(\omega)$, which is then given by[9.17],

$$G(\omega) = (2/\pi) \int g(\omega)(\hbar/\Gamma_{iv})[(E_{ck} - \hbar\omega)^2 + (\hbar/\Gamma_{iv})^2]^{-1}dE_{ck}, \qquad (9.23)$$

where Γ_{iv} is the intraband scattering time. The effect of this broadening is to smooth out the gain curve as shown by the dotted line in Fig. 9.4. It also causes a small shift in the position of the peak gain. However, these effects are not large enough to cause a major change in the gain characteristics.

The gain frequency curves for quantum wells are quite distinct from those for the bulk materials. In bulk materials, the gain starts from a value of zero for $\hbar\omega = E_g$, increases with increase in $\hbar\omega$, reaches a maximum value and then decreases again. The frequency of the maximum gain for bulk materials varies with the carrier density. On the other hand, in quantum wells, the maximum gain occurs for all carrier densities when $\hbar\omega = E_g + E_{c1} + E_{v1}$, i.e., at the modified band edge. There is only a minor change of ω due to band gap renormalization[9.16] with the carrier density and due to intraband scattering. The difference in the character of the quantum well and the bulk gain curves arises from the step density of states in quantum wells and parabolic density of states in bulk materials. It is also because of this difference that the injected carriers contribute more to the emission in quantum wells. In bulk materials, as the current is increased to cause the carrier density to increase, the maximum gain shifts to higher energies, and the number of carriers contributing to the emission does not increase proportionately. On the other hand, in quantum wells, the injected carriers increase the carrier concentration at the band edge and contribute more effectively to emission.

The gain spectrum illustrated in Fig. 9.4 is also found to be sharper in quantum wells, as the density of states does not change with the energy. This character of the gain curve leads to better spectral purity in the laser output.

The maximum value of the gain coefficient has been calculated[9.18,19] for GaAs and GaInAs for different injected carrier densities assuming the width of the well to be 200 Å. The calculated curves for GaAs[9.18] at different temperatures are illustrated in Fig. 9.5. The gain initially increases rapidly with the carrier

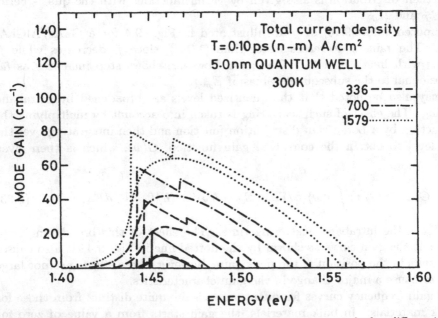

Figure 9.4. Gain of a GRIN-SCH-SQW laser for different phonon energy for four different injection levels. The upper and the lower curve for each injection level give respectively the unconvolved and the convolved gain vs photon energy .[After S. R. Chinn, P. S. Zory and A. R. Reisinger, *IEEE J. Quantum Electron.* **24**,2191 (1988); Copyright (=A9 1988=IEEE)].

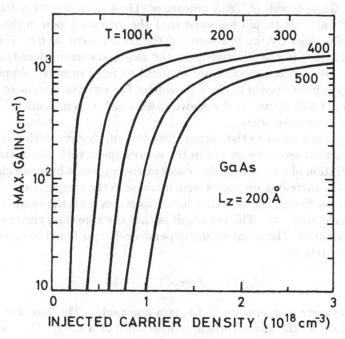

Figure 9.5. Calculated maximum gain for heavy-hole-1 to conduction-band-1 level transitions as a function of injected carrier density in GaAs single quantum wells of 200 Å thickness with $Al_{0.52}Ga_{0.48}As$ barriers. The temperature is indicated by the numbers on the curves.[After N. K. Dutta, *J. Appl. Phys.* **53**, 7211 (1982); Copyright: American Institute of Physics].

density but finally saturates. The saturation is characteristic of the quantum wells and is due to the energy- independent density of states and unchanging frequency of maximum gain. These states get filled up for large injected carrier density and cause thereby saturation of the gain. In contrast, as the maximum gain shifts to higher energies in bulk materials with increase in the carrier density and since the density of states increases continuously with energy no saturation in gain is observed.

9.4.3. ESTIMATION OF THRESHOLD CURRENT

The threshold current for lasing may be calculated by using the gain curves, and the estimated values of α_i, α_m and Eq. (14) as illustrated in Reference 9.18.

For example, the estimated values of Γ, α_i, and R are 0.04, 10 cm^{-1} and 0.3 for a GaAs/ Al$_{0.52}$Ga$_{0.48}$As well of 200 Å thickness. These data give for a 380 μm long well, $g_{th} \approx 10^3$ cm^{-1}. It should be noted that the increased gain in the quantum wells is partially offset by the low value of the confinement factor. However, in spite of this decrease in the effective gain, the required current density turns out to be lower as the injected carriers are required to fill a smaller volume and the current density is proportional to the well width. The estimated value of threshold current density is 550 A/cm^2 in the above case, which is significantly lower than the value for a conventional laser.

The gain, being related to the carrier distribution, changes with the temperature. As the carriers are more spread in the energy space with rise in the temperature, concentration of the carriers decreases at energy levels from which emission occurs. The gain, therefore, decreases with increase in the temperature. The internal absorption coefficient, on the other hand, increases with the temperature due to increased scattering rate. The net result is that the threshold current increases with the temperature. The temperature dependence[9.8] is found to be expressible by the following relation,

$$J_{th}(T) = J_0 \exp(T/T_0 - 1) \qquad (9.24)$$

around an operating temperature T (J_0 is a constant). The constant temperature T_0 is defined as the characteristic temperature of a laser. The value of the threshold current may be estimated by carrying out the analysis, outlined above, for different temperatures and the value of T_0 may be derived therefrom. It may, however, be noted that the gain for QWL's is expected to be less dependent on temperature, since the Fermi level for the carriers in quantum wells is almost temperature independent for degenerate concentration. Detailed calculations also indicate that the characteristic temperature for QWL's is higher than that of conventional double-heterostructure lasers.

9.4.4. EQUIVALENT CIRCUIT MODEL

The design of a cicuit for using a laser for a specific purpose requires an equivalent circuit model of the laser. Several model circuits have been proposed [9.20]. Consideration of these models is beyond the scope of this book. We may, however, note that it is possible to predict the turn-on delay, relaxation oscillation frequency and 0n/0ff aspect ratio resulting from the application of a pulse voltage by using these models. Variation of the parameters may also be predicted . It has been generally concluded that the bias dependence of QWL's are significantly different from that of the coventional double-heterostructure(DH) lasers.

9.5. Experimental Results

The work on QWL's has been so extensive that it is not possible to discuss even a fraction of this work within the limited scope of the book. Interested reader may consult books solely devoted to this topic[9.21]. Some salient features of this work are, however, discussed below.

9.5.1. GaAs/AlGaAs QWL

Experimental work on QWL's was started with the GaAs/AlGaAs system, as the crystal growth techniques for this lattice-matched system was already developed. The first QWL[9.22] , using a single 200 Å well was reported in 1977; it required a threshold current density J_{th} of 3 kA/cm^2 at 300 K. The performance characteristics were radically improved in later years by improving the quality of the crystals, particularly the interface, so as to reduce the interface scattering. The structures were also optimized by choosing the proper dimension of the active layer quantum well and number of wells in MQW systems. Aluminum content and thickness of the barrier layers were optimized to facilitate greater carrier injection and the outer cladding layers were designed to ensure confinement of the optical energy. The number of wells was also optimized to minimize absorption and maximize confinement. An optimized MQWL with four 120 Å wells and 380 μm length was reported[9.23] to have a threshold current density J_{th} of 250 A/cm^2 . The value of J_{th} of a single well laser was also reduced[9.24] to 160 A/cm^2 by using a 1125 μm long cavity. The increased cavity length reduced the effect of facet loss and the carrier absorption loss was reduced by reducing the doping of the cladding layers. Greater optical confinement was ensured by using the GRIN-SCH structure.

Work on GaAs/GaAlAs lasers is being continued and development of such lasers have been reported even in recent years. In one report[9.25], the laser had a threshold current,I_{th}, of 18 mA, characteristic temperature T_0 of 240 K and external quantum efficiency η_d of 70 $-$ 74%. The power output, P_{out} , was 20 mW at 840 nm and operated up to 70°C. In a second report[9.26], the power output was 500 mW at 50°C. The value of J_{th} at 300 K has also been reduced[9.27] to 52 A/cm^2 and of I_{th} to 0.35 mA[9.28]. GaAs/AlGaAs lasers have also been developed to have low temperature sensitivity[9.29], high modulation bandwidth[9.30] and reduced spectral band width[9.31] and chirp (initial wave length variation).

9.5.2. InGaAsP QWL

The development of optical fibers for communication[9.32] stimulated work on diode lasers for the zero dispersion wavelength 1.1 μm and lowest attenuation wavelength 1.55 μm. It was recognized that the quaternary $In_x Ga_{1-x}As_yP_{1-y}$, with suitably chosen values of x and y has the required band gap values for the generation of these wavelengths and may be lattice matched to InP. Conventional heterostructure lasers using this material system grown by LPE are well-established. Considering the expected advantages of QW lasers, attempts have been made to realize QWL's with this material. However, good quality phosphides being difficult to grow by MBE or MOCVD, first few lasers were made by LPE[9.33,34] or hydride-transport VPE technique[9.35].

It may be recalled that the confinement factor for the MQW being larger in comparison to that for SQW, the threshold carrier density is smaller in the former. Auger non-radiative recombination[9.36,37] being more effective in these lasers because of the smaller band gap, the carrier density is tried to be kept as low as possible since Auger recombination increases as the cube of the carrier density. The InGaAsP QWL's are, therefore, based on MQW systems. Initial experiments[9.34] on 1.3 μ m lasers yielded I_{th} of 15-20 mA, J_{th} of 1.2 kA/cm^2 and T_0 of 70-145 K. The threshold current has been further reduced[9.38] to 5-8 mA by reflective coating of one facet by Si/SiO$_2$.

MQL's for 1.55 μm wavelength have been fabricated by MBE, but using the $In_{0.53}Ga_{0.47}As/Al_{0.48}In_{0.52}As$ system[9.39] and by CBE[9.40], using the $In_{0.53}Ga_{0.47}As/InP$ systems. The value of J_{th} is about 2.4 kA/cm^2, while T_0 is about 60-100 K. A 1.5 μm QWL using InGaAs/InGaAsP has been reported[9.41], which has J_{th} of 170 A/cm^2, η_i of 83%, α of 3.8 cm^{-1} and T_0 of 45 K.

It appears from the published work that the performance characteristics of InGaAsP MQL's are not yet as good as those of GaAs MQL's. Values of threshold currents have been reduced but the characteristic temperature has still a low value.

9.5.3. STRAINED-LAYER LASERS

Heterostructures may be grown with materials having small lattice mismatch. The layers grown on a host material of a different lattice constant gets strained, but the strain may be accommodated[9.42] unless the layer is thicker than about 30-40 Å. The strain, however, alters the energy band structure[9.43]. The light hole band and the heavy hole band are decoupled and as a result the upper most band has a reduced value of effective mass for fairly large values of hole energy as illustrated schematically in Fig. 4.11. The valence band Fermi level is therefore deeper in the band for the same hole concentration. The difference between the electron quasi-Fermi level and the hole quasi-Fermi level is, therefore, larger for the same injected carrier density in strained layers in comparison to that in the unstrained layers. The gain in QWL's using strained layers is, therefore, expected to be higher for the same carrier density.

A phenomenological analysis has been given [9.44] , which describes the high-power high-temperature operation of a compressively strained InGaAsP MQW ridged waveguide laser.It has also been experimentally verified that it is the crystal heating which defines the optical power saturation.

Strained-layer lasers are also of interest for generating radiation of wavelengths of 0.8-1.1 μm, required for E_r^{3+} - doped optical fiber amplifiers[9.45]. The ternary material $In_xGa_{1-x}As$ of the required band gap has a lattice constant, different from the readily available substrates and is required to be grown as a strained layer.

Work on the strained-layer QWL's is being carried out mostly by using the $In_xGa_{1-x}As/GaAs$ system. The early reports[9.46] on InGaAs/GaAs lasers emitting at 1.0 μm gave J_{th} of 1.2 kA/cm^2 . The performance of these lasers, however improved with the years and 0.98-1 μm lasers have been reported with I_{th} ranging between 3 mA and 85 mA[9.47-51] and J_{th} ranging between 56 A/cm^2 and 212 A/cm^2[9.52-56]. The threshold current has been further reduced to 0.35 mA by using reflective coating on a facet[9.57].

Recently, a 1×12 monolithically integrated array has been demonstrated[9.58] with AlGaInAs/InP GRIN-SCH strained MQW 1.5 μm lasers. The diodes on the array had a characteristic temperature of 88 K, slope efficiency drop of 0.9 dB between 20 and 80^{0}C, threshold current of 5.91 mA, slope efficiency of 0.281 W/A at 20^{0}C. The lasing wavelength was 1510 nm at 20^{0}C and the 3 dB modulation band width was 12 GHz.

The effect of strain on the performance characteristucs of $In_xGa_{1-x}P/In_{0.5}(Al_{0.7}Ga_{0.3})_{0.5}P$ strained-layer lasers have been analysed by Ahn [9.59]. The threshold current density of compressive as well as tensile-strained quantum wells were found to be lower than that in unstrined wells. Lowest threshold current density of about 385 A/cm^2 was obtained for a laser with a 500 μm long cavity, which agreed with experiments.

9.5.4. VISIBLE MQWL

Lasers for visible light are required for different applications, e.g., in bar-code scanners and erasable optical disks. The AlGaInP system lattice-matched to GaAs can give radiation of wavelengths around 650 nm. The room temperature AlGaInP laser was first reported in 1986[9.60]. The performance characteristics have been improved with time. However, I_{th} in these lasers are comparatively larger because of the larger carrier effective mass, which effectively reduces the quasi-Fermi levels and increases the transparency concentrations. The band offsets, being also small in these structures the leakage current is larger. Characteristic temperatures for these lasers are in the range between 80 and 140 K. The development of AlGaInP lasers with improved performance characteristics is therefore engaging great attention. A strained-layer single QW AlGaInP/GaInP laser has been reported[9.61] to have J_{th} of 215 A/cm^2 . In another report[9.62] 60 mW light output has been obtained at 100°C with a I_{th} of 36 mA. Light output at 634 nm has also been reported[9.63] for a laser with I_{th} of 59 mA and T_o of 46 K.

There has been great interest in the development of Light-Emitting Diodes (LED's) and lasers from the ultraviolet to the yellow region . Such LED's have been developed by using InGaN single quantum wells and these are commercially available [9.64,65]. An InGaN MQWL has been demonstrated [9.66] , which produced 215 mW of power at a forward current of 2.3 A with sharp peak light output at 417 nm and had a FWHM of 1.6 nm under pulsed condition at room temperature. The threshold current density fior the laser was 4 kA/cm^2.

9.5.5. SURFACE-EMITTING LASERS

In many applications. e.g., parallel light-wave communication systems, quick-access optical disks, optical computing, optical interconnects, monolithic integration and two-dimensional arrays, light is required to be emitted from the surface. Surface-emitting lasers[9.67-72] are realized either by using a vertical cavity or by using a horizontal cavity and a built-in reflector.

A vertical-cavity surface-emitting laser (VCSEL) is shown schematically in Fig. 9.6(a). It has the usual laser structure, but the top and the bottom surfaces are covered by Bragg reflectors. For example[9.67], for an InGaAs VCSEL, a Bragg reflector consisting of quarter wave layers of $Al_{0.67}Ga_{0.33}As$ and GaAs is first grown on a GaAs substrate. Then the active layers consisting of three 80 Å $In_{0.2}Ga_{0.8}As$ wells with 100 Å GaAs barriers are grown, separated by $Al_{0.33}Ga_{0.67}As$ spacers. On top of the active structure another Bragg reflector is grown. The total vertical length of the wells, barriers and spacers is one wavelength. As the reflectivity of the Bragg reflectors is very close to unity (99.75-99.95 %), the edge reflection loss in the vertical direction is much smaller than that in the lateral direction. Hence the lasing occurs in the vertical mode to produce emission through the substrate GaAs, which is transparent to the generated radiation at 930 nm. The lateral extent of the carriers is also limited by the contact dimensions to concentrate the injected carriers over a small diameter. VCSEL's have been constructed with InGaAs and GaAs[9.69] quantum wells, which have I_{th} of 1-4 mA, J_{th} of 0.8-1.4 kA/cm^2 , P_{out} of 1-9 mW and η_d of 10- 25 %.

Horizontal cavity surface-emitting lasers (HCSEL) are essentially the edge-emitting lasers with in-built 90°/45° mirrors, as illustrated schematically in Fig. 10.6(b,c). To one end of the laser consisting of the n and p-type GaAlAs layers and a GRINSCH active layer, a $90^o S_3N_4$ mirror is constructed by reactive ion etching and to the other end is constructed a $45^o S_3N_4$ mirror with ion-beam etching. The laser works as an edge- emitting laser with output at one end, which is reflected upwards to be emitted through the surface. Such lasers have been constructed with GaAs[9.70] and InGaAs[9.71,72] active layers to give P_{out} of 100-250 mW, J_{th} of 300-500 A/cm^2 and η_d of 10-22 %.

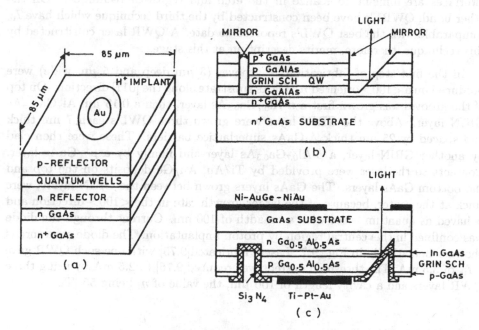

Figure 9.6. Schematic diagram of surface-emitting quantum well laser. (a) Vertical cavity SE-QWL. The cavity is formed by the top p-type and the bottom n-type Bragg reflectors. Current is confined to the central region by H$^+$ implants around the contact. The quantum well system is similar to the structures used in other lasers. (b) GaAs/GaAlAs horizontal cavity SEQWL. The cavity is formed by the two mirrors at the end faces. Light emitted from the edge is reflected vertically upwards by the 45°-mirror. (c) InGaAs/GaAs horizontal SEQWL. Light is emitted vertically upwards after reflection from the end mirror.

10.5.6. QUANTUM WIRE LASER

The improvement in the performance of diode lasers by the one-dimensional confinement is expected to be further enhanced by two- or three-dimensional confinement. Attempts are, therefore, being made to realize structures in which the carrier momentum will be quantized in two directions, the so called quantum wires (QWR). Various techniques, e.g., etching and regrowth[9.73], growth on vicinal substrates[9.74], growth on patterned non- planar substrates[9.73-78], have been tried for realizing QWR's.

Quantum wire lasers (QWRL) have been constructed by the first technique but the expected low J_{th} could not be realized, apparently because damage free interfaces are difficult to realize in the etch and regrowth technique. On the other hand, QWRL's have been constructed by the third technique which have I_{th} comparable with the best QWL's reported to date. A QWR laser constructed by this technique, therefore, merits description at this stage.

In the first step of fabrication, V-grooves (3 μm deep and 5 μm wide) were produced on a <100> oriented n^+ GaAs substrate along the [011] direction. On top of the grooves was grown first a n-Al$_{0.7}$Ga$_{0.3}$As layer, then a 0.15 μm Al$_x$Ga$_{1-x}$As GRIN layer. Above the GRIN layer were grown three QWL's each 7 nm thick and spaced by 25 nm thick AlGaAs superlattice barriers. These were then clad by another GRIN layer, a p-Al$_{0.7}$Ga$_{0.3}$As layer and finally by a p^+ GaAs layer. Contacts to the laser were provided by Ti/Au, Au/Ge deposits on the top and the bottom GaAs layers. The GaAs layers grown between the GRIN layers, were thick at the center, because of the larger growth rate in the <100> direction and behaved as quantum wires with a breadth of 100 nm. Current through the diode was confined in this central region by proton implantation. The diode structure is shown schematically in Fig. 9.7. Lasers fabricated[9.75] with one such QWR gave a I_{th} of 3.5 mA. The threshold current was reduced[9.76] to 2.5 mA by using three QWR layers and a cavity length of 100 μm, the value of η_d being 55 |The

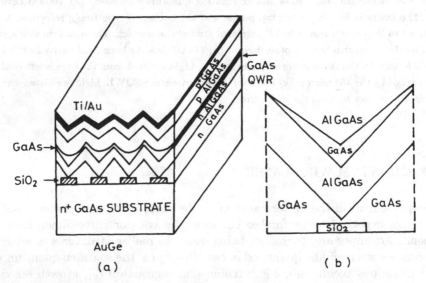

Figure 9.7. Schematic diagram of a quantum-wire laser. Quantum wires are formed by the GaAs layers deposited on the AlGaAs layers with corrugated surface. The diode is formed by the top p-type AlGaAs and the bottom n-type AlGaAs. Carriers are confined mostly to the GaAs wires with triangular cross-section.

threshold current was further reduced to 0.6 mA by applying high-reflection coating (R= 0.95) and using a length of 135 μm.

Superlattice QWR structures, constructed by SiO_2 masks on GaAs substrates, as described in Chapter 2 have also been tested for laser action by optical pumping[9.77,78]. It has been predicted from these results that it may be possible to realize lasers with I_{th} <100 μA by suitable design of the cavity of a vertical-cavity laser.

Quantum wire lasers have also been produced by a single step molecular epitaxy method[9.79]. It has been found that QWR lasers offer the benefits of higher gain, reduced temperature sensivity , higher modulation band width and narrower spectral linewidth. In these structures a short-period superlattice is grown with lateral modulation via the strain-induced lateral ordering [9.80]. A quantum wire is realized by using this structure as the quantum well region of a conventional quantum well as shown in Fig. 9.8. The band structure of such a laser has been analysed by Li and Chang [9.81], who concluded that the strain increases the quantum confinement and enhances the anisotropy of the optical transitions.

9.5.7 QUANTUM DOT LASER

Quantum dots are also being studied for use in the active region of injection lasers[9.82,83]. Lasers based on a single sheet of quantum dots (QD's) lase via the ground state in the low temperature range (150-180 K) and exhibit the expected low threshold current density and ultrahigh temperature stability [9.84]. However, the electrons and the holes escape fronm the QD's at higher temperatures, which causes a decrease in gain and increase in the threshold current. The problem has been solved by using vertically coupled QD's(VCQD's). The threshold current has been found to decrease dramatically at 300 K [9.85].A 1 W CW laser has been

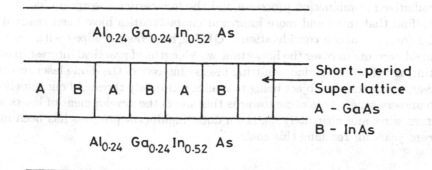

Figure 9.8. A schematic diagram of a quantum wire laser using $Ga_xIn_{1-x}As$ with strained lateral ordering.

demonstrated by using InGaAs VCQD's in an AlGaAs matrix [9.86].

QD lasers have also been realized [9.87] with CdSe/ZnSSe strain-compensated quantum dot superlattices. A threshold current density of 55 kW/cm^2 and 0.8 kW/cm^2 were realized respectively for T(temperature) = 300 and < 75 K. The characteristic temperature was as high as 750 K.

In spite of the lower values of the threshold current densities and higher characteristic temperatures so far realized, QD lasers cannot be considered as fully developed . The main difficulty arises from the size variation of the QD's, which leads to a large number of lasing wavelengths. It may be expected that with further improvement in the QD growth techniques , this problem will be minimised and the full potential of QD lasers will be realized.

9.6. Conclusion

The physics underlying the operation of quantum well lasers have been discussed in this chapter. Different kinds of QWL's realized to-date have also been briefly described. The basic idea behind the improvement of the performance characteristics is the confinement of carriers in a smaller and smaller physical space so that greater concentration of carriers is obtained for the same total carrier injection. Confinement by reducing the physical size, confines the electron also in the energy space by quantization. However, the small physical size has the attendant problem of small optical confinement. Steps are required to be taken for greater optical confinement. GRIN and MMQ systems have been evolved for this purpose.

Improvement of the performance characteristics also requires minimization of the loss and the non-radiative processes. The effective edge loss may be reduced by increasing the length of the cavity and by high-reflection coating of the edges. Crystal growth techniques are also required to be improved so as to reduce the defects which may cause non-radiative recombination. The carrier concentration is also required to be kept low for the threshold conditions to minimize the Auger non-radiative recombination processes and the free-carrier absorption loss.

We find that more and more improved characteristics have been realized for QWL's from the above considerations. QWL's have been realized with very low threshold currents to cover the important wavelengths of practical interest in edge-emitting as well as the surface-emitting lasers. Interest in the diode laser research, however, continues, the object being to realize ultimately threshold currents in the micro ampere range. One route towards this aim is the development of lasers with quantum wires and ultimately quantum dots. Significant progress has been made in recent years for reaching this goal.

CHAPTER 10

QUANTUM WELL DETECTOR, MODULATOR
AND SWITCH

Quantum well devices discussed in Chapter 7-9 have replaced conventional devices in many electronic equipment. Extensive work is also being carried out on quantum well detectors, modulators and switches. These devices are not, however, fully developed yet and several varieties of each device are being experimented on. In order to keep the discussion within limits we shall consider only the basic operating principles of these devices and describe one or two forms of each device as example.

10.1. Quantum Well Detector

Detectors for infrared or long wavelength radiation are of interest for night vision, space surveillance, space exploration, remote monitoring of environment and such other applications. Such detectors have been made since the sixties by using mercury cadmium telluride, the band gap of which may be varied from near-zero value to 1.44 eV by varying the composition. The detectors developed by using the absorption phenomena in quantum wells provide a challenging alternative to these detectors for some of the applications.

Quantum well infrared photo detector (QWIP)'s have been realized by using intersubband absorption in $GaAs/Ga_xAl_{1-x}As$, $InGaAs/Al_xGa_{1-x}As$ and Si_xGe_{1-x} / Si MQW systems, by using interband absorption in $InAs/Ga_{1-x}In_xSb$ superlattices and also by using $HgTe/CdTe$ superlattices. The intersubband absorption detectors are comparatively well developed and have reached the application stage. We shall discuss only this detector. Interested reader may consult Reference 10.1 and 10.2 for a more detailed discussion on the QWIP's.

10.1.1. PRINCIPLE OF OPERATION

The width of the wells and the composition of the barrier material are so chosen for a QW intersubband detector that the wells have two energy levels separated by the energy of a photon to be detected. The upper energy level is arranged to be either a resonant level in the continuum or a level just below the barrier level in the presence of the applied voltage as shown in Fig. 10.1. The wells are heavily doped n-type to a concentration of about $10^{18}/cm^3$. The incident radiation excites electrons from the lower ground state to the upper level and the electrons so excited are transported out of the well freely or by tunneling due to the applied field.

The performance of the detector depends on three factors: the absorption of the incident radiation, the current produced by the excited carriers, dark current and noise. These factors are discussed below.

Absorption of incident radiation

Intersubband absorption has been discussed in Section 5.1.4 for quantum wells with barriers of infinite height. The barrier height for the wells in experimental detectors are, however, finite, which cause extension of the envelope functions into the barrier region. The effective mass m^* is a function of z as a consequence and the matrix element for transitions are modified. The modification may be worked out by using the envelope function equation after including the vector potential operator $\mathbf{A}_{op}$ arising from the incident radiationThe modified equation is given below

$$\{(1/2)(-i\hbar\nabla + |e|\mathbf{A}_{op})[1/m(z)](-i\hbar\nabla + |e|\mathbf{A}_{op}) + V_p(z)\}F(\mathbf{r}) = EF(\mathbf{r}). \quad (10.1)$$

Proceeding as in Section 5.1 and neglecting the sqare of $\mathbf{A}_{op}$ we get for the matrix element for transitions from level 1 to level 2,

$$M_{12} = -i|e|A_0 a_z (E_2 - E_1) < z > /2\hbar, \quad (10.2)$$

where $\mathbf{A}_{op}$ has been replaced by $a_z A_0 \exp(i\kappa.\mathbf{r})$. E_2 and E_1 are the energies corresponding to the two levels and a_z is the component of the polarization vector

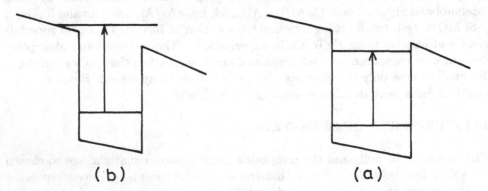

Figure 10.1.   Energy levels in quantum well infrared photodetector (QWIP). (a) Bound-to-bound QWIP. Both the lower and the upper levels are bound. (b) Bound-to- continuum QWIP. The lower level is bound, but the u pper level is in the contnuum.

in the z-direction. Using this matrix element in the expression for the absorption coefficient we get

$$\alpha = C_{\text{abis}} f_0 a_z^2 (\Gamma/2\pi)[(\omega - \omega_0)^2 + (\Gamma/2)^2]^{-1}, \qquad (10.3)$$

where

$$C_{\text{abis}} = (\pi e^2 \hbar / 2 L_w \epsilon_0 n_r c m_{cw}^*)(N_{s1} - N_{s2}), \quad f_0 = (2m_{cw}^* \omega/\hbar) < z >^2 . \qquad (10.4)$$

L_w is the width and m_{cw}^* is the electron effective mass of the well. All other symbols are the same as defined for Eq. (5.). The oscillator strength for the transitions from E_1 to E_2 is f_0 . It should be mentioned that when the matrix element is expressed in terms of the dipole moment $|e| < z >$, the effective mass is eliminated from C_{abis} . However, it reappears when α is expressed in terms of f_0 . If the mass in f_0 is taken[10.3] as m_0 , then C_{abis} has the factor $1/m_0$. On the other hand, when m_{cw}^* is used to define f_0 , as is usually done[10.2] , C_{abis} has the factor $1/m_{cw}^*$. Values of f_0 quoted in the literature are required to be interpreted accordingly. We note that the half-width Γ in the above expression is required to be chosen to fit the experimental curves. However, with properly chosen values of Γ, experimental curves are explained well as shown in Fig. 10.2

The oscillator strength f_0 depends on the structure of the detector. Its value is 0.96 for transitions from the ground level to the next higher level in a well with infinite barrier height. For wells with finite barrier height, $< z >$ is required to be evaluated by using the solutions $F_1(z)$ and $F_2(z)$. The solution for $F_n(z)$ may be generally written as,

$$\begin{aligned}
F_n(z) &= A \sin(n\pi/2 - k_w z), \quad \text{for } -(L/2) < z < (L/2) \\
&= A \sin(n\pi/2 + k_w L_w/2) \sin\{k_b[(L_b + L_w)/2 + z]\} \\
&\quad \times [\sin(k_b L_b/2)]^{-1}, \quad \text{for } -(L_b + L_w)/2 < z < -L_w/2 \\
&= A \sin(n\pi/2 - k_w L_w/2) \sin\{k_b[(L_b + L_w)/2 - z]\} \\
&\quad \times [\sin(k_b L_b/2)]^{-1}, \quad \text{for } (L_w/2) < z < (L_b + L_w)/2. \qquad (10.5)
\end{aligned}$$

$$\begin{aligned}
A &= (2/L_w)^{1/2}\{1 + (-1)^{n-1} \sin(k_w L_w)/k_w L_w \\
&\quad + [\sin^2(n\pi/2 - k_w L_w/2)/(\sin^2 k_b L_b/2)][L_b/L_w - \sin(k_b L_b)/k_b L_w]\}^{-1} \\
k_w^2 &= (2m_{cw}^*/\hbar^2)(E - E_{cw}), \quad k_b^2 = (2m_{cb}^*/\hbar^2)(E - E_{cb}) \qquad (10.6)
\end{aligned}$$

where E_{cw} and E_{cb} are the bandedges respectively of the well and the barrier layer. The width of the well is L_w and the total width of the barrier layer is L_b ($L_b/2$ being the width on each side). The energy eigenvalues are obtained from the equation,

$$\tan[k_w L_w/2 + (1 + (-1)^n)\pi/4] = r \cot(k_b L_b/2), \qquad (10.7)$$

$$r = k_b m_{cw}^* / k_w m_{cb}^*. \qquad (10.8)$$

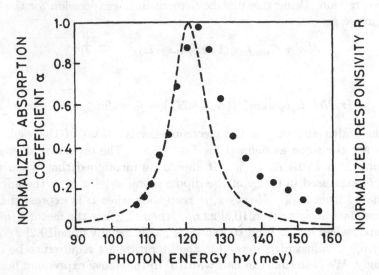

Figure 10.2. Absorption spectrum for a bound-to-bound detector. The dashed line is a Lorentzian fit to the experimental curve. [After B. F. Levine, *J. Appl. Phys.* **74**, R1 (1993); Copyright: American Institute of Physics].

We note that the level corresponding to E_n is bound if $E_n < E_{cb}$, since k_b is then imaginary and $F(z)$ decays in the barrier. Equation (10.7) giving the energy levels then reduces to Eq. (4.8). The transition is from a bound level to a bound level $(b - b)$, when both E_1 and E_2 are smaller than E_{cb}. The absorption is given in this case by the delta function or the Lorentzian function as in wells with infinite barrier. But, the value of α is reduced due to the reduced value of f_0. Values of f_0 have been computed[10.2] for the GaAs/GaAlAs, GaInAs/AlInAs systems. The value is approximately 0.4.

For well widths less than a critical value there is only one bound level E_1. Transition may, however, occur to levels in the continuum ,$(b - c)$, for which $E_n > E_{cb}$. The absorption spectrum[10.3] for $(b - c)$ transition is illustrated in Fig. 11.3. Absorption commences when the photon energy is equal to the difference between the barrier band-edge energy, E_{cb}, and the bound level energy, E_1. As the photon energy is increased, it sharply rises to a peak value and then slowly falls to near-zero value for large photon energies. The nature of the absorption spectrum is essentially determined by the variation of f_0 or $< z >$ with the frequency, ω, of radiation. Computed curves obtained by using the expressions of Eq. (10.3) agree with the experimental results[10.4]. The nature of the curve may also be explained as follows.

Absorption is zero for photon energies $\hbar\omega < (E_{cb} - E_1)$, since there are no levels available for transition. It starts for $\hbar\omega = E_{cb} - E_1$. As ω increases

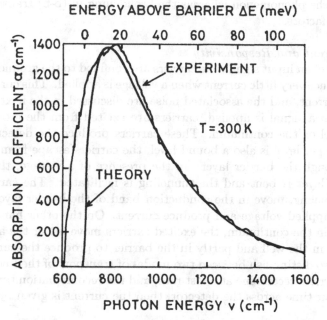

Figure 10.3. Absorption spectrum for a bound-to-continuum detector. [After B. F. Levine, C. G. Bethea, K. K. Cho, J. Walker and R. J. Malik, *J. Appl. Phys.* **64**, 1591 (1988); Copyright: American Institute of Physics].

the absorption increases, since the probability density for the electron is more and more concentrated within the well. The probability density within the well reaches the maximum value when $k_w L$ or the total phase shift is a multiple of π. The corresponding energy levels are referred as virtual levels, the energy for which are given by,

$$E_{vn} = (\hbar^2/2m^*_{cw})(n\pi/L)^2. \qquad (10.9)$$

It may be expected that f_0 will have the largest value when $\hbar\omega = E_{vn} - E_1$. The first such value corresponds to $n = 2$. In reality, however, the peak value occurs for lower energies[10.5] Since the envelope function $F_1(z)$ corresponding to E_1 extends in narrow wells into the barrier layer, $< z >$ has the largest value when $F_2(z)$ also has non-zero value in the barrier layer. The peak absorption, therefore, occurs for photon energies less than $E_{v2} - E_1$. Experimentally, it has been found to be larger than $E_{cb} - E_1$ by a few milli electron volts, about 20 meV. The absorption, therefore, sharply rises to the peak value within about 20 meV of the cut-off photon energy. After reaching the peak value it decreases as the envelope function for the second level gradually diffuses into the barrier layer and becomes uniform like that of a free electron.

We note that the broadening of the absorption spectrum for (b-c) transitions is inherent in the phenomenon whereas the broadening in (b-b) transitions is due to extraneous factors.

Detector current and Responsivity

In the absence of an input signal, the carriers are confined to the ground level in the wells and produce very little current when a voltage is applied. This current, the so-called dark current, and the associated noise are discussed in the next subsection. However, when a signal is applied, carriers are excited from the ground level to the upper level or the continuum. These carriers produce the detected current. In case, the upper level is also a bound level, the carriers escape from the well by tunneling through the barrier layer. In the presence of a voltage, the bandedge of the barrier layer is bent and the tunneling is facilitated. The carriers, which escape by tunneling, move in the conduction band of the barrier layer under the action of the applied voltage and produce current. On the other hand, when the upper level is in the continuum, the excited carriers move under the action of the voltage partly in the well and partly in the barrier to produce the current.

We may also distinguish between two modes of transport of the free carriers. If the excited carriers recombine at a fast rate, and the recombination time is smaller than the transit time across the detector, then the current is given by,

$$I_p = A_r |e| n_p, \tag{10.10}$$

where n_p is the number of photoexcited carriers exiting per unit time per unit area at the output end. A_r is the surface area of the detector. We note that for signal power P, incident per unit area on the detector, the number of photo excited carriers per unit time per unit area in one well is

$$n_{ex} = (1/\hbar\nu)P\cos\theta(1 - R)(1/2)[1 - \exp(-p\alpha L_w)], \tag{10.11}$$

where ν is the incident signal frequency, θ is the angle of the direction of propagation of the signal with the normal to the surface of the well inside the well, R is the reflection coefficient of the detector surface, p is the number of passes through the well, α is the absorption coefficient given by Eq. (10.3) and L_w is the width of a single well. The factor of $(1/2)$ is introduced to account for the fact that for unpolarized input signal only the component in the plane of incidence is absorbed. If now, p_e be the probability of escape from the well and there be N_w wells, then the number of electrons reaching the output end per unit area

$$n_p = \sum_{n=1}^{N_w} n_{ex} p_e \exp(-n\tau_{tr}/\tau_r), \tag{10.12}$$

where p_e is the probability of escape of the excited electrons from the well, τ_{tr} is the transit time of the escaped electrons across one period, $L_p = (L_w + L_b)$ and τ_r is the recapture or recombination lifetime. Wells are counted from the exit end.

As electrons exit from the output end, other electrons exit from the opposite end contact to maintain charge neutrality of the detector. If τ_{tr} be small or the end contact be made directly to the well in single-well detector these injected carriers recombine before travelling to any significant distance to contribute to the current. The current is contributed in this case by the primary excited electrons given by Eq. (10.10). and it is given by

$$I_p = A_r|e|p_e\eta_a P \cos\theta \sum_{n=1}^{N} \exp(-n\tau_{tr}/\tau_r), \qquad (10.13)$$

where $p\alpha L_w$ has been assumed to be much smaller than τ_r and

$$\eta_a = (1/h\nu)(1 - R)(1/2)p\alpha L_w. \qquad (10.14)$$

η_a is defined as the quantum efficiency. Performance of a detector is characterized by responsivity which is defined as

$$R_p = I_p/A_r P \cos\theta. \qquad (10.15)$$

Responsivity for this mode of transport is

$$R_p = |e|\eta_a p_e \sum_{n=1}^{N_w} \exp(-n\tau_{tr}/\tau_r). \qquad (10.16)$$

It should be noted that τ_{tr} may be taken to be inversely proportional to the drift velocity v of the carriers, which for a voltage V is given by,

$$v = \mu(V/L)[1 + (\mu V/Lv_s)^2]^{-1/2}, \qquad (10.17)$$

v_s being the saturation velocity, and μ the mobility. The responsivity, therefore, increases with the increase in voltage through the exponential factor.

The other mode of carrier transport is effective for $\tau_r > N_w\tau_{tr}$. For this case, the injected electrons also contribute to the current. In effect, the excited carriers travel around the detector a number of times before being recaptured. The steady- state free carrier concentration n is given by the condition that the rate of recapture is equal to the rate of excitation and escape per unit volume

$$n/\tau_r = n_{ex}p_e/L_w, \qquad (10.18)$$

n_{ex} being the number of electrons excited from each well per unit area. The current in this case is given by

$$I = A_r n|e|v = A_r(N_w n_{ex}\tau_r p_e/N_w L_w)|e|v, \qquad (10.19)$$

v being the drift velocity of the free electron.

The factor $v\tau_r/N_w L_w$ is defined as the optical gain g; it gives the increase in current due to the circulation of the excited carriers. The value of g for some designs[10.6] may be as large as 8.1. The gain is often expressed as

$$g = v\tau_r/N_w L_w = (1/N_w)(\tau_r/\tau_{\text{tr}}) = (1/N_w)(1/p_c), \qquad (10.20)$$

where p_c is the capture time, being equal to τ_{tr}/τ_r , i.e., the ratio between the transit time across well and the recapture life time, Replacing v by (17), we get for the responsivity is in this case,

$$R = |e|\eta_a p_e \tau_r (1/L_w)(\mu V/L)[1 + (\mu V/Lv_s)^2]^{-1/2}. \qquad (10.21)$$

Expression (10.21) gives the dependence of the performance characteristics on the different factors. The variation of responsivity with the bias voltage V is illustrated in Fig. 10.4 for two QWIP's[10.7]. The curve E is for a GaAs/Ga$_{0.74}$Al$_{0.26}$As MQWIP with 50 Å wells and 500 Å barriers, while curve F is for a MQWIP in which L_w =50 Å, but the barrier consists of a 50 Å Ga$_{0.7}$Al$_{0.3}$As layer followed by a 500 Å Ga$_{0.74}$Al$_{0.26}$As layer. The responsivity for the detector E is initially near zero, then increases almost linearly and finally saturates. The near-zero

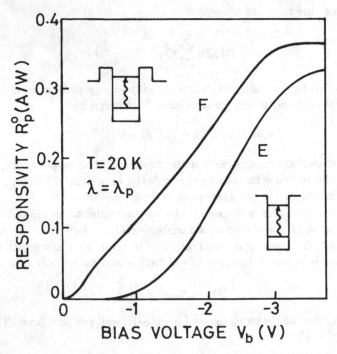

Figure 10.4. Responsivity vs. bias voltage for two QWIP's. (a) Bound-to-bound. (b) Bound-to-quasicontinuum. Insert shows the energy levels in the QWIP's.[After B. F. Levine, *J. Appl. Phys.* **74**, R1 (1993); Copyright: American Institute of Physics].

resonsivity for low voltages is due to the small value of the tunneling probability and hence 0f p_e. The rest of the curve is in accordance with the variation of the carrier velocity with the voltage as given by (10.17). The responsivity of the detector F increases almost linearly from low voltages and finally saturates. The barrier being very thin, excited electrons escape from very low voltages to produce current and the current varies with the voltage like V.

Expression (10.21) for the responsivity also shows that its value will increase with increase in the value of p_e and η_a . Since α is directly proportional to the carrier density the responsivity should increase with increase in the doping. Also, in bound-to-bound QWIP's, R_p should increase with decrease in the barrier width, which enhances the value of p_e . Optimization of these two factors is controlled by the dark current and the associated noise, discussed below.

Dark current and Noise

Dark current, as mentioned earlier, is the detector current in the absence of the infrared radiation . It is required to be as low as possible as its magnitude determines the sensitivity of the detector. The current is produced by the tunneling of the carriers out of the ground state in the well. It is given by

$$I_D = n^* |e| v A_r,\qquad(10.22)$$

where v is the velocity of the escaped electrons and the density of such electrons is for bound-to-continuum QWIP,

$$n^* = (m^*/\pi\hbar^2 L_p) \int_{E_1}^{\infty} f(E,T) T_u(E,V) dE,\qquad(10.23)$$

where m^* is the effective mass in the well, $f(E)$ is Fermi distribution function and $T_u(E_1,V)$ is the tunneling probability for an applied voltage V and electron energy E_1. For bound-to- bound QWIP an additional integral extending from E_2 to ∞ is to be included.

We may conclude from Eq. (10.23) that the dark current will be strongly temperature dependent. For example, if we idealize $T_u(E,V)$ as $T_u = 0$ for $E < (E_b - E_1)$ (E_b is the barrier layer bandedge) and $T_u = 1$ for $E > (E_b - E_1)$, we get

$$n^* = (m^* k_B T/\pi\hbar^2 L_p) \exp[-(E_b - E_1 - E_F)/k_B T],\qquad(10.24)$$

E_F being the Fermi-energy corresponding to the doping density N_D . Calculations[10.8] show that the current changes by a factor of about 10^8 as the temperature is raised from 4 K to 77 K. The current will also depend strongly on the width of the barrier layer, since the tunneling probability decreases exponentially with the width of the layer. The thickness of the barrier layer in bound-to-bound QWIP is made about 100-150 Å, so that the escape probability for the excited carriers is not too small. On the other hand, the thickness of the barrier layer may be made

500 Å or more in bound-to-continuum QWIP without reducing significantly the value of p_e or g. The five-fold increase in the barrier layer thickness reduces the dark current for these QWIP's by a factor of 10^4 -10^{11} over the temperature range 4-80 K. The difference is accentuated as the operating voltage is also much smaller in these detectors, the excited electrons being not required to tunnel out.

The current flow through the detector is associated with shot noise. The root mean square shot-noise current is

$$i_n = (\overline{i_n^2})^{1/2} = (4|e|I_D g\Delta f)^{1/2}, \tag{10.25}$$

where I_D is the dark current.

The current is magnified by the optical gain,g. The bandwidth Δf is taken as 1 Hz for characterization, so that the unit of i_n is $A/Hz^{1/2}$.

The net performance of the detector is specified by detectivity, defined as

$$D = R_p(A_r\Delta f)^{1/2}/i_n, \tag{10.26}$$

in units of cm.$Hz^{1/2}$ /W, unit of R_p being A/W.

Optimization of a detector structure is aimed at maximizing R_p and D. The effect of barrier width on R_p and I_D has been discussed. R_p may be increased also by increasing the absorption coefficient α, which is proportional to the carrier density and hence to the doping of the well. The increased doping enhances also the the dark current and thereby mitigates a part of the increase in D. The doping dependence of D may be understood by considering the bound-to-continuum QWIP. The dark current is proportional to n^* ,which may be expressed in terms of the doping concentration N_D as follows. Assuming all the donors to be ionized, we may express N_D as,

$$N_D = (m^* k_B T/\pi\hbar^2 L_w)\ln[1 + \exp(E_F/k_B T)] \tag{10.27}$$

Hence,

$$n^* = (n_0 L_w/L_p)[\exp(E_1 - E_v)/k_B T].[\exp(N_D/n_0) - 1], \tag{10.28}$$

where

$$n_0 = m^* k_B T/\pi\hbar^2 L_w \tag{10.29}$$

The dark current, I_D , therefore, increases exponentially with N_D and $D \propto \rho/[\exp(\rho) - 1]$, where $\rho = N_D/n_0$. Consequently, the detectivity initially increases with N_D , attains a peak value and then decreases with further increase of N_D . The peak is, however, very broad. Computations show that the detectivity changes by a factor of 2 for a change of N_D by a factor of 30.

It is also of interest to note that the detectivity is relatively insensitive to the magnitude of the bias voltage V, since for low voltages, I_p as well as $(I_D g)^{1/2}$ vary as V. However, at high biases I_D increases faster than V. On the other hand, the reduction in the value of p_e for very low voltages reduces the value of D. The detectivity as a result has a broad maximum for moderate operating voltages.

10.1.2. EXPERIMENTAL RESULTS

The first QWIP was demonstrated for a wavelength of 10.9 μm by Levine et al[10.9] following the exploratory work on intersubband absorption by West and Eglash[10.3]. A multiquantum well system consisting of fifty 65 Å GaAs wells, doped 1.4×10^{18} cm $^{-3}$, and 95 Å $Al_{0.25}Ga_{0.75}As$ barriers was used. The detector was constructed by providing ohmic contacts to the top and bottom n GaAs covering layers as shown in Fig. 11.5. Also, to ensure optical coupling, radiation was made incident from the back by polishing the substrate at 45° angle. The responsivity of the detector was 0.52 A/W. The value of p_e and $v\tau_r$ were estimated to be 60 % and 2500 Å. The dark current of the device was however, very large Subsequently the performance of the detector was improved[10.10] by increasing the barrier thickness to 95-140 Å and the barrier height by using $Al_{0.36}Ga_{0.69}As$. The value of R_p was 1-9 A/W and I_D was 2×10^{-4} A at 15 K.

Further improvement in the performance characteristics was achieved[10.8] by using thinner quantum wells, so that the transitions occurred from the bound state to the continuum. Fifty GaAs quantum wells 40 Å thick, (doped 1.4×10^{18} cm^{-3}) and 305 Å $Al_{0.29}Ga_{0.71}As$ barriers were used. The cut-off wavelength was 8.4 μm . The dark current was about 10 A at 77 K and less than 10^{-12} A at 15 K. Another structure with increased barrier width of 480 Å and composition of $Al_{0.25}Ga_{0.75}As$, had a cut-off wavelength of 10.7 μm and gave dark current of about 10^{-5} A at 77 K and $<10^{-12}$ A at 15 K.

Extensive work[10.1] has been reported, in which the well width was varied

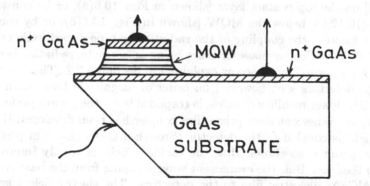

Figure 10.5. Structure of an AlGaAs/GaAs QWIP. The shaded region is n^+ GaAs. The GaAs substrate is polished at an angle of 45° so that normally incident light is incident on the detector at 45°. The hatched area is the MQW.[After B. F. Levine, *J. Appl. Phys.* **74**, R1 (1993); Copyright: American Institute of Physics].

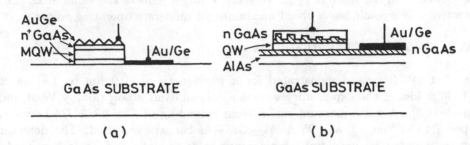

Figure 10.6. QWIP's with grating and waveguide. (a) Schematic diagram showing a detector with a grating on top. The normally incident signal is reflected at angles with the normal to the grating, which is then optically coupled to the well. [After G. Hasnain, B. F. Levine, C. G. Bethea, R. A. Logan and J. Walker, *Appl. Phys. Lett.* **54**,2515 (1989)]. (b) Schematic diagram showing a detector with a waveguide formed by the QW's and the bottom and the top n-doped GaAs layer. [After J. Y. Andersson, L. Lundqvist and Z. F. Paska, *Appl. Phys. Lett.* **58**, 2264 (1991)]. Copyright: American Institute of Physics.

from 40-70 Å, the barrier width was about 500 Å, the doping was raised between 0.3 -2×10^{18} cm^{-3}, and the value of x in Al$_x$Ga$_{1-x}$As was also varied between 0.1-0.3. These covered the wavelength range 7.5 to 16.6 μm. The peak responsivity at 20 K was in the range 0.4-0.7 A/W. The detectivity varied from 10^{10} cm.Hz$^{1/2}$/W at 77 K to 10^{13}cm.Hz$^{1/2}$/W at 35 K for a 10.7 μm detector and form 10^9 cm.Hz$^{1/2}$ /W at 50 K to 10^{12}cm.Hz$^{1/2}$/W at 20 K for a 19 μm detector.

Experiments were conducted on various kinds of structures, which are described in Reference 10.2. Optical coupling has been improved by using either a grating[10.10,11] on the top contact layer [shown in Fig. 10.6(a)] or by constructing a waveguide[10.12,13] below the MQW [shown in Fig. 10.6(b)] or by constructing both. To improve the coupling of the radiation, p-type doped wells were also used[10.14,15]. The effective mass for holes being anisotropic, radiation is coupled to holes even for normal incidence as explained in Section 5.2.2. The responsivity of the p-type detectors was, however, an order of magnitude lower than that of n-QWIP's. The lower mobility of holes, is responsible for the poorer performance.

Recently, an evanescent wave prism film coupler has been developed [10.16] in which the light is coupled to the detector through a GaAs thin film prism .The angle of the prism is so chosen that the incident light is totally internally reflected from the base. But, the evanescent wave escaping from the base is coupled through a AlGaAs dielectric film to the detectors. The electric field component of the evanescent wave being normal to the bottom surface of the prism , light interacts effectivelyt with the electrons in the QWIP. It is reported that a tenfold increasein electron photon coupling may be realized by this technique in comparison to optimal grating coupling systems.

Table 10.1 Characteristics of Quantum Well Infrared Photo detectors (QWIP).
λ-Wavelength.$\Delta\lambda$-Spectral width.

Matarial	Well width(Å)	Barrier width(Å)	Number of wells	Wave-length(μm)	$\Delta\lambda/\lambda$ %	Responsivity(A/W)
GaAs	50	500	25	7.5	11	0.35
GaAlAs[a]	40	500	50	9.8	20	0.5
	60	500	50	13.2	19	0.7
	70	500	50	16.6	28	0.7
InGaAs/ InAlAs[b]	50	300	50	4	33	0.025
InGaAs/ InP[c]	50	500	20	8	-	6.5
GaAs/ GaInP[d]	40	300	10	8	53	0.34
GaAs/ AlInP[e]	30	500	20	3	12	-
InGaAs/	80	500	5	15.3	50	0.63
GaAs[f]	50	600	10	16.7	18	0.29

a - Reference 10.12, b - Reference 10.17, c - Reference 10.18,
d - Reference 10.19, e,f - Reference 10.1

Materials other than GaAs/GaAlAs have also been used [10.17-20] to construct QWIP's of different wavelengths. Typical results are summarized in Table 10.1. Some results for the GaAs/AlGaAs system are also included for comparison.

10.1.3 QUANTUM DETECTOR SYSTEMS UNDER EXPERIMENTATION

Quantum well infrared photodetectors (QWIP's) have been realized by using both p and n type $Si_{1-x}Ge_x$/Si strained-layer MQW structures. Because of the anisotropic electron effective mass, n-type as well as p-type detectors work on normal incidence. The estimated value of detectivity is 1×10 cm.Hz$^{1/2}$/W at 77 K for 9.5 μm wavelength in the structures on which experiments have been conducted[10.21].

The $InAs$/$Ga_xIn_{1-x}Sb$ heterojunctions have also been considered for detectors. The junctions are of type II; the conduction band and the valence band edges of InAs are both lower than those of $Ga_xIn_{1-x}Sb$. Consequently electrons are localized in the InAs layer, while holes are localized in the $Ga_xIn_{1-x}Sb$ layers. Separation between the conduction band electron energy levels of InAs and the valence band hole energy levels may be made to be such as to absorb 10 μm radiation. Also, by properly choosing the thickness of the barrier layers the electron and hole wave functions may be made to overlap so that electrons may be excited from the valence band to the conduction band by absorbing photons. Some preliminary studies

have been made[10.2] on this structure, which show that the optical properties of
$InAs/Ga_xIn_{1-x}Sb$ detectors may be comparable to those of HgCdTe.

Another quantum well structure currently being studied for QWIP is the
HgTe/CdTe superlattice. It is expected that some of the drawbacks of HgCdTe
detectors, e.g., requirement of stringent uniformity, large tunneling currents, may
be eliminated by using the superlattice. Theory predicts that in HgTe/CdTe superlattice detector, wavelength will be easier to control, tunneling current will be
reduced and operating temperatures may be reasonable. In comparison to other
QWIP's, this material has the advantage of polarization-independent absorption
and large absorption coefficient. QWIP using HgTe/CdTe superlattice has been
demonstrated[10.22] for the 4.5 μm wavelength. Work is also being continued for
extending the wavelength to longer values. However, problems associated with the
growth of good quality structures are not yet fully resolved.

Extensive work has been done on the so-called double-barrier QWIP (DBQWIP),
in which a barrier layer is introduced within the well layer to divide it into a broad
and a narrow well. There are two energy levels below the barrier in the broad
well and one level in the narrow well , close to the top level of the broad well
(see Fig. 10.7). The broad well is doped while the narrow well is undoped. In
the absence of light electrons tunnel sequentially through the nartrow well, while
under illumination the excited electrons tunnel coherently through the narrow
well. As a result, the DBQWIP's have good responsivity but low dark current.
Such detectors have been constructed with GaAs/AlAs/AlGaAs [10.1] and Al-
GaAs/AlAs/InGaAs [10.21-23].

The tchnology of GaAs/AlGaAs QWIP's have been so perfected that focal
plane arrays may be realized to detect light of wavelengths 6 to 25 μm [10.26-
30]. Arrays having dimensions of 128×128, 256×256 and 640×486 have been
constructed and used in the focal plane of snap-shot cameras. Efforts have been
made to reduce the dark current by using bound to quasi-continuum (the upper
level is just below the continuum) structure in place of the bound-to- continuum
structure.

Although the technology of QWIP's have matured for commercial exploitation , work is being continued for further improvement[10.25,31,32].

10.2. Quantum Well Modulator

Quantum well modulators have been realized by utilizing the field-dependence of
the excitonic absorption in quantum wells. We have presented in Fig. 10.7 the
absorption spectrum for a GaAs/AlGaAs quantum well system for a voltage V,
applied across it. The spectrum has a peak for $V = 0$ near about 990 nm. This
peak shifts to higher wavelengths, i.e., it is red- shifted by the applied voltage.
The magnitude of the peak also decreases with increase in the value of V.

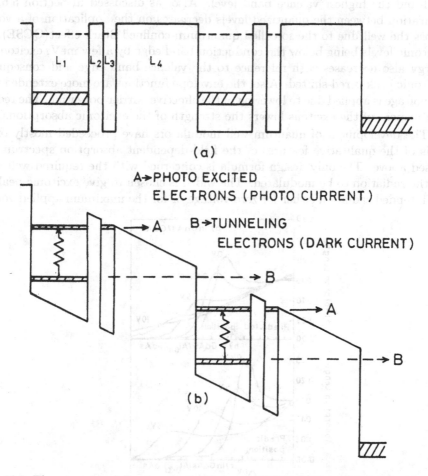

Figure 10.7. The energy band diagram of a double barrier QWIP. (a) Potential distribution with no bias.(b) Potential distribution under forward bias.L_1 - 72 Å GaAs well doped to $1.0 \times 10^{18} \text{cm}^{-3}$. L_2- 39 Å $Al_{0.33}Ga_{0.67}As$ undoped barrier.L_3-18 Å undoped GaAs well. L_4- 152 Å undoped $Al_{0.433}Ga_{0.67}As$ barrier.

Excitons and excitonic absorption have been discussed in Section 5.3 and 5.5.2. The nature of the experimental excitonic absorption curves may be explained on the basis of the theory presented in these sections. We recall that the binding energy of excitons in quantum wells is ideally eight times that for bulk materials (see Section 5.3). Excitons are, therefore, excited at room temperature and produce the peaks in the absorption below the band edge, which in the case of quantum wells correspond to the separation between the lowest quantized conduction band level and the highest valence band level. Also, as discussed in Section 5.6, this separation between the quantized levels decreases on the application of a voltage across the well due to the so-called quantumm-confined Stark effect(QCSE). The excitonic levels being below the conduction band edge by a few meV, exciton level energy also decreases with reference to the valence band edge and consequently excitonic peak is red-shifted. Also, the envelope functions are more extended when the voltage is applied due to lowering of the effective barrier potential. The reduced confinement of the excitons lowers the strength of the excitonic absorption.

The development of quantum well modulators have proceeded mostly on the basis of the qualitative features of the field-dependent absorption spectrum, discussed above. The only design formula is concerned with the required well width for the radiation to be modulated. The width is chosen to give excitonic peaks for the intended radiation either for zero voltage or for the maximum applied voltage.

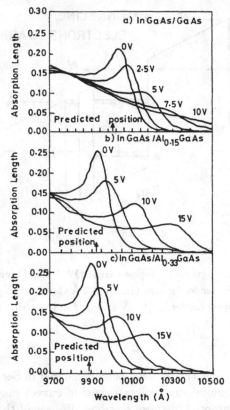

Figure 10.8. Absorption spectrum for InGaAs/GaAlAs quantum wells. Samples consisted of twenty 75 Å $In_{0.2}Ga_{0.8}As$ wells separated by 200 Å $Ga_{1-x}Al_xAs$ in $p-i-n$ diode configuration. [After B. Pezeski, S. M. Lord and J. S. Harris,*Jr. Appl. Phys. Lett.* **59**, 888 (1991); Copyright: American Institute of Physics].

The barrier height and width are also chosen to ensure maximum confinement and maximum interaction with the incident radiation. We review below the characteristics and structures of modulators that have been developed. Three basic forms of modulators have been developed as described in the following subsections.

10.2.1. TRANSVERSE TRANSMISSION MODULATOR

A transverse transmission modulator was first demonstrated[10.33] using the structure shown schematically in Fig. 10.9(a). An MQW structure consisting of 50 intrinsic 90 Å GaAs wells, was sandwiched between two intrinsic superlattice buffer layers in order to improve the quality of the quantum wells. The buffer layers were clad by p and n type conducting superlattices, followed by p and n-$Al_{0.32}Ga_{0.68}As$ layers. The complete structure was grown on GaAs substrate. Contacts were provided on the top and the bottom side by metallization. Holes were etched through the metal contacts and the GaAs substrate for the input and output light. Dimensions of different layers are indicated in the figure.

Light of wavelength matching the exciton lines is modulated by applying a reverse bias voltage to the diode. The field appearing across the MQW, causes red shift of the exciton lines and the absorption of the incident light is altered

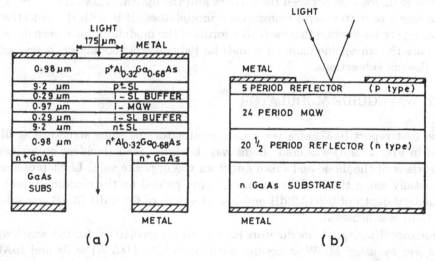

(a)   (b)

Figure 10.9. Schematic diagram of a quantum well modulator. (a) Transverse transmission modulator. Radiation is incident through the top hole and is transmitted through the MQW system and the bottom hole in the substrate. Voltage is applied between the nmetallized contacts on the top p^+ AlGaAs and the bottom n^+ AlGaAs to modulate the transmission. (b) Fabry-Perot modulator. Light is incident on the top through the hole in the metallized contact. The reflected light is modulated by applying a voltage between the top p-type reflector and the bottom n-type reflector through the surface.

as a consequence. The intensity of the emergent light can thus be modulated by varying the bias voltage. The characteristic of the modulator is often indicated by the contrast ratio, which is defined as the ratio of the maximum and the minimum intensity of the emergent light. The particular modulator described above had a contrast ratio of 2.

The structure was made simpler in later versions of the modulator [10.34] by using GaAs/AlGaAs. The diode consisted of an n-type GaAs substrate, a 1 μm thick n-type $Al_{0.3}Ga_{0.7}As$ layer, a 60-well GaAs/AlGaAs MQW structure, a 0.5 μm thick p-type $Al_{0.3}Ga_{0.7}As$ layer and a 50 Å p^+ GaAs layer, grown successively. Experiments have been carried out to optimize the composition and the thickness of the barrier layers and it has been concluded that optimum normally-on operation is obtained by using 20-30 Å AlAs barriers with 83-91 Å GaAs wells.

Transverse transmission modulators have been realized[10.35,36] also by using $In_xGa_{1-x}As/GaAs$ and $In_xGa_{1-x}As/Al_{0.15}Ga_{0.85}As$ MQW structures. These are grown on GaAs substrates with p^+ and n^+ GaAs cladding layers, and operate at wavelengths of 1.02 to 1.07 μm and 980 nm. The structures do not require removal of the GaAs substrate as it is transparent to these radiations. This is of advantage in using these modulators in many-device systems.

Performance characteristics of the modulators may be improved by arranging for greater interaction between the carriers and the optical signal, than is possible in the single pass transverse transmission modulators. It is with the objective of increasing the interaction that two other forms of the modulator have been devised. These are the waveguide modulator and the Fabry-Perot modulator, described in the following subsections.

10.2.2. WAVEGUIDE MODULATOR

In the first report[10.37] of a waveguide modulator the same structure as illustrated in Fig. 10.9(a) was used. Light was, however, made incident on a cleaved side surface of the diode and taken out from the opposite side. Light propagates horizontally along the MQW structures, being guided by the cladding layers. A modulation depth of 9.2-10.2 dB and insertion loss of 6± 2 dB (3 dB was due to reflection) was achieved.

Horizontal waveguide modulators have been realized[10.38] for the wavelength of 1.5 μm by using MQW structures with InGaAlAs (186 Å) wells and InAlAs (50 Å) barriers. The device consisted of an n-InP substrate, a p-InAlAs layer, a p-InGaAs layer grown consecutively. Modulation was done by applying a reverse bias to the metal contacts on the two surfaces. The modulator had contrast ratio of 20 dB and a modulation bandwidth of 40 GHz.

In a further development of the device[10.39] an integrated modulator and amplifier has been realized on InP to operate at 1.55 μm, with a contrast ratio of 10.5 dB and effectively no insertion loss.

10.2.3. FABRY-PEROT MODULATOR

A Fabry-Perot(FP) modulator is also essentially a $p - i - n$ diode, the intrinsic region being made of MQW's. The MQW layers are, however, sandwiched between two Bragg reflectors made of alternate quarter wave GaAs and AlGaAs layers. For example, a FP modulator has been reported[10.40] in which the MQW structure consists of 100 Å /100 Å GaAs/Al$_{0.2}$Ga$_{0.8}$As layers. This structures is sandwiched between a 5-period quarter-wave Bragg reflector at the top and a $20\frac{1}{2}$ period Bragg reflector at the bottom. Reflectors are made of 725 Å AlAs and 625 Å Al$_{0.2}$Ga$_{0.8}$As layers. The structure is schematically illustrated in Fig. 10.9(b).

The total reflectivity R_T, of a Fabry-Perot cavity may be expressed as[10.41]

$$R_T = \{[R_f^{1/2} - R_b^{1/2}\exp(-\alpha)]^2 + 4R_f^{1/2}R_b^{1/2}\exp(-\alpha)\sin^2\phi\}$$
$$\times\{[1 - R_f^{1/2}R_b^{1/2}\exp(-\alpha)]^2 + 4R_f^{1/2}R_b^{1/2}\exp(-\alpha)\sin^2\phi\}^{-1}, \quad (10.30)$$

where R_f and R_b are the reflectivities of the front and the backside mirrors respectively, α is the absorption and ϕ is the phase shift of the light beam in one pass.

The resonant frequency of the cavity corresponds to $\sin\phi = 0$; it is determined by the length of the cavity and the permittivity of the material of the cavity. The reflectivity depends on the relative values of R_f, R_b and $\alpha.R_f$ is made small, of the order of 50 %, while R_b is made as close to unity as possible. The front quarter-wave Bragg mirror is, therefore, constructed with 5 or fewer periods and in some designs the semiconductor/air interface constitutes the front mirror. On the other hand, the back Bragg mirror is made of 20 or more periods. The device may be made to work as normally-on [10.42] or normally-off[10.43] by choosing properly the operating wavelength in relation to the exciton line. Optimal performance is obtained[10.44] when the cavity is designed to operate as normally-on and the reflectivity is arranged to be maximum when the absorption, α, is minimum. As the bias voltage is increased, the exciton line shifts towards the incident signal wavelength and α increases. $R_b^{1/2}\exp(-\alpha)$ therefore, approaches $R_f^{1/2}$ and R_T decreases. As a consequence, the output light decreases in intensity as the bias voltage is increased. On the other hand, in a normally-off device, initial absorption is so arranged that $R_b^{1/2}\exp(-\alpha)$ is close to $R_f^{1/2}$ for zero bias. As the bias is increased the exciton line shifts away from the signal wavelength, α decreases and R_T increases. The signal, therefore, increases in intensity as the bias voltage is increased. An experimental reflectivity spectrum is given in Fig. 10.10 for a normally-on modulator.

It should be noted that the realization of large contrast ratios require larger interactions with the carriers, but such interaction increases the insertion loss. The design of the cavity is therefore aimed at a compromise between the values of the contrast ratio and the insertion loss.

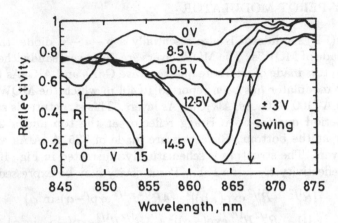

Figure 10.10. Reflection spectra for a normally-on Fabry-Perot modulator. Reflectivity is maximum for zero applied voltage and decreases with increase in the modulating voltage. [After C. C. Barron, C. J. Mahon, B. J. Thibeault, G. Wang, W. Jiang, L. A. Coldren and J. E. Bowers, *IEEE J. Quantum Electron.* **31**, 1484 (1995);Copyright (=A9 1995=IEEE)].

Fabry-Perot modulators have been realized[10.44] with GaAs/GaAlAs MQW's for the wavelength of 829 nm with an insertion loss of 1.2 dB and contrast ratio of 7.5 using an optimum value of 35 Å for the well thickness. A few other[10.45,46] GaAs/GaAlAs F-P modulators have been reported in which the integrated Bragg mirrors act as the p and n regions for the $p-i-n$ diode, the i- region being formed by the MQW structure. Contrast ratios ranging between 15 and 100 have been realized by suitably designing the mirrors and the MQW structure. Higher values of contrast ratio, 60-130, have also been realized[10.47,48] in GaAs/GaAlAs F-P modulators by replacing the MQW's by superlattice structures. The exciton line is blue shifted in superlattices due to the so- called Wannier-Stark localization[10.49-51]. In this case, due to band bending electron states change from the extended states in minibands of the superlattice to localized states, the energy gap between the hole and the electron levels increase and blue shift results. The absorption coefficient being also high and the cavity lengths relatively short, the contrast ratio has a higher value in superlattice modulators.

Fabry-Perot modulators have also been realized by using GaAs/InGaAs strained layer multi-quantum wells for the operating wavelength of 938-960 nm. The structures were grown on n^+GaAs substrate and had two n and p-doped Bragg mirrors at the bottom and the top side. The top mirror had a few periods and low reflectivity, while the bottom mirror had a large number of periods and a reflectivity larger than 99 %. The MQW structure consisting of 50 or more periods of $In_xGa_{1-x}As$/GaAs wells was sandwiched between the mirrors.

Contrast ratio of 37-66 and modulation of 8-9 dB with insertion loss of of 5.2 dB have been realized[10.52-54].

An inverted cavity GaAs/InGaAs strained layer F-P modulator has also been reported[10.41] for the wavelength of 956 Å. One side of the cavity was formed by a 5.5 period GaAs(654 Å)/ AlAs(784 Å) Bragg reflector, grown on n^+GaAs buffer layer on an insulating GaAs substrate. The other side of the cavity was formed by an undoped 90 Å GaAs layer and $p - p^+$ doped 570 Å GaAs layer covered by a metallic or dielectric mirror. The MQW structure consisted of 50 periods of $In_{0.15}Ga_{0.85}As$ (100 Å)/GaAs(125 Å) layers. As GaAs is transparent to the operating wavelength, light could be made incident through the substrate and the reflected light could also be collected from the same side. The structure therefore has the potential for quasi-hermetic hybridization with protected surfaces. The reflectors being also simpler, the processing of the modulator is less complex.

10.3. Quantum Well Switch

Switches for optic signals have been realized[10.36,54-58] by using the quantum well modulators described in the preceding section. Three different kinds of circuit have been used. These are the resistance coupled self electro-optic effect device (R-SEED)[10.54], the diode-coupled self electro-optic device (D-SEED)[10.58] and symmetric self electro- optic device (S-SEED)[10.57]. The basic device is a QW modulator, but the circuit operation becomes different because of the different loads used in the three forms. Operating principles of the devices are described below.

10.3.1. R-SEED

A resistive load with a resistance of the order of 1 MΩ is connected[10.57] in an R-SEED in series with a QW modulator and the voltage supply, as shown in the insert of Fig. 10.11(a). The modulator may be transmission type or reflection type. It is, however, so designed that for zero operating bias the absorption is maximum and the absorption decreases with increasing bias. The diode is also initially biased with a large voltage. The QW modulator therefore transmits (in case of a transmission modulator) or reflects (in case of a reflection modulator) low-intensity incident signal. It should be understood that the absorbed signal causes photocurrent to flow through the device and this current causes a drop across the load resistance. Hence, as the signal intensity is increased, the photocurrent increases, the voltage drop increases and the voltage across the diode decreases. The decreased voltage causes enhancement of absorption and at some point, this positive feedback in absorption becomes cumulative and the voltage across the diode reaches a small value. Absorption of the incident signal is then maximum,

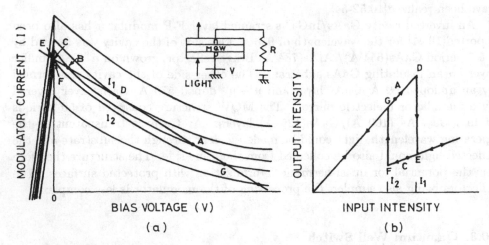

Figure 10.11. Schematic diagram of the current-voltage and input-output characteristic of an R-SEED. (a) Current-voltage characteristic. Insert shows the circuit arrangement. R is the load resistance. (b) Input-output characteristic.

giving rise to a characteristic shown in Fig. 10.11(b). The device thus acts as a self-activated switch. It is also found that, as the signal intensity is reduced after switching, the device remains in the low-output state down to a minimum input intensity. At this minimum, the reverse transition occurs and the device switches to the large-output state.

The hysteresis may be explained by drawing the load line on the current-voltage($I - V$) characteristic of the modulator for different input, as shown in Fig. 10.10(a). The current through the QW-modulator, which acts as a photo-diode, initially increases from zero for a positive voltage across it. The diode current at this voltage cancels the photocurrent. It attains the maximum value for $V = 0$ and then decreases with increasing value of V, since the absorption in the modulator decreases with increasing field as a result of the shift in exciton line due to QCS effect. The load line for a resistive load is given by ABC. For low as well as high input, the load line cuts the $I - V$ curve of the modulator at one point. The intersection point corresponds to a high value of V for low light input. The modulator is, therefore, in a high-output state for low input. For large input the intersection point corresponds to a low voltage. the modulator is then in a low-output state. But, for input of some intermediate value, there are three intersection points, A,B,C, and the circuit may settle to either A, corresponding to a high-output state or to C, corresponding to a low-output state. Hence, as the intensity of the input is increased to I_1 , a point D is reached beyond which only the low-output intersection point exists. The modulator therefore switches from the high-output state to the low-output state as the input intensity is increased beyond I_1 . On the other hand, as the input is reduced to I_2 , a point E is reached below which the load-line intersects the current-voltage characteristic at high voltage or the high-output point. Reverse switching from the low-output

to the high-output state therefore occurs as input intensity goes below I_2 . It is evident that the input-output characteristic will exhibit hysteresis as $I_2 < I_1$. The extent of hysteresis depends on the bending of the $I - V$ characteristic of the modulator, which may be enhanced by suitable design of the modulator.

It may also be noted that the presence of hysteresis in the characteristic makes the R-SEED usable as an optical bistable device.

It is also evident from Fig. 10.11(a) that values of the input switching power, I_1 and I_2, decrease with increase in the value of the load resistance R. On the other hand, the switching speed is ultimately controlled by the value of the time constant RC, C being the capacitance of the modulator. Hence, as the switching power is reduced by increasing R the switching time is increased. Experimentally,[10.54] the switching power has been varied from 250 μW to 0.5 μW by varying the load resistance from 100 kΩ to 100 MΩ; switching time was about 20 μs for 1 MΩ load resistance for which the threshold power was about 30 μW.

10.3.2. D-SEED

A diode-coupled SEED (D-SEED) is realized[11.58] by connecting a photo-diode in series with the QW-modulator and the bias voltage V as shown in the insert of Fig. 10.12(a). The modulator is so arranged that it is normally-off, e.g., the absorption is maximum for zero bias. The absorption decreases as for the R-SEED with increasing bias voltage. The current through the modulator, therefore, changes with the voltage V across it, as shown in Fig. 10.12(b) for a fixed input optic-signal. The voltage across the photo-diode is $V_{abs} - V$ and the diode being also negatively biased (the anode connected to the ground side), its current varies with V as shown by the dotted lines for different input signal. The currents of the QW modulator and the photo-diode are required to be the same in the steady state as they are in series. It is evident that this requirement may be satisfied for low and also for high input power to the photo diode by a unique voltage across the modulator indicated respectively by A and B. For point A the voltage across the modulator is small and its output is small. On the other hand, for point B, voltage across the modulator is large and its output is large. Hence the modulator remains in the low-output state when input to the photo diode is small and it remains in the high-output state when the input to the photo diode is large. The low-output state of the modulator continues up to an input I_1 to the photo-diode, for which its current touches the point D of the modulator current. The modulator switches to the high-output state as photo-diode input is increased beyond I_1 . Similarly, when the photo-diode input is reduced from a high value the high-output state of the modulator continues down to an input I_2 to the photo-diode for which its current touches the point E of the modulator current. The modulator switches back to the low-output state as the photo- diode input is lowered below I_2 . The output characteristic of the modulator therefore varies with the photo-diode input as shown in Fig. 10.12(b). The hysteresis in the characteristic extends from I_2 to I_1 . The device acts like the R-SEED as a switch and also an optical bistable device. It is also evident from Fig. 10.12(a) that the values of I_1 and I_2 depend on the power input to the modulator and the values increase as the modulator input power increases. So, larger input power will be required to control larger power in the modulator.

10.3.3. S-SEED

The working of symmetric self electro-optic device (S- SEED)[10.57] is similar to the D-SEED. In a S-SEED the photo-diode is replaced by a QW-modulator and one of the two modulators may be switched by changing the input to the other modulator. The switching modulator acts as the photo-diode and the switched modulator acts as a switch and a bistable device. The two devices, the controlled and the controller, being identical switching power is required to be the same as the controlled power.

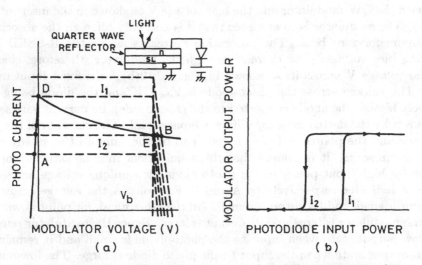

Figure 10.12. Schematic diagram of the current-voltage and the input-output characteristic of a D-SEED. (a) Current-voltage characteristic. Solid line - QW modulator. Dashed line - Photo-diode. Insert shows the circuit arrangement. (b) Photo-diode input-output characteristic.

SEEDs of different forms have been realized by using GaAs/AlGaAs, $In_{0.1}Ga_{0.9}As$/GaAs and $In_xGa_{1-x}As$/$Al_{0.15}Ga_{0.85}As$ modulators based on MQW's or SL's. Contrast ratios have been realized from 5:1 to 130:1. The switching-input powers are of the order of a few microwatts and the switching energy is of the order of a few $fJ/\mu m^2$. The switching speed varies from a few microseconds to a few nanoseconds.

10.4. Optical Bistable Device(OBD)

The SEED's described in Section 10.3 provide two output levels , switching between the two levels being controlled by the intensity of an input optical signal. Switching in these devices, however, occur via the current flow through the device. The change in current through the device with change in the input optic signal provide the feedback for switching. These devices are therefore considered as hybrid.

All-optic bistable devices are realized by using the optical nonlinearity associated with the excitonic absorption, which is very significant and continues up to the room temperature in quantum wells as discussed in Chapter 5. The absorption increases sharply as the wavelength of the incident signal approaches the exciton wavelength, reaches a maximum and then falls again to merge with the absorption for band-to-band transitions. The change in the absorption coefficient causes also a change in the refractive index of the composite structure, which may be worked out by using the Krammers-Kronig relations. These changes are illustrated schematically in Fig. 10.13. The absorption α, as well as the change in refractive index, Δn_r , depend on the intensity of the input signal since the excitons are affected by the generated carriers through various mechanisms, such as screening, phase-space filling, saturation and band renormalization. The net effect is a reduction in the magnitude of α and Δn_r with increasing intensity of the optical signal. Bistable devices are realized by utilizing these input-signal-induced changes.

The effect of the changes are enhanced by enclosing the MQW structure in a Fabry-Perot(FP) etalon. The transmitted intensity I_T , for an input intensity I_i is given for a FP etalon by the following expression,

$$I_T = I_i(1 - R)[1 - R\exp(-2\alpha L)]\{[1 - R\exp(-\alpha L)]^2 + 4R\exp(-\alpha L)\sin^2\phi\}^{-1},$$
(10.31)

where R is the reflectivity of the mirrors of the etalon, α is the absorption coefficient of the enclosed material, ϕ is the total phase shift of the signal in traversing the length L of the cavity and in the reflection. The total phase shift ϕ may be expressed as

$$\phi = (2\pi/\lambda_0)(n_r + \Delta n_r)L + \phi_r,$$
(10.32)

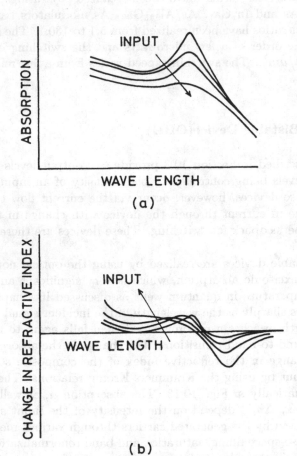

Figure 10.13. Schematic diagram showing the variation of the absorption coefficient and the refractive index with the radiation intensity near the exciton resonance. (a) Absorption coefficient. (b) Refractive index.

where λ_0 is the free-space wavelength of the incident signal, n_r is the refractive index of the material for low input, Δn_r is the change in n_r caused by the input signal and ϕ_r is the phase change introduced by the mirrors. Equation (10.31) may also be written as

$$I_T = I_i T_M / (1 + F \sin^2 \Delta\phi), \qquad (10.33)$$

$$\text{where} \quad T_M = (1 - R)[1 - R\exp(-2\alpha L)]/[1 - R\exp(-\alpha L)]^2, \qquad (10.34)$$

$$F = 4R\exp(-\alpha L)/[1 - R\exp(-\alpha L)]^2. \tag{10.35}$$

$$\Delta\phi = \phi - l\pi, \tag{10.36}$$

l being the integer nearest to ϕ/π.

The transmission coefficient of the etalon may, therefore, be controlled by the input signal through the changes in α and Δn_r or $\Delta\phi$. Two types of OBD's may be devised. These are the absorption-based and the dispersion-based OBD.

10.4.1. ABSORPTION-BASED OBD

These devices are realized by designing the MQW and the FP etalon so as to make identical the wavelength of the input signal, the exciton wavelength of the MQW and the resonant wavelength of the FP etalon, i.e., the wavelength for which D=0. The absorption being very high for this condition, the output for low input signal is small. However, as the intensity of the input increases, the absorption decreases and the transmission coefficient I_T/I_0 also incre ases. The reduction in absorption also causes the intensity of the signal, I_{et} , inside the etalon to be high and the enhancement causes further reduction of the absorption. This feedback mechanism leads to a drastic reduction in the absorption at a particular input level I_1 . The output then switches to a high value and continues to be high for further increase of the input. However, when the input is decreased, the output remains high below I_1 and switches back to the low level for an input level I_2 lower than I_1 . The hysteresis is due to the fact that initiation of the reverse process of desaturation and destruction feedback requires a lowering of I_{et} from the value corresponding to I_1 . Evaluation of the criterion for absorptive bistability however, is somewhat involved and is not therefore, discussed.

Designing of the MQW's and FP's for satisfying the conditions so that the device acts as an OBD being fairly critical, such devices are not much reported.

10.4.2. DISPERSION-BASED OBD

The basic structure of dispersion-based OBD's are the same as that of the absorption based OBD. But, the FP resonant wavelength is made larger than the wavelength of the input signal which, however, is close to the exciton wavelength. The resonator is hence in a low-transmission mode. As the input intensity increases, Δn_r which is negative increases in magnitude and causes the resonant wavelength of FP etalon to approach the wavelength of the input signal. The intensity inside the cavity therefore increases, which causes further increases in $|\Delta n_r|$. This positive feedback causes the cavity to switch to the high- transmission mode when the input intensity reaches a level I_1 . The high-transmission mode continues for further increase in the input. However, as the input is decreased, the device switches back to the low-transmission mode for an intensity I_2 which is lower than I_1 under certain conditions. The transmission curves have been worked out theoretically for determining the required conditions. The conditions may be visualized also graphically as follows.

It may be noted that the change in refractive index Δn_r is caused by the radiation I_{et} , inside the etalon. For the sake of simplicity, Δn_r and therefore $\Delta \phi$ may be taken to be proportional to I_{et} . On the other hand, the transmitted signal I_T is proportional to I_{et} and hence $\Delta \phi$ may be taken to be proportional to I_T . Equation (10.33) may now be written as

$$[1 + F \sin^2(\Delta \phi - \gamma I_T)]^{-1} = I_T/T_M I_i, \tag{10.37}$$

where γ has been introduced as the proportionality constant and $\Delta \phi_0$ is the low-input value of $\Delta \phi$. The solution of the above equation gives I_T for an input I_i . The basic nature of the characteristics may be explained by plotting the functions on the two sides of Eq. (10.37) against I_T , as shown schematically in Fig. 10.14(a). The curves for the functions intersect at a single point for small values of I_i (e.g., A) and also for high values of I_i (e.g.,J). But, for input intensities lying between I_1 and I_2 the curves intersect at three points (e.g., D,E,F). Hence as the input intensity I_i is increased from a low value, output continues to be low till I_i equals I_1 . For larger inputs, the output jumps to H or the large output point. Similarly, when the input is reduced from a high value, the high-transmission state

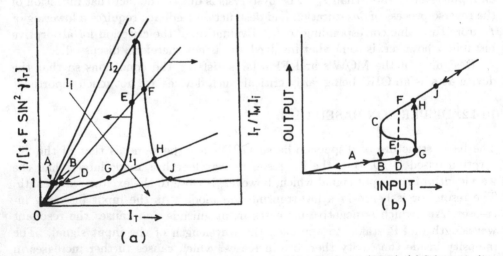

Figure 10.14. Characteristic of dispersion-based optical bistable device (OBD). (a) Schematic diagram showing the variation of $[1 + F\sin^2(\Delta \phi)]^{-1}$ and $I_T/T_M I_i$ [of Eq. (11.33)] with I_T. (b) Input-output characteristic.

continues till I_i reaches I_2 for which the two curves touch at C. For further reduction of I_i , the device switches to the low- transmission mode at B. The resultant input-output characteristic is shown in Fig. 10.14(b). It may be noted that points on the left side of the $[(1 + F \sin^2(\Delta\phi - \gamma I_T)]^{-1}$ vs I_T curves or the etalon resonance curve are not stable from G to C. For these points, a reduction in I_T causes an increase in $\Delta\phi$ and further reduction in I_T . Similarly, for an increase in I_T , $\Delta\phi$ is decreased which causes a further increase in I_T.

It is evident from the above discussion that the characteristic of the device is critically dependent on the choice of the etalon wavelength and the exciton wavelength. The operating wavelength should be close to the exciton wavelength to get significant change in Δn_r with I_i . It should not also be too far away or too close to the FP etalon resonant wavelength. When it is too far away, the intersection point of the two curves remain on the low transmission side and when it is too close, the intersection points lie on the right side of the etalon resonance curve and the input-output curve remains single valued. We also find that the separation between I_1 and I_2 is determined by the width of the etalon resonance curve. It is necessary to make the resonance curve not-too sharp for getting a significant difference between I_1 and I_2 . A value of R, not-too close to unity and some absorption are favorable from these considerations. The switching time of the device is controlled by the photon life time in the FP cavity and can be made small by shortening its length.

An OBD using a semiconductor, namely GaAs was first demonstrated[10.59] in 1979. The sensitivity of the device was much improved by using a MQW structure in place of the bulk sample[10.60,61]. The MQW structure had 300 periods of 66 Å GaAs and 64 Å $Al_{0.3}Ga_{0.7}As$ grown by MBE. It was sandwiched between mirrors of reflectivity 0.9 and 0.98. The device could be switched by using 6mW power from a laser diode operating at 830 nm.

An OBD has been realized[10.62] also by using InGaAs/InAlAs MQW's. The mirror for the Fabry-Perot etalon was provided by high-reflection dielectric mirror coating on the two surfaces of the MQW structures and the structure was cemented to a sapphire slab. The device was operated at 1.5 μ m wavelength with 20 mW power and had a switching time of 10 ns.

10.5. Waveguide All-optic Switch

The enhanced optical nonlinearity of MQW's have been utilized also to realize waveguide all-optic switches. It uses the property of two coupled waveguides, in which the power fed into one switches back and forth with distance[10.54]. The power fed to one waveguide is fully transferred to the other waveguide in a distance L_c , called the beat length. This length is determined by the guide wavelength and may, therefore, be controlled by varying the refractive index of the material of the waveguide.

An waveguide all-optic switch is realized by using a MQW structure, covered by two cladding layers, in which two waveguides are formed. The separation between the two waveguides is arranged so that they work as coupled waveguides due to leaking of optic energy through the sides. Power in waveguide 1, to which the signal is applied, may be expressed[10.63] as

$$P_1 = P(1/2)\{1 + cn[\pi(L/L_c)|(P^2/P_c^2)]\}, \tag{10.38}$$

where L is the coupling length of the waveguides, L_c is a characteristic length, the so-called beat length, $cn(\phi|m)$ is the Jacobi elliptic function[10.51], P is the input power and P_1 is the output power from waveguide 1, P_c is the critical power, given by

$$P_c = \lambda A_{\text{eff}}/n_2 L, \tag{10.39}$$

for which the incremental phase shift in a waveguide is 2π. A_{eff} is the effective cross-sectional mode area of a waveguide, λ is the wavelength of the incident radiation and n_2 is the coefficient of nonlinear refractive index, i.e.,

$$n_2 = [n(I) - n(0)]/I, \tag{10.40}$$

$n(I)$ being the refractive index for an intensity I of radiation.

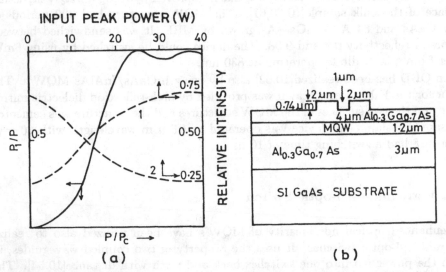

Figure 10.15. Schematic diagram of the structure and input-output characteristic of waveguide all-optic switch. (a) Normalized input-output characteristic. Solid line - Theoretical. Dashed line - Experimental. (b) Structure of the device.

The beat length L_c is determined by the waveguide material and the device structure. The guide coupling length is made equal to L_c .

The variation of power at the output of the waveguides for different inputs to waveguides 1 is illustrated schematically in Fig. 10.15(a). For small values of incident power, $P, P/P_c$ is small; the Jacobi elliptic function behaves as a cosine function and P, varies cosinusoidally with the coupling length. The power at the output of waveguide 1 is, therefore, zero for $L = L_c$ and low intensity input. The input power P then comes out of the coupled waveguide i.e., waveguide 2. As the output is increased the output of waveguide 1 increases and that of waveguide 2 decreases. For $P = P_c$, $cn[\pi|(P^2/P_c^2)] \approx 0.08$. The output from the two waveguides are then nearly equal. As the input power is further increased $cn[\pi|(P^2/P_c^2)]$ approaches unity and the output of waveguide 1 is nearly equal to the input and consequently that of guide 2 goes to zero. The ideal variation as descrived above is illustrated schematically in Fig. 10.15(a).

The waveguide all-optic switch [10.65-72], also called nonlinear directional coupler (NLDC) has been realized by using GaAs/ AlGaAs MQW's with AlGaAs cladding layers. The waveguides were realized either by depositing metal strips or by etching ridges[10.65]. Vertical confinement is produced in these structures by the cladding layers, while the horizontal confinement is caused by the strains at the edges of deposited metal strips[10.63-65] or by the ridges[10.68,69]. The structure used with ridge waveguides is illustrated in Fig. 10.15(b). Experimental curves are also illustrated schematically by dotted lines in Fig. 10.15(a). The cross-coupling for low inputs was found to be 1:3-1:5, while for high inputs (about 400 pJ) it changed to a value higher than 1:1/3. It should be noted that the NLDC described above has associated with it a large absorption and the transmission is a few percent. It was, however, surmised that the transmission could be increased to about 15 % by suitable design.

10.6. Conclusion

Hybrid and all-optic devices have been discussed in this chapter. The principles of operation and the underlying physics have been explained. Reported experimental results have also been quoted. There are various other devices, based on quantum wells, discussion of which is not however, possible within the limited scope of the book.

These devices use either the electric field dependent or the signal intensity dependent absorption or change in refractive index for realizing the intended function. MQW structures are used because of large excitonic effects in these structures, which persist up to room temperature. The physics of these devices are, therefore, related mostly to electron-photon interaction and optical absorption

phenomena discussed in Chapter 5 . Work has been mainly concerned so long
with proving ideas and improving the device characteristics from the knowledge of
the qualitative features of the involved phenomena. However, developments have
now reached a stage at which detailed study of the devices and the underlying
phenomena have become relevant.

REFERENCES

CHAPTER 1

1.1. L. Esaki and R. Tsu, *IBM Internal Res. Rep. RC 2418, Mar.26* (1969)

1.2. L. Esaki and R. Tsu , *IBMJ. Res. Dev.* **14**, 61 (1970)

1.3. L. Esaki, L. L. Chang, W. E. Howard and V. L. Rideout, *Proc. 11th Int. Conf. Phys. Semicond.*, Warsaw, Poland, 1972, Polish Academy of Sciences, ed., PWN - Polish Scientific Publishers, Warsaw, Poland, 1972, p.431

1.4. R. O. Grondin, W. Porod, J. Ho, D. K. Ferry and G. J. Iafrate, *Superlattices Microstruct.* **1**, 183 (1985)

1.5. L. L. Chang, L. Esaki and R. Tsu, *Appl. Phys. Lett.* **24**, 593 (1974)

1.6. A. Y. Cho and J. R. Arthur, *Prog. Solid-State Chemistry*, Vol.10, Pt.3, 157 (1975)

1.7. R. Dingle, W. Wiegmann and C. H. Henry, *Phys. Rev. Lett.* **33**, 827 (1974)

1.8. H. L. Störmer, R. Dingle, A. C. Gossard, W. Wiegmann and M. D. Sturge, *Solid State Commun.* **29**, 705 (1979)

1.9. T. Mimura, S. Hiyamizu, T. Fuzi and K. Nanbu,*Jpn. J. Appl. Phys.* **19** L225 (1980)

1.10. J. H. English, A. C. Gossard, H. L. Störmer and K. W. Baldwin, *Appl. Phys. Lett .* **50**, 1826 (1987)

1.11. L. Pfeiffer, K. W. West, H. L. Störmer and K. W. Baldwin, *Appl. Phys. Lett.* **55**, 1888 (1989)

1.12. J. R. Schrieffer, in *Semiconductor Surface Physics*, R. H. Kingston, ed., University of Pennsylvania Press, Philadelphia, 1957, p.55

1.13. T. Ando, A. B. Fowler and F. Stern, *Rev. Mod. Phys.* **54**, 437 (1982)

1.14. P. Ho, M. Y. Kao, P. C. Chao, K. H. G. Duh, J. M. Ballingall, S. T. Allen, A. T. Tessmer and P. M. Smith, *Electron. Lett.* **27**, 325 (1991)

1.15. J. P. Van der Ziel, R. Dingle, R. C. Miller, W. Weigmann and W. A. Nordland, Jr., *Appl. Phys. Lett.* **26**, 463 (1975)

1.16. R. D. Dupuis, P. D. Dapkus, N. Holonyak, Jr., E. A. Rezek and R. Chin, *Appl. Phys. Lett.* **32**, 295 (1978)

1.17. R. D. Dupuis, P. D. Dapkus, R. Chin, N. Holonyak, Jr., S. W. Kirchoefer, *Appl. Phys. Lett.* **34**, 265 (1979)

1.18. R. D. Dupuis, P. D. Dapkus, N. Holonyak, Jr. and R. M. Kolbas, *Appl. Phys. Lett.* **35**, 487 (1979)

1.19. W. T. Tsang, C. Weisbuch, R. C. Miller and R. Dingle, *Appl. Phys. Lett.* **35**, 673 (1979)

1.20. H. Temkin, K. Alave, W. R. Wagner, T. P. Pearsall and A. Y. Cho, *Appl. Phys. Lett.* **42**, 845 (1983)

1.21. S. Nakamura, M. Senoh, S. Nagahama, N. Iwasa, T. Yamada, T. Matsushita, H. Kiyoku and Y. Sugimoto, *Jpn. J. Appl. Phys.* 35, L74 (1996)

1.22. E. Kapon, S. Simhony, J. P. Harbison, L. T. Florez and P. Worland, *Appl. Phys. Lett.* **56**, 1825 (1990)

1.23. Sei-Ichi Miyazawa, Y. Sekiguch and M. Okuda, *Appl. Phys. Lett.* **63**, 3583 (1993)

1.24. T. C. L. G. Sollner, W. D. Goodhue, P. E. Tannenwald, C. D. Parker and D. D. Peck, *Appl. Phys. Lett.* **43**, 588 (1983)

1.25. V. K. Reddy, A. J. Tsao, D. P. Neikirk, *Electron. Lett.* **26**, 1742 (1990)

1.26. T. P. E. Broeckaert, W. Lee and C. G. Fonstad, *Appl. Phys. Lett.* **53**, 1545 (1988)

1.27. E. R. Brown, J. R. Söderström, C. D. Parker, L. J. Mahoney, K. M. Molver and T. C. McGill, *Appl. Phys. Lett.* **58**, 229 (1991)

1.28. R. H. Davis and H. H. Hosack, *J. Appl. Phys.* **34**, 864 (1963)

1.29. N. Yokoyama, K. Imamura, S. Muto, S. Hiyamizu and H. Nishi, *Jpn. J. Appl. Phys.* **24**, L853 (1985)

1.30. F. Capasso, S. Sen, A. C. Gossard, A. L. Hutchinson and J. H. English, *IEEE Electron Dev. Lett.* **EDL-7**, 573 (1986)

1.31. F. Capasso, S. Sen, F. Beltram and A. Y. Cho, in *Physics of Quantum Electron Devices.* F. Capasso, ed., Springer Verlag, Berlin, 1990, p.181

1.32. L. Esaki and H. Sakaki, *IBM Tech. Disc. Bull.* **20**, 2456 (1977)

1.33. S. Smith, L. C. Chiu, S. Margalit, A. Yariv and A. Y. Cho, *Infrared Phys.* **23**, 93 (1983)

1.34. L. C. West and S. J. Eglash, *Appl. Phys. Lett.* **46**, 1156 (1985)

1.35. B. F. Levine, K. K. Choi, C. G. Bethea, J. Walker and R. J. Malik, *Appl. Phys. Lett.* **50**, 1092 (1987)

1.36. L. I. Kozlowski, G. M. Williams, G. J. Sullivan, C. W. Farley, R. J. Anderson, J. Chen, D. T. Cheung, W. E. Tennant and R. E. DeWames, *IEEE Trans. Electron. Dev.* **ED-38**, 1124 (1991)

1.37. S. D. Gunapala, K. M. S. V. Bandara, B. F. Levine, G. Sarusi, J. S. Park, T. L. Lin, W. T. Pike and J. K. Liu, *Appl. Phys. Lett.* **64**, 3431 (1994)

1.38. M. O. Manasreh, ed., *Semiconductor Quantum Wells and Superlattices for Long-Wave length Infrared Detectors*, Artech House, Boston, 1993

1.39. D. A. B. Miller, D. S. Chemla, T. C. Damen, A. C. Gossard, W. Wiegmann, T. H. Wood and C. A. Burrus, *Phys. Rev. B*, **32**, 1043 (1985)

1.40. T. K. Woodward, Theodore Sizer II, D. L. Sivco and A. Y. Cho, *Appl. Phys. Lett.* **57**, 548 (1990)

1.41. R. H. Yan, R. J. Simes, L. A. Coldren and A. C. Gossard, *Appl. Phys. Lett.* **56**, 1626 (1990)

1.42. M. Whitehead, A. Rivers, G. Parry, J. S. Roberts and C. Button, *Electron. Lett.* **25**, 984 (1989)

1.43. C. C. Barron, C. J. Mahon, B. J. Thibeault, G. Wang, W. Jiang, L. A. Coldren and J. E. Bowers, *IEEE J. Quantum Electron.* **31**, 1484 (1995)

1.44. H. M. Gibbs, *Optical Bistability : Controlling Light with Light*, Academic Press, New York, 1985

1.45. D. A. B. Miller, D. S. Chemla, D. J. Eilenberger, P. W. Smith, A. C. Gossard and W. Wiegmann, *Appl. Phys. Lett.* **42**, 925 (1983)

1.46. R. Jin, C. L. Chuang, H. M. Gibbs, S. W. Koch, J. N. Polky and G. A. Rubenz, *Appl. Phys. Lett.* **53**, 1791 (1988)

CHAPTER 2

2.1. H. Kressel and H. Nelson, *RCA Rev.* **30**, 106 (1969)

2.2. D. C. Tsui, R. A. Logan, *Appl. Phys. Lett.* **35**, 99 (1979)

2.3. Y. Sasai, J. Ohya. M. Ogura, *IEEE. J. Quantum Electron.* **QE-25**, 662 (1989)

2.4. I. Hayashi, M. B. Panish and P. W. Foy, *IEEE J. Quantum Electron.* **QE-5**, 211 (1969)

2.5. Zh. I. Alferov, V. M. Andreev, E. L. Portnoi and M. K. Trakan, Sov. Phys. Semicond. **3**, 1107 (1970)

2.6. H. Kroemer, *Proc. IEEE*, **70**, 13 (1982)

2.7. H. Nelson, U.S. Patent 3565, 702 (1971)

2.8. I. Hayashi, M. B. Panish, P. W. Foy and S. Sumski, *Appl. Phys. Lett.* **17**, 109 (1970)

2.9. G. H. Olsen, in *GaInAsP Alloy Semiconductors*, T. P. Pearsall, ed., John Wiley & Sons, New York, 1982, Chap. 1

2.10. J. R. Arthur, *J. Appl. Phys.* **39**, 4032 (1968)

2.11. A. Y. Cho and J. R. Arthur, *Prog. Solid State Chem.* **10**, 157 (1975)

2.12. H. M. Manasevitt, *Appl. Phys. Lett.* **12,** 156 (1968)

2.13. M. J. Ludowise, *J. Appl. Phys.* **58**, R31 (1985)

2.14. H. M. Manasevitt and W. I. Simpson, *J. Electrochem. Soc.* **116**, 1725 (1969)

2.15. R. D. Dupuis, P. D. Dapkus, R. D. Yingling and L. A. Moudy, *Appl. Phys. Lett.* **31**, 201 (1978)

2.16. W. T. Tsang, *Appl. Phys. Lett.* **45**, 1234 (1984)

2.17. M. B. Panish, H. Temkin and S. Sumski, *J. Vac. Sc. Technol.* B 3, 657 (1985)

2.18. W. I. Wang, E. E. Mendez and F. Stern, *Appl. Phys. Lett.* **45,** 639 (1984)

2.19. I. Lahiri, D. D. Nolte, J. C. P. Chang, J. M. Woodall and M.R. Melloch,*Appl. Phys. Lett.* **67**, 1244 (1995)

2.20. D. V. Lang, M. B. Panish, F. Capasso, J. Allam, R. A. Hamm and A. M. Sergent, *Appl. Phys. Lett.* **50**, 736 (1987)

2.21. F. Lobentanzer, W. König, W. Stolz, K. Ploog, T. Elsaesser and R. J. Baürle, *Appl. Phys. Lett.* **53**, 571 (1988)

2.22. I. J. Fritz, T. J. Drummond and G. C. Osbourn, J. E. Schirber and E. D. Jones, *Appl. Phys. Lett.* **48**, 1678 (1985)

2.23. T. Inata, S. Muto, Y. Nakata, S. Sasa, T. Fujii and S. Hiyamizu, *Jpn. J.Appl. Phys.* **26**, L1332 (1987)

2.24. Y. B. Li, R. A. Stradling, L. Artus, S. J. Webb, R. Cusco, S. J. Chung and A. G. Norman, *Semicond. Sci. Technol.* **11**, 868 (1993)

2.25. C. R. Bolognese, I. Sela, J. Ibbetson, B. Brar, H. Kroemer and J. H. English, *J. Vac. Sci. Technol. B* **11**, 868 (1993)

2.26. D. Yang, Y. C. Chen, T. Brock and P. K. Bhattacharya, *IEEE Electron Dev. Lett.* **15**, 350 (1992)

2.27. H. Mani, A. Joullie, A. M. Joullie, B. Girault and C. Alibert, *J. Appl. Phys.* **61**, 2101 (1987)

2.28. T. Fujii, Y. Nakata, Y. Sugiyama and S. Hiyamizu, *Jpn. J. Appl. Phys.* **25**, L254 (1986)

2.29. M. Haines, T. Kerr, S. Newstead and P. B. Kirby, *J. Appl. Phys.* **65**, 1942 (1989)

2.30. H. Lee, P. K. York, R. J. Menna, R. U. Martinelle, D. Garbuzov and S. Y. Narayan, *J. Cryst. Growth* **150**, 1354 (1995)

2.31. B. F. Levine, A. Y. Cho, J. Walker, R. J. Malik, D. A. Kleinman and D. L. Sivco, *Appl. Phys. Lett.* **52**, 1481 (1988)

2.32. H. Asahi, S. Emura, S. Gonda, Y. Kawamara and H. Tanaka, *J. Appl. Phys.* **65**, 5007 (1989)

2.33. S. Emura, T. Nakagawa, S. Gonda and S. Shimizu, *J. Appl. Phys.* **62**, 4632 (1987)

2.34. F. Schaffler and H. Jorke, *Appl. Phys. Lett.* **58**, 397 (1991)

2.35. Y. Hefetz, J. Nakahara, A. V. Nurimiko, L. A. Kolodzieskii, R. L. Gunshor and S. Datta, *Appl. Phys. Lett.* **47**, 989 (1985)

2.36. R. H. Miles, G. Y. Wu, M. B. Johnson, T. C. McGill, J. P. Faurie and S. Sivanathan, *Appl. Phys. Lett.* **48**, 1383 (1986)

2.37. M. Schultz, F. Heinrichs, U. Merkt, T. Colin, T. Skauli and S. Lovoid, *Semicond. Sci. Technol.* **11**, 1168 (1996)

2.38. A. N. Baranov, Y. Cuminal, G. Boissier, J. C. Nicolas, J. L. Lazzari, C. Alibert and A. Joullie, *Semicond. Sci. Technol.* **11**, 1185 (1996)

2.39. *Proceedings of Molecular Beam Epitaxy,* 1994, Eighth International Conference on Molecular Beam Epitaxy, Osaka, Japan, 29 Aug. - 2 Sept., 1994, published in *J. Cryst. Growth* **150**, No. 1-4 (May 1995)

2.40. P. D. Dapkus, *Ann. Rev. Mater. Sci,* **12**, 243 (1982)

2.41. M. J. Ludowise, D. Biswas and P. K. Bhattacharya, *Appl. Phys. Lett.* **56,** 958 (1990)

2.42. D. Biswas, N. Debbar, P. Bhattacharya, M. Razeghi, M. Defour and F. Omnes, *Appl. Phys. Lett.* **56,** 833 (1990)

2.43. T. Y. Wang and G. B. Stringfellow, *J. Appl. Phys.* **67,** 344 (1990)

2.44. R. P. Schneider, Jr. and B. W. Wessels, *J.Appl. Phys.* **70,** 405 (1991)

2.45. P. A. Anderson, R. F. Kazarinov, N. A. Olsson, T. Tanbun-Ek and R. A. Logan, *IEEE J. Quantum Electron.* **30,** 219 (1994)

2.46. A. Asif Khan, Q. Chen, C. J. Sun, M. Shur and B. Gelmont, *Appl. Phys. Lett.* **67,** 1429 (1995)

2.47. M. Razeghi, J. P. Hirtz, U. O. Ziemeliś, C. Delalande, B. Etienne and M. Voos, *Appl. Phys. Lett.* **43,** 585 (1983)

2.48. A. P. Roth, M. Sacilotti, R. A. Masut, P. J. D'Arcy, B. Watt, G. I. Sproule and D. F. Mitchell, *Appl. Phys. Lett.* **48,** 1452 (1986)

2.49. K. Uomi, S. Sasaki, T. Tsuchya and N. Chinone, *J. Appl. Phys.* **67,** 904 (1990)

2.50. C. A. Larsen, C. H. Chen, M. Kitamura, G. B. Stringfellow, D. W. Brown and A. J. Robertson, *Appl. Phys. Lett.* **48,** 1531 (1986)

2.51. A. H. Cowley, B. L. Benac, J. G. Ekerdt, R. A. Jones, K. B. Kedd, J. Y. Lee and J. E. Miller, *J. Ann. Chem. Soc.* **110,** 628 (1988)

2.52. W. T. Tsang, *Appl. Phys. Lett.* **45,** 1234 (1986)

2.53. W. T. Tsang, in *Semiconductors and Semimetals*, Vol. 24, R. Dingle, ed., Academic Press, New York, 1987, Chapt. 7

2.54. H. Kinoshita, H. Fujiyasu, *J. Appl. Phys.* **51,** 5845 (1980)

2.55. K. Shinohara, Y. Nishijima, H. Ebe, A. Ishida and H. Fujiyasu, *Appl. Phys. Lett.* **47,** 1184 (1985)

2.56. M. B. Panish, H. Temkin, R. A. Hamm and S. N. G. Chu, *Appl. Phys. Lett.* **49,** 164 (1986)

2.57. C. L. Goodman and M. V. Pessa, *J. Appl. Phys.* **60,** R65 (1986)

2.58. S. J. Pearton, ed., *GaN and Related Materials II* , (Gordon Breach Science Publishers, Asian Edition, 2000) p. 118

2.59. R. N. Bicknell, N. C. Giles and J. F. Schetzina, *Appl. Phys. Lett.* **49,** 1095 (1986)

2.60. J. T. Cheung, G. Niizawa, J. Moyle, N. P. Ong, B. M. Paine and T. Vreeland, Jr., *J. Vac. Sci. Technol.* A **4,** 2086 (1986)

2.61. K. Kodama, M. Ozeki and J. Komeno, *J. Vac. Sci. Technol.* B **1,** 696 (1983)

2.62. L. J. Ketelsen and R. F. Kazarinov, *IEEE J. Quantum Electron.* **31,** 811 (1995)

2.63. M. Notomi, M. Naganuma and T. Nishida, T. Tamamura, H. Iwamura, S. Nojima and M. Okamoto, *Appl. Phys. Lett.* **58,** 720 (1991)

2.64. P. Ills, M. Michel, A. Forchel, I. Gyuro, M. Klenk and E. Zielinski, *Appl. Phys. Lett.* **64**, 496 (1994)

2.65. B. Hubuer, B. Jacobs, C. Greus, R. Zengerb and A. Forched, *J. Vac. Sci. Technol.* *B* **12**, 3658 (1994)

2.66. T. Fukui and H. Saito, *Jpn. J. Appl. Phys.* **29**, L731 (1990)

2.67. S. Hara, J. Ishizaki, J. Motohisa, T. Fukui and H. Hasegawa, *J. Cryst. Growth* **145**, 692 (1994)

2.68. K. Inoue, Hu Kun Huang, M. Takeuchi, K. Kimura, H. Nakashima, M. Iwane, O. Matsuda and K. Muruse, *Jpn. J. Appl. Phys.* **34**, 1342 (1995)

2.69. J. Cibert, P. M. Petroff, G. J. Dolan, S. J. Pearton, A. C. Gossard and J. H. English, *Appl. Phys. Lett.* **49**, 1275 (1986)

2.70. H. A. Zarem, P. C. Sercel, M. E. Hoenk, J. A. Lebens and K. J. Vahala, *Appl. Phys. Lett.* **54**, 2692 (1989)

2.71. R. K. Kupka, Y. Chen, R. Planel and H. Latnois, *J. Appl. Phys.* **77**, 1990 (1995)

2.72. E. Kapon, M. C. Tamargo and D. M. Hwang, *Appl. Phys. Lett.* **50**, 347 (1987)

2.73. Y. Hasegawa, T. Egawa, T. Junbo and M. Umeno, *J. Cryst. Growth* **145**, 728 (1994)

2.74. K. Shimoyama, S. Nagai, Y. Inove, K. Kojomi, N. Horoi, K. Fujii and H. Gotoh, *J. Cryst. Growth* **145**, 734 (1995)

2.75. S. Tsukamoto, Y. Nagamune, M. Nishioka and Y. Arakawa, *J. Appl. Phys.* **71**, 533 (1992)

2.76. M. Dilger, M. Hohenstein, F. Phillipp, K. Eberl, A. Kurtenbach, P. Grambow, A. Lehmann, D. Heilman and K. Von Klitzing, *Semicond. Sci. Tecchnol.* **9**, 2258 (1994)

2.77. Y. Lin, Y. Nishimoto, S. Seimomura, K. Garmo, K. Munuse, N. Sano, A. Adachi, K. Kanamoto, T. Isu, K. Fujita, T. Watanabe and S. Hiyamizu, *Jpn. J. Appl. Phys.* **33**, 719 (1994)

2.78. S. Tsukamoto, Y. Nagamune, M. Nishioka and Y. Arakawa, *Appl. Phys. Lett.* **62**, 49 (1993)

2.79. A. Hartmann, L. Vescan, C. Dieker and H. Luth, *J. Appl. Phys.* **77**, 1959 (1995)

2.80. H. Temkin, G. J. Dolan, M. B. Panish and S. N. G. Chu, *Appl. Phys. Lett.* **50**, 413 (1987)

2.81. Y. Nagamune, S. Tsukamoto, N. Nishioka and Y. Arakawa, *J. Cryst. Growth* **126**, 707 (1992)

2.82. Y. Arakawa, *Solid-St. Electron.* **37**, 523 (1993)

2.83. J. Oshimowo, M. Nishioka, S. Ishida, Y. Arakawa, *Jpn. J. Appl. Phys.* **33**, L1634 (1994)

2.84. J. Y. Marzin, J. M. Gérerd, A. Izraël and D. Barrier, *Phys. Rev. Lett.* **73**, 716 (1994)

2.85. D. Leonard, M. Krishnamurthy, C. M. Reves, S. P. Denbaars and P. M. Petroff, *Appl. Phys. Lett.* **64**, 2727 (1994)

2.86. A. Madhukar, Q. Xie, P. Chen and A. Konkar, *Appl. Phys. Lett.* **64**, 2727 (1994)

2.87. R. Nötzel, *Semicond. Sci. Technol.* **11**, 1365 (1996)

2.88. J. Ahopelto, A. A. Yamaguchi, K. Nishi, A. Usui and H. Sakaki, *Jpn. J. Appl. Phys.* **32**, L32 (1993)

2.89. R. Apetz, L. Vescan, A. Hartman, C. Dieker and H. Luth, *Appl. Phys. Lett.* **66**, 445 (1995)

2.90. A. Kurtenbach, K. Eberl and T. Shitara, *Appl. Phys. Lett.* **66**, 361 (1995)

2.91. S. C. Moss, D. Ila, H. W. H. Lee and D. J. Norris, eds., *Semiconductor Quantum Dots: Mater. Res. Soc. Symp. Proc.* **571** (1999)

2.92. E. L. Hue, *FED Journal* **10**, Suppl. 2, 5 (1999)

2.93. M. K. Saker, D. M. Whittaker, M. S. Skolnick, C. F. McConville, C. R. Whitehouse, S. J. Barnett, A. D. Pitt, A. G. Cullis and G. M. Williams, *Appl. Phys. Lett.* **65**, 1118 (1994)

2.94. H. Hillmer, R. Losch, W. Schlapp and H. Burkhard, *Phys. Rev.* B **52**, R17025 (1995)

CHAPTER 3

3.1. O. Madelung, ed., *Semiconductors*, Springer-Verlag, Berlin, 1991

3.2. J. Van Vechten, in *Handbook of Semiconductors*. Vol.3, S. P. Keller, ed., North Holland, Amsterdam, 1980, p.33

3.3. L. Esaki, *IEEE J. Quantum Electron.* **QE-27**, 1611 (1986)

3.4. J. Singh, *Physics of Semiconductors and their Heterostructures*, McGraw-Hill, New York, 1993, p.192

3.5. R. L. Anderson, *Solid-St. Electron.* **5**, 341 (1962)

3.6. D. W. Niles and G. Margaritondo, *Phys. Rev.* B **34**, 2923 (1986)

3.7. J. O. McCaldin, T. C. McGill and C. A. Mead, *Phys. Rev. Lett.* **36**, 56 (1976)

3.8. Su-Huai Wei and A. Zunger, *Phys. Rev. Lett.* **59**, 144 (1987)

3.9. W. A. Harrison, *J. Vac. Sci. Technol.* **14**, 1016 (1977)

3.10. W. R. Frensley and H. Kroemer, *Phys. Rev.* B, **16**, 2642 (1977)

3.11. J. Tersoff, *Phys. Rev.* B **30**, 4874 (1984)

3.12. G. Margaritondo, *Phys. Rev.* B **31**, 2526 (1985)

3.13. C. G. Van de Walle and R. M. Martin, *Phys. Rev.* B **35**, 8154 (1987)

3.14. N. E. Christensen, *Phys. Rev.* B **37**, 4528 (1988)

3.15. N. E. Christensen, I. Gorczyca, O. B. Christensen, U. Schmid, and M. Cardona, *J. Cryst. Growth* **101**, 318 (1990)

3.16. C. G. Van de Walle and J. Neugebaurer, *Mater. Res. Soc. Symp. Proc.* **v. 449**, 861 (1997)

3.17. J. A. Majeworky, M. Stadele and P. Vogel, *ibid*, p.917

3.18. P. Bernardine, V. Fiorentini and D. Vanderbilt, *ibid*, p.923

3.19. R. Dingle, W. Wiegmann and C. H. Henry, *Phys. Rev. Lett.* **33**, 827 (1974)

3.20. R. C. Miller, D. A. Kleinman and A. C. Gossard *Phys. Rev. B* **29**, 7085 (1984)

3.21. B. R. Nag and S. Mukhopadhyay, *Appl. Phys. Lett.* **58**, 1056 (1991)

3.22. K. Uomi, S. Sasaki, T. Tsuchiya and N. Chinone, *J. Appl. Phys.* **67**, 904 (1990)

3.23. M. A. Khan, R. A. Skogman, J. M. Van Hove, S. Krishnankutty and R. M. Kolbas, *Appl. Phys. Lett.* **56**, 1257 (1990)

3.24. K. Itoh, T. Kawamoto, H. Awano, K. Hiramatsu and I. Akasaki, *Jpn. J. Appl. Phys.* **30**, 1924 (1991)

3.25. M. Heiblum, M. I. Nathan and M. Eizenberg, *Appl. Phys. Lett.* **47**, 503 (1985)

3.26. M. A. Haase, M. A. Emanuel, S. C. Smith, J. J. Coleman and G. E. Stillman, *Appl. Phys. Lett.* **50**, 404 (1987)

3.27. M. K. Kelly, D. W. Niles, E. Colavita, G. Margaritondo and M. Henzler, *Appl. Phys. Lett.* **46**, 768 (1985)

3.28. D. W. Niles and G. Margaritondo, *Phys. Rev. B* **34**, 2923 (1986)

3.29. M. Heiblum and H. V. Fischetti, in *Physics of Quantum Electron Devices*, F. Capasso, ed., Springer-Verlag, Heidelberg, 1990, p. 275

3.30. S. R. Forrest and O. K. Kim, *J. Appl. Phys.* **53**, 5738 (1982)

3.31. H. Okumura, S. Misawa, S. Yoshida and S. Gonda, *Appl. Phys. Lett.* **46**, 377 (1985)

3.32. W. I. Wang, E. E. Mendez and F. Stern, *Appl. Phys. Lett.* **45**, 639 (1984)

3.33. E. A. Kraut, P. W. Grant, J. R. Waldrop and S. P. Kowalczyk, *Phys. Rev. Lett.* **44**, 1620 (1980)

3.34. G. J. Gualtieri, G. P. Schwartz, R. G. Nuzzo and W. A. Sunder, *Appl. phys. Lett.* **49**, 1037 (1986)

3.35. G. J. Gualtieri, G. P. Schwartz, R. G. Nuzzo, R. G. Malik and J. F. Walker, *J. Appl. Phys.* **61**, 5337 (1987)

3.36. P. M. Hui, H. Ehrenrich and N. F. Johnson, *J. Vac. Sci. Technol. A* **7**, 424 (1989)

3.37. G. Martin, S. Strite, A. Botchkarev, A. Agarwal, A. Rockett, H. Morcoc, W. R. L. Lambrecht and B. Segall, *Appl. Phys. Lett,* **65**, 610 (1994)

3.38. T. M. Duc, C. Hsu and J. P. Faurie, *Phys. Rev. Lett.* **58**, 1127 (1987)

3.39. E. T. Yu, D. H. Chow and T. C. McGill, *Phys. Rev. B* **38**, 12764 (1988)

3.40. G. Martin, A. Botchkarev, A. Rockett and H. Morcoc, *Appl. Phys. Lett.* **68**, 2541 (1996)

3.41. P. K. Bhattacharya, ed., *Indium Gallium Arsenide*, INSPEC, London, 1993, p.86

3.42. D. Biswas, N. Debbar, P. Bhattacharya, M. Razeghi, M. Defour andF. Omnes, *Appl. Phys. Lett.* **56**, 833 (1990)

3.43. S. H. Feng, J. Krynicki, V. Donchev, J. C. Bourgoin, M. Di, Fort-Poisson, C. Brylinski, S. Delage, H. Blanck and S. Alaya, *Semicond. Sci. Technol.* **8**, 2092, (1993)

3.44. J. J. O'Shea, C. M. Reaves, S. P. Den Baars, M. A. Chin and V. Narayanmurti, *Appl. Phys. Lett.* **69**, 3022 (1996)

3.45. T. Kobayashi, K. Taira, F. Nakamura and H. Kawai, *J. Appl. Phys.* **65**, 4898 (1989)

3.46. K. Kodama, M. Hoshino, K. kilahara, M. Takikawa and M. Ozeki, *Jpn. J. Appl. Phys.* **25**, L127 (1986)

3.47. M. S. Faleh, J. Tasselli, J. P. Braille and A. Marty, *Appl. Phys. Lett.* **69**, 1288 (1996)

3.48 Il Jong Kim, Yong-Hoon Cho, Kwan-Shik Kim, Byung-Doo Choe and A. Lim, *Appl. Phys. Lett.* **68**, 3489, (1996)

3.49 S. H. Kwok, P. Y. Yu, K. Uchida and T. Arai, *Appl. Phys. Lett.* **71**, 1110, (1997)

3.50. W. I. Wang, *Solid-St. Electron.* **29**, 133 (1986)

3.51. A. R. Bonnefoi, T. C. Mc Gill, R. D. Burnham and G. B. Anderson, *Appl. Phys. Lett.* **50**, 344 (1987)

CHAPTER 4

4.1. W. Kohn, in *Solid State Physics*, Vol. 5, F. Seitz and D. Turnbull, eds., Academic Press, New York, 1957, p.274

4.2. J. M. Luttinger and W. H. Kohn, *Phys. Rev.* **97**, 869 (1955)

4.3. M. Altereli, in *Heterojunctions and Superlattices*, G. Allen, G. Bastard, N. Bocorra, M. Lannoo and M. Voos, eds., Springer Verlag, Berlin, 1986, p.14

4.4. E. O. Kane, in *Semiconductors and Semimetals*, Vol.1, R. K. Willardson and A. C. Beer, Academic Press, New York, 1966, p.75

4.5. J. M. Luttinger, *Phys. Rev.* **102**, 1030 (1956)

4.6. D. A. Broido and L. J. Sham, *Phys. Rev. B*, **31**, 888 (1985)

4.7. D. Ahn and Shun-Lien Chuang, *IEEE J. Quantum Electron.* **24**, 2400 (1988)

4.8. B. R. Nag, *J. Appl. Phys.* **77**, 4148 (1995)

4.9. R. Eppenga, M. F. H. Schuurmans and S. Colak, *Phys. Rev. B*, **36**, 1554 (1987)

4.10. T. Hiroshima and R. Lang, *Appl. Phys. Lett.* **49**, 456 (1986)

4.11. G. T. Einevoll, P. C. Hemmer and J. Thomsen, *Phys. Rev. B* **42**, 3485 (1990)

4.12. R. A. Morrow and K. R. Brownstein, *Phys. Rev. B* **30**, 678 (1984)

4.13. D. J. Ben Daniel and C. B. Duke, *Phys. Rev.* **152**, 683 (1966)

4.14. B. R. Nag, *Appl. Phys. Lett.* **59**, 1620 (1991)

4.15. B. R. Nag, *J. Appl. Phys.* **70**, 4623 (1991)

4.16. D. F. Nelson, R. C. Miller and D. A. Kleinman, *Phys. Rev. B* **35**, 7770 (1987)

4.17. U. Ekenberg, *Phys. Rev.* B **36**, 6152 (1987)

4.18. R. Lassing, *Phys. Rev.* B **31**, 8076 (1985)

4.19. B. R. Nag and S. Mukhopadhyay, *J. Phys:Condens.Matter*, **3**, 3757 (1991)

4.20. U. Wiesner, J. Pillath, W. Bauhofer, A. Kohl, A. Mesquida Kiisters, S. Brittner and K. Heime, *Appl. Phys. Lett.* **64**, 2520 (1994)

4.21. B. R. Nag, S. Mukhopadhyay, *Phys. Stat. Sol.* **175**, 103 (1993)

4.22. B. R. Nag, *Electron Transport in Compound Semiconductors*, Springer-Verlag, Berlin, 1980, p.60

4.23. U. Rössler, *Solid-St. Commun.* **49**, 943 (1984)

4.24. W. Zawadzki and P. Pfeffer, *Semicond. Sci. Technol.* **5**, S179 (1990)

4.25. J. W. Conley and G. D. Mahan, *Phys. Rev.* **161**, 681 (1967)

4.26. F. A. Padovani and R. Stratton, *Phys. Rev. Lett.* **16**, 1202 (1966)

4.27. D. Mukherje and B. R. Nag, *Phys. Rev.* B, **12**, 4338 (1975)

4.28. G. Bastard, *Acta Electron.* **25**, 147 (1983)

4.29. J. R. Söderstörm, D. H. Chow and T. C. McGill, *Electron. Dev. Lett.* **11**, 27 (1990)

4.30. T. Ando, *J. Phys. Soc. Jpn.* **51**, 3893 (1982)

4.31. T. Ando, A. B. Fowler and F. Stern, *Rev. Mod. Phys.* **54**, 437 (1982)

4.32. F. Stern and S. Das Sarma, *Phys. Rev.* B, **30**, 840 (1984)

4.33. M. Abramowitz and I. A. Stegun, eds., *Handbook of Mathematical Functions*, National Bureau of Standards Applied Mathematics Series, No. 55, U.S.Goverment Printing Office, Washington, 1964.

4.34. F. Stern, *Phys. Rev.* B **5**, 4891 (1972)

4.35. F. F. Fang and W. E. Howard, *Phys. Rev. Lett.* **16**, 797 (1966)

4.36. F. Stern and W. E. Howard, *Phys. Rev.* **163**, 816 (1967)

4.37. T. Ando, *J. Phys. Soc. Jpn.* **51**, 3900 (1982)

4.38. G. Bastard, *Surf. Sci*, **142**, 284 (1984)

4.39. K. S. Yoon, G. B. Stringfellow and R. J. Huber, *J. Appl. Phys.* **62**, 1931 (1987)

4.40. D. Gershoni, H. Temkin, G. J. Dolan, J. Dunsmuir, S. N. G. Chu and M. B. Panish, *Appl. Phys. Lett.* **53**, 995 (1988)

4.41. B. R. Nag and S. Gangopadhyay, *Phys. Stat, Sol.(a)* **179**, 463 (1993)

4.42. J. Shertzer and L. R. RamMohan, *Phys. Rev.* B, **41**, 9994 (1990)

4.43. S. Gangopadhyay and B. R. Nag, *Phys. Stat. Sol.* **195**, 123 (1996)

4.44. S. Gangopadhyay and B. R. Nag, *J. Appl. Phys.* **81**, 7885 (1997)

4.45. M. Notomi, M. Naganuma, T. Nishida, T. Tamamura, H. Iwamarn, S. Nojima and M. Okamoto, *Appl. Phys. Lett.* **58**, 720 (1991)

4.46. S. Tsukamoto, Y. Nagamune, M. Niskioko, and Y. Arakawa, *Appl. Phys. Lett.* **62**, 49 (1993)

4.47. Y. Arakawa, *Solid-St. Electron.* **37**, 523 (1994)

4.48. P. Ils, M. Michel, A. Forchel, I. Gynro, M. Klenk and E. Zielinski, *Appl. Phys. Lett.* **64**, 496 (1994)

4.49. S. Koshiba, H. Noge, H. Akiyama, T. Inoshita, Y. Nakamura, A. Shimizu, Y. Nagamune, M. Tsuchiya, H. Kano, H. Sakaki and K. Nada, *Appl. Phys. Lett.* **64**, 363 (1994)

4.50. H. Temkin, G. J. Dolan, M. B. Panish and S. N. G. Chu, *Appl. Phys. Lett.* **50**, 413 (1987)

4.51. Y. Nagamune, M. Nishioka, S. Tsukamoto and Y. Arakawa, *Solid-St. Electron.* **37**, 579 (1994)

4.52. S. S. Nedorezov, *Sov. Phys. Solid-St.* **12**, 1815 (1977)

4.53. J. C. Maan, A. Fasolino, G. Belle, M. Altarelli and K. Ploog, *Physica*, **127 B**, 426 (1984)

4.54. J. A. Brum, L. L. Chang and L. Esaki, *Phys. Rev. B*, **38**, 12977 (1988)

4.55. G. C. Osbourn, *J. Appl. Phys.* **53**, 1586 (1982)

4.56. J. C. Bean, *J. Cryst. Growth*, **81**, 411 (1987)

4.57. B. S. Meyerson, *Appl. Phys. Lett.* **48**, 797 (1986)

4.58. R. People, *IEEE J. Quantum Electron.* **QE-22**, 1696 (1986)

4.59. G. E. Pikus and G. L. Bir, *Sov. Phys. Solid-St.* **1**, 1502 (1958)

4.60. C. Kittel, *Introduction to Solid State Physics*, 2nd Ed. John Wiley & Sons, New York, 1953, p.85

4.61. M. Cardona and N. E. Chistensen, *Phys. Rev. B* **35**, 6182 (1987)

4.62. D. D. Nolte, W. Walukiewicz and E. E. Haller, *Phys. Rev. Lett.* **59** 501 (1987)

4.63. S-C Hong, G. P. Kothiyal, N. Debbar, P. Bhattacharya and J. Singh, *Phys. Rev. B*, **37**, 878 (1988)

4.64. K. Ismail, B. S. Meyerson and J. Wang, *Appl. Phys. Lett.* **58**, 2197 (1991)

4.65. F. Schaffler and H. Jorke, *Appl. Phys. Lett.* **58**, 398 (1991)

4.66. G. Abstreiter, H. Brugger, T. Wolf, H. Jorke and H. J. Herzog, *Phys. Rev. Lett.* **54**, 2441 (1985)

4.67. H. Hertle, G. Schuberth, E. Gornik, G. Abstreiter and F. Schäffler, *Appl. Phys. Lett.* **59**, 2977 (1991)

4.68. R. P. G. Karunasiri, J. S. Park and K. L. Wang, *Appl. Phys. Lett.* **59**, 2588 (1991)

4.69. J. N. Schulman and Y. C. Chang, *Phys. Rev. B* **24**, 4445 (1981)

4.70. G. Edwards and J. C. Inkson, *Semicond. Sci. Technol.* **5**, 1023 (1990)

4.71. Yia-Chung Chang, *Phys. Rev. B*, **37**, 8215 (1988)

CHAPTER 5

5.1. H. M. Gibbs, S. L. McCall, T. N. C. Venkatesan, A. C. Gossard, A. Passner and W. Weigmann, *Appl. Phys. Lett.* **35**, 451 (1979)

5.2. S. S. Tarng, H. M. Gibbs, J. L. Jewell, N. Peyghambarian, A. C. Gossard, T. Venkatesan and W. Weigmann, *Appl. Phys. Lett.* **44**, 360 (1984)

5.3. D. A. B. Miller, D. S. Chemla, P. W. Smith, A. C. Gossard and W. Wiegmann, *Optics Lett.* **8**, 477 (1983)

5.4. D. A. B. Miller, D. S. Chemla, T. C. Damen, A. C. Gossard, W. Wiegmann, T. H. Wood and C. A. Barrus, *Phys. Rev. B* **32**, 1043 (1985)

5.5. L. C. West and S. J. Eglash, *Appl. Phys. Lett.* **46**, 1156 (1985)

5.6. B. F. Levine, S. D. Gunapala, J. M. Kuo, S. S. Pei and S. Hui, *Appl. Phys. Lett.* **59**, 1864 (1991)

5.7. B. R. Nag, *Theory of Electrical Transport in Semiconductors*, Pergamon Press, Oxford, 1972, p.52

5.8. B. R. Nag, *Electron Transport in Compound Semiconductors*, Springer-Verlag, Berlin, 1980, p.100

5.9. J. L. Powell and B. Crasemann, *Quantum Mechanics* Addison-Wesley, Reading,1961, p.416

5.10. M. Voisin, in *Heterojunction and Semiconductor Superlattices*, G. Allan, G. Bastard, N. Boccara, M. Lannoo and M. Voos, eds., Springer-Verlag, Berlin, 1986, p.77.

5.11. E. O. Kane, in *Semiconductors and Semimetals*, Vol.1, R. K. Willardson and A. C. Beer, eds., Academic Press, New York, 1966, p.80

5.12. E. J. Johnson, in *Semiconductors and Semimetals*, Vol.3, R. K. Willardson and A. C. Beer, eds., Academic Press, New York, 1967, p.166

5.13. P. A. Lindsay, *Introduction to Quantum Mechanics for Electrical Engineers*, McGraw-Hill, New York, 1967, p.182.

5.14. M. Asada, A. Kameyama and Y. Suematsu, *IEEE J. Quantum Electron.* **QE-20**, 745 (1984)

5.15. H. Barry Bebb and E. W. Williams, in *Semiconductors and Semimetals*, Vol. 3, R. K. Willardson and A. C. Beer, Academic Press, New York, 1967, p.181

5.16. H. C. Casey, Jr., and M. B. Panish, *Heterostructure Lasers*, Academic Press, New York, 1978, Part A, Chapter 3.

5.17. C. Hermann and C. Weisbuch, *Phys. Rev. B* **15**, 823 (1977)

5.18. H. Xie, J. Piao, J. Katz and W. I. Wang, *J. Appl. Phys.* **70**, 3152 (1991)

5.19. U. Rössler, *Solid-St. Commun.* **49**, 943 (1984)

5.20. F. M. Peeters, A. Matulis, M. Helm, T. Fromherz and W. Hilber, *Phys. Rev. B* **48**, 12008 (1993)

5.21. H. M. Gibbs, S. S. Tarng, J. L. Jewell, D. A. Weinberger, K. Tai, A. C. Gossard, S. L. McCall, A. Passner and W. Weigmann, *Appl. Phys. Lett.* **41**, 221 (1982)

5.22. A. Migus, A. Antonetti, D. Hulin, A. Mysyrowicz, H. M. Gibbs, N. Peyghambrian and J. L. Jewell, *Appl. Phys. Lett.* **46**, 70 (1985)

5.23. G. H. Wannier, *Phys. Rev.* **52**, 191 (1937)

5.24. J. O. Dimmock, in *Semiconductors and Semimetals*, Vol. 3, R. K. Willardson and A. C. Beer, eds., Academic Press, New York, 1967, p.278

5.25. G. Bastard, E. E. Mendez, L. L. Chang and L. Esaki, *Phys. Rev. B*, **26**, 1974 (1982)

5.26. Y. Shinozuka and M. Matsuura, *Phys. Rev. B*, **28**, 4878 (1983)

5.27. R. L. Greene and K. K. Bajaj, *Phys. Rev. B*. **31**, 6498 (1985)

5.28. H. Mathieu, P. Lefevre and P. Christol, *J. Appl. Phys.* **72**, 300 (1992)

5.29. R. L. Greene, K. K. Bajaj and D. E. Phelps, *Phys. Rev. B*, **29**, 1807 (1984)

5.30. M. Shinada and S. Sugano, *J. Phys. Soc. Japan*, **2**, 1936 (1966)

5.31. J. J. Hopfield, in *Proc. 7th Int. Conf. Physics of Semiconductors*, Dunod, Paris and Academic Press, New York, 1964, p.725

5.32. R. R. Sharma and S. Rodriguez, *Phys. Rev.* **153**, 823 (1967); **159**, 649 (1967)

5.33. J. R. Haynes, *Phys. Rev. Lett.* **4**, 361 (1960)

5.34. M. E. Pistol and X. Liu, *Phys. Rev. B* **45**, 4312 (1992)

5.35. O. Brandt, H. Lage and K. Ploog, *Phys. Phys. Rev. B* **45**, 4217 (1992)

5.36. P. W. Yu, S. Chaudhuri, D. C. Reynolds, K. K. Bajaj, C. W. Litton, W. T. Masselink, R. Fischer and H. Morkoc, *Solid-St. Commun.* **54**, 159 (1985)

5.37. M. S. Skolnick, *Semicond. Sci. Technol.* **1**, 29 (1986)

5.38. E. Fermi, *Nuclear Physics*, University of Chicago Press, Chicago, 1950, p. 142

5.39. B. K. Ridley, *Quantum Processes in Semiconductors*, Clarendon Press, Oxford, 1988, p. 196

5.40. N. Peyghambarian, S. W. Koch and A. Mysyrowicz, *Introduction to Semiconductor Optics*, Prentice Hall, Englewood Cliffs, 1993, p. 123

5.41. G. Lasher and P. Stern, *Phys. Rev.* **133**, A553 (1964)

5.42. G. W. Taylor, *J. Appl. Phys.* **70**, 2508 (1991)

5.43. Bose, *Zeits. Physik.* **26**, 178 (1924)

5.44. S. Schmitt-Rink, D. S. Chemla and D. A. B. Miller, *Adv. Phys.* **38**, 89 (1989)

5.45. R. C. Miller, A. C. Gossard, D. A. Kleinmann and O. Munteanu, *Phys. Rev. B* **29**, 3740 (1984)

5.46. M. H. Meynandier, C. Delalande, G. Bastard, M. Voos, F. Alexandre and J. L. Liéven, *Phys. Rev. B* **31**, 5539 (1985)

5.47. W. Heitler, *The Quantum Theory of Radiation*, Clarendon Press, Oxford, 1954, p. 32

5.48. L. C. West and S. J. Eglash, *Appl. Phys. Lett.* **46**, 1156(1985)

5.49. B. F. Levine, R. J. Malik, J. Walker, K. K. Choi, C. G. Bethea, D. A. Kleinman and J. M. Vandenberg, *Appl. Phys. Lett.* **50**, 273 (1987)

5.50. F. H. Julien, J. M. Lourtioz, N. Herschkorn, D. Delacourt, J. P. Pocholle, M. Papuchon, R. Planel and G. Le Roux, *Appl. Phys. Lett.* **53**, 116 (1988)

5.51. Prince J. S. A. Adelabu, B. K. Ridley, E. G. Scott and G. J. Davies, *Semicond. Sci. Technol*, **3**, 873 (1988)

5.52. B. F. Levine, A. Y. Cho, J. Walker, R. J. Malik, D. A. Kleinmann and D. L. Sivco, *Appl. Phys. Lett.* **52**, 1482 (1988)

5.53. H. Lobentanzer, W. König, W. Stolz, K. Ploog, T. Elsaesser and R. J. Bäuerle, *Appl. Phys. Lett.* **53**, 571 (1988)

5.54. H. Asai and Y. Kawamura, *Phys. Rev. B* **43**, 4748 (1991)

5.55. S. D. Gunapala, B. F. Levine, D. Ritter, R. Hamn and M. B. Panish, *Appl. Phys. Lett.* **58**, 2024 (1991)

5.56. Y. Shakuda and H. Katahama, *Jpn. J. Appl. Phys.* **29**, L552 (1990)

5.57. H. Hertle, G. Schuberth, E. Gornik, G. Abstreiter and F. Schaffler, *Appl. Phys. Lett.* **59**, 2977 (1991)

5.58. C. Jelen, S. Slivken, J. Hoff, M. Razeghi and G. J. Brown, *Appl. Phys. Lett.* **70**, 360 (1991)

5.59. D. J. Newson and A. Kurobe, *Semicond. Sci. Technol.* **3**, 786 (1988)

5.60. J. Y. Anderson, L. Lundquist and R. F. Pasha, *Appl. Phys. Lett.* **58**, 2264 (1991)

5.61. G. Hasnain, B. F. Levine, C. G. Bethea, R. A. Logan, J. Walker and R. J. Malik, *Appl. Phys. Lett.* **54**, 2515 (1989)

5.62. K. W. Goossen, S. A. Lyon and K. Alavi, *Appl. Phys. Lett.* **53**, 1027 (1988)

5.63. H. Xie, J. Piao, J. Katz and W. I. Wang, *J. Appl. Phys.* **70**, 3152 (1991)

5.64. B. F. Levine, S. D. Gunapala, J. M. Kuo, S. S. Pei and S. Hui, *Appl. Phys. Lett* **.59**, 1864 (1991)

5.65. S. D. Gunapala, K. M. S. V. Bandara, B. F. Levine, G. Sarusi, J. S. Park, T. L. Lin, W. I. Pike and J. K. Liu, *Appl. Phys. Lett.* **64**, 3431 (1994)

5.66. W. Franz, *Z. Naturf.(a)*, **13**, 484 (1958)

5.67. L. V. Keldysh, *Soviet Phys. JETP*, **34**, 788 (1958)

5.68. R. A. Houston, *A Treatise on Light*, Longmans Green & Co. London, 1952, p. 316

5.69. D. A. B. Miller, D. S. Chemla, T. C. Damen, A. C. Gossard, W. Wiegmann, T. H. Wood and C. A. Burrus, *Phys. Rev. B* **32**, 1043 (1985)

5.70. G. Bastard, E. E. Mendez, L. L. Chang and L. Esaki, *Phys. Rev.* B **28**, 3241 (1983)

5.71. D. Ahn and S. L. Chung, *Appl. Phys. Lett.* **49**, 1450 (1986)

5.72. E. J. Austin and M. Jaros, *Phys. Rev.* B **31**, 5569 (1985)

5.73. T. Hiroshima and R. Lang, *Appl. Phys. Lett.* **49**, 639 (1986)

5.74. Der-San Chuu and Yu-Tai Shih, *Phys. Rev.* B **44**, 8054 (1991)

5.75. R. K. Jain, *Optical Engineering*, **21**, 199 (1982)

5.76. C. K. N. Patel, R. E. Slusher and P. A. Fleury, *Phys. Rev. Lett.* **17**, 1011 (1966)

5.77. F. Bassani, G. Pastori Parravicine and R. A. Ballinger, *Electron States and Optical Transitions in Solids*, Pergamon, Oxford, 1975, p.154

5.78. D. H. Auston, S. McAfee, C. V. Shank, E. P. Ippen and O. Teschke, *Solid-St. Electron.* **21**, 147 (1978)

5.79. D. S. Chemla, D. A. B. Miller, P. W. Smith, A. C. Gossard and W. Weigmann, *IEEE J. Quantum Electron.* **QE-20**, 265 (1984)

5.80. D. A. B. Miller, D. S. Chemla, D. J. Eilenberger, P. W. Smith, A. C. Gossard and W. Wiegmann, *Appl. Phys. Lett.* **42**, 925 (1983)

5.81. K. Tai, J. Hegarty and W. T. Tsang, *Appl. Phys. Lett.* **51**, 86 (1987)

5.82. J. S. Weiner, D. B. Pearson, D. A. B. Miller, D. S. Chemla, D. Sivco and A. Y. Cho, *Appl. Phys. Lett.* **49**, 531 (1986)

5.83. S. Schmitt-Rink, D. S. Chemla and D. A. B. Miller, *Adv. Phys.* **38**, 89 (1989)

5.84. H. Barry Bebb and E. W. Williams, in *Semiconductors and Semimetals*, Vol.8, R. K. Willardson and A. C. Beer, eds. Academic Press, New York, 1972, p.184

5.85. H. Hillmer, A. Forchel, C. W. Tu and R. Sauer, *Semicond. Sci. Technol.* **7**, B235 (1992)

5.86. D. F. Weltch, G. W. Wicks and L. F. Eastman, *Appl. Phys. Lett.* **46**, 991 (1985)

5.87. J. Lee, E. S. Koteles and M. O. Vassell, *Phys. Rev.* B **33**, 5512 (1986)

5.88. P. K. Basu, *Phys. Rev. B.* **44**, 8798 (1991)

5.89. R. A. Stradling and P. C. Klipstein, eds. *Growth and Charcterisation of Semiconductors*, Adam-Hilger, Bristol, 1990, p.135

5.90. K. Brunner, G. Abstreiter, G. Böhm, G. Tränkle and G. Weimann, *Appl. Phys. Lett.* **64**, 3320 (1994)

5.91. R. C. Miller, A. C. Gossard, W. T. Tsang and O. Munteanu, *Solid-St. Commun.* **43**, 519 (1982)

5.92. C. Weisbuch, R. C. Miller, R. Dingle, A. C. Gossard and W. Weigmann, *Solid-St. Commun.* **37**, 219 (1981)

5.93. F. Yang, M. Wilkinson, E. J. Austin and K. P. O'Donnell, *Phys. Rev. Lett.* **70**, 323 (1993)

5.94 H. Wang, M. Jiang and D. G. Steele, *Phys. Rev. Lett.* **65**, 1225 (1990)

5.95. L. Munoz, L. Vina, N. Mestres and W. I. Wang, *Solid-St. Electron.* **37**, 877 (1994)

5.96. S. Lutgen, T. F. Albrecht, T. Marschner, W. Stolz and E. O. Gobel, *Solid-St. Electron.* **37**, 899 (1994)

5.97. D. K. Nayak, N. Usami, H. Sunamura, S. Fukatsu and Y. Shiraki, *Solid-St. Electron.* **37**, 937 (1994)

5.98. P. D. Colbourne and D. T. Cassidy, *IEEE J. Quantum Electron.*, **29**, 62 (1993)

5.99. H. Hillmer, S. Hansmann, A. Forchel, M. Morohashi, E. Lopez, H. P. Meier and K. Ploog, *Appl. Phys. Lett.* **53**, 1937 (1988)

5.100. H. Hillmer, A. Forchel, S. Hansmann, M. Morohashi, E. Lopez, H. P. Meier and K. Ploog, *Phys. Rev. B.* **39**, 10901 (1989)

5.101. P. K. Basu and P. Ray, *Phys. Rev. B* **44**, 1844 (1991)

5..102 H. Hillmer, A. Forchel and C. W. Tu, *J. Phys: Condens. Matter*, **5**, 5563 (1993)

CHAPTER 6

6.1. B. R. Nag, *Theory of Electrical Transport in Semiconductors*, Pergamon, Oxford, 1972

6.2. B. R. Nag, *Electron Transport in Compound Semiconductors*, Springer-Verlag, Berlin, 1980

6.3. S. Mukhopadhyay and B. R. Nag, *Phys. Rev. B* **48**, 17960 (1993)

6.4. N. Mori and T. Ando, *Phys. Rev. B* **40**, 6175 (1989)

6.5. J. Lee, H. N. Spector, V. K. Arora, *J. Appl. Phys.* **54**, 6995 (1983)

6.6. T. Ando, *J. Phys. Soc. Jpn.* **51**, 3900 (1982)

6.7. F. Stern and W. E. Howard, *Phys. Rev.* **163**, 816 (1967)

6.8. Ching-Yuan Wu and G. Thomas, *Phys. Rev. B* **9**, 1724 (1974)

6.9. K. Hess, *Appl. Phys. Lett.* **35**, 484 (1979)

6.10. W. Walukiewicz, H. E. Ruda, J. Lagowski and H.C. Gatos, *Phys. Rev. B* **30**, 4571 (1984)

6.11. F. Stern, *Phys. Rev. Lett.* **18**, 546 (1967)

6.12. T. Ando, A. B. Fowler and F. Stern, *Rev. Mod. Phys.* **54**, 437 (1982)

6.13. A. Katz, *Principles of Statistical Mechanics*, W. H. Freeman, San Francisco, 1967, p. 162

6.14. A. Hönig, *Phys. Rev. Lett.* **17**, 186 (1966)

6.15. D. S. Tang, *Phys. Rev. B* **36**, 2757 (1987)

6.16. R. Dingle, H. L. Störmer, A. C. Gossard and W. Wiegmann, *Appl. Phys. Lett.* **33**, 665 (1978)

6.17. H. L. Störmer and W. T. Tsang, *Appl. Phys. Lett.* **36**, 685 (1980)

6.18. P. J. Price, *Ann. Phys.* **133**, 217 (1981)

6.19. H. L. Störmer, A. Pinczuk, A. C. Gossard and W. Wiegmann, *Appl. Phys. Lett.* **38**, 691 (1981)

6.20. K. Y. Cheng, A. Y. Cho, T. J. Drummond and H. Morkoc, *Appl. Phys. Lett.* **40**, 147 (1982)

6.21. Y. Guldner, J. P. Vieren, P. Voisin, M. Voos, M. Razeghi and M. A. Poisson, *Appl. Phys. Lett.* **40**, 877 (1982)

6.22. A. Kastalsky, R. Dingle, K. Y. Cheng and A. Y. Cho, *Appl. Phys. Lett.* **41**, 274 (1982)

6.23. K. T. Chan, L. D. Zhu and J. M. Ballantyne, *Appl. Phys. Lett.* **47**, 44 (1985)

6.24. P. J. Price, *Surf. Sc.* **113**, 199 (1982)

6.25. S. Hiyamizu, J. Saito, K. Nanbu and T. Ishikawa, *Jpn. J. Appl. Phys.* **22**, L609 (1983)

6.26. G. Bastard, *Appl. Phys. Lett.* **43**, 591 (1983)

6.27. P. K. Basu and B. R. Nag, *Appl. Phys. Lett.* **43**, 689 (1983)

6.28. N. Sano, H. Kato and S. Chika, *Solid- St. Commun.* **49**, 123 (1984)

6.29. J. C. M. Hwang, A. Kastalsky, H. L. Störmer and V. G. Keramidas, *Appl. Phys. Lett.* **44**, 802 (1984)

6.30. H. L. Störmer, A. C. Gossard, W. Wiegmann, R. Bondell and K. Baldwin, *Appl. Phys. Lett.* **44**, 139 (1984)

6.31. M. Heiblum, E. E. Mendez and F. Stern, *Appl. Phys. Lett.* **44**, 1064 (1984)

6.32. P. J. Price, *Surf. Sci.* **143**, 145 (1984)

6.33. E. E. Mendez, P. J. Price and M. Heiblum, *Appl. Phys. Lett.* **45**, 294 (1984)

6.34. B. Vinter, *Appl. Phys. Lett.* **45**, 581 (1984)

6.35. B. J. F. Lin, D. C. Tsui, M. A. Paalanen and A. C. Gossard, *Appl. Phys. Lett.* **45**, 695 (1984)

6.36. K. Inoue, H. Sakaki and J. Yoshino, *Jpn. J. Appl. Phys.* **23**, L767 (1984)

6.37. E. E. Mendez and W. I. Wang, *Appl. Phys. Lett.* **46**, 1159(1985)

6.38. B. J. F. Lin, D. C. Tsui and G. Weimann, *Solid- St. Commun.* **56**, 287 (1985)

6.39. K. Inoue, H. Sakaki and J. Yoshino, *Appl. Phys. Lett.* **47**, 614 (1985)

6.40. S. Sasa, J. Saito, K. Nanbu, T. Ishikawa, S. Hiyamizu and M. Inoue, *Jpn. J. Appl. Phys.* **24**, L281 (1985)

6.41. B. Vinter, *Phys. Rev. B* **33**, 5904 (1986)

6.42. B. Vinter, *Surf. Sci.* **170**, 445 (1986)

6.43. W. Walukiewicz, *Phys. Rev. B* **31**, 5557 (1985)

6.44. K. Hirakawa and H. Sakaki, *Phys. Rev. B* **33**, 8291 (1986)

6.45. C. Guillemot, M. Badet, M. Gauneau, A. Regreny and J. C. Portal, *Phys. Rev.* B **35**, 2799 (1987)

6.46. M. Artaki and K. Hess, *Phys. Rev.* B **37**, 2933 (1988)

6.47. M. Tomizowa, T. Furutu, K. Yokoyama and A. Yoshi, *IEEE Trans. Electron. Dev.* **36**, 2380 (1989)

6.48. L. Pfeiffer, K. W. West, H. L. Stormer and K. W. Baldwin, *Appl. Phys. Lett.* **55**, 1888 (1989)

6.49. B. Laikhtman, *Appl. Phys. Lett.* **59**, 3021 (1991)

6.50. W. Ted Masselink, *Phys. Rev. Lett.* **66**, 1513 (1991)

6.51. T. Kawamura and S. Das Sarma, *Phys. Rev.* B **45**, 3612 (1992)

6.52. K. Griepel and U. Rossler, *Semicond. Sci. Technol.* **7**, 487 (1992)

6.53. K. Hess, H. Morkoç, H. Shichijo and B. G. Streetman, *Appl. Phys. Lett.* **35**, 469 (1979)

6.54. I. C. Kizilyalli and K. Hess, *J. Appl. Phys.* **65**, 2005 (1989)

6.55. K. S. Yoon, G. B. Stringfellow and R. J. Huber, *J. Appl. Phys.* **62**, 1931 (1987)

6.56. K. Yokayama, *J. Appl. Phys.* **63**, 938 (1988)

6.57. M. Tomizawa, T. Furutu, K. Yokoyama and A. Yoshi, *IEEE Trans. Electron Dev* **36**, 2380 (1989)

6.58. D. Bose and B. R. Nag, *Semicond. Sci. Technol.* **6**, 1135 (1991)

6.59. M. P. Chamberlain, D. Hoare, R. W. Kelsall and R. A. Abram, *Semicond. Sci. Technol.* **7**, B45 (1992)

6.60. S. M. Goodnick and J. E. Lary, *Semicond Sci. Technol.* **7**, B109 (1992)

6.61. W. P. Hong and P. K. Bhattacharya, *IEEE Trans. Electron Dev.* **ED-34**, 1491 (1989)

6.62. D. Yang, P. K. Bhattacharya, W. P. Hong, R. Bhattacharya and J. R. Hayes, *J. Appl. Phys.* **72**, 174 (1992)

6.63. W. T. Masselink, *Semicond. Sci. Technol.* **4**, 503 (1989)

6.64. T. Takagahara, *Phys. Rev.* B **31**, 6552 (1985)

6.65. S. Rudin and R. L. Reinecke, *Phys. Rev.* B **41**, 3017 (1990)

6.66. D. S. Chemla, D. A. B. Miller, P. W. Smith, A. C. Gossarel and W. Wiegmann, *IEEE J. Quantum Electron.* **QE-20,** 265 (1984)

6.67. W. T. Tsang and E. F. Schubere, *Appl. Phys. Lett.* **49,** 220 (1986)

6.68. M. Razeghi, J. P. Hirtz, U. O. Ziemelis, C. Delalande, B. Etienne and M. Voos, *Appl. Phys. Lett.* **43**, 585 (1983)

6.69. C. P. Kuo, K. L. Fry and G. B. Stringfellow, *Appl. Phys. Lett.* **47**, 855 (1985)

6.70. J. H. Marsh, J. S. Roberts and P. A. Claxton, *Appl. Phys. Lett.* **46**, 1161 (1985)

6.71. H. Temkin, M. B. Panish, P. M. Petroff, R. A. Hamn, J. M. Vandenberg and S. Sumski, *Appl. Phys. Lett.* **47**, 394 (1985)

6.72. D. F. Welch, G. W. Wicks and L. F. Eastman, *Appl. Phys. Lett.* **46**, 991 (1985)

6.73. K. Kawamara, K. Wakita and H. Asaki, *Electron. Lett.* **21**, 1169 (1985)

6.74. B. Deveaud, J. Y. Emery, A. Chomette, B. Lambert and M. Baudet, *Appl. Phys. Lett.* **45**, 1078 (1984)

6.75. H. Kawai, K. Kaneko and N. Watanabe, *J. Appl. Phys.* **56**, 463 (1984)

6.76. K. Ohta, H. Funbshi, T. Sakamoto, T. Nakagowa, N. J. Kawai, T. Kojima and M. Kawashima, *J. Electron. Mater.* **15**, 97 (1986)

6.77. B. R. Nag and S. Mukhopadhyay, *Jpn. J. Appl. Phys.* **31**, 3287 (1992)

6.78. D. Grützmacher, K. Wolter, H. Jürgensen, P. Balk and C. W. T. Bulle Lieuwma, *Appl. Phys. Lett.* **52**, 872 (1986)

6.79. H. Sakaki, T. Noda, K. Hirakawa, M. Tanaka and T. Matsue, *Appl. Phys. Lett.* **51**, 1934 (1987)

6.80. W. C. Mitchel, G. J. Brown. I. Lo, S. Elhamri, M. Razeghi and X. He, *Appl. Phys. Lett.* **65**, 1578 (1994)

6.81. B. R. Nag, S. Mukhopadhyay and M. Das, *J. Appl. Phys.* **86**, 459 (1999)

6.82. J. M. Redwing, M. A. Tishler, J. S. Flynn, S. Elhamri, M. Aouja, R. S. NewRock and W. C. Mitchell, *Appl. Phys. Lett.* **69**, 963 (1996)

6.83. F. Stengel, S. N. Mohammad and H. Morkoç, *J. Appl. Phys.* **80**, 3031 (1996)

6.84. M. A. Khar, J. M. Van Hove, J. N. Kunzia and D. T. Olson, *Appl. Phys. Lett.* **58**, 2408 (1991)

6.85. A. Ozgur, W. Kim, Z. Fan, A. Botchkarev, A. Sawador, S. N. Mohammad, B. Sverdlov and H. Morkoc, *Electron. Lett.* **31**, 1389 (1995)

6.86. S. J. Pearton, ed.,*GaN and Related Materials II*, Gordon and Breach Science Publishers, Asian Edition, 2000, p. 220

6.87. M. S. Shur and L. F. Eastman, *IEEE Electron. Dev.* **ED-26**, 1677 (1979)

6.88. M. S. Shur and L. F. Eastman, *Solid-st. Electron.* **24**, 11 (1981)

6.89. R. Landauer, *Phil. Mag.* **21**, 863 (1970)

6.90. M. Buttiker, Y. Imry, R. Landauer and S. Pinhas, *Phys. Rev. B* **31**, 6207 (1985)

6.91. R. Landauer, *IBM J. Res. Dev.* **1**, 223 (1957)

6.92. B. J. Van Wees, H. Van Houten, C. W. J. Beenakker, J. G. Williamson, L. P. Kouwenhoven, D. Van der Marel and C. T. Foxon, *Phys, Rev. Lett.* **60**, 848 (1988)

6.93. Y. Hirayama and T. Saku, *Appl. Phys. Lett.* **54**, 2556 (1989)

6.94. G. L. Snider, M. S. Miller, M. J. Rook and E. L. Hu, *Appl. Phys. Lett.* **59**, 2727 (1991)

CHAPTER 7

7.1. T. Mimura, S. Hiyamizu, T. Fujii and K. Nanbu, *Jpn. J. Appl. Phys.* **19**, L225 (1980)

7.2. D. Delagebeaudeuf, M. Laviron, P. Delescluse, Pham N. Tung, J. Chalplart and N. T. Linh, *Electron. Lett.* **18**, 103 (1982)

7.3. T. J. Drummond, W. Kopp, R. E. Thorne, R. Fischer and Morkoç, *Appl. Phys. Lett.* **40**, 879 (1982)

7.4. R. Fischer, T. J. Drummond, J. Klem, W. Kopp, T. S. Henderson, D. Perrachione and H. Morkoç, *IEEE Trans. Electron Dev.* **ED-31**, 1028 (1984)

7.5. K. Inoue and H. Sakaki, *Jpn. J. Appl. Phys.* **23**, L61 (1984)

7.6. H. Daembkes, ed., *Modulation Doped Field-Effect Transistors*, IEEE Press, New York, 1991.

7.7. T. J. Drummond, H. Morkoc, K. Lee and M. Shur, *IEEE Electron. Dev. Lett.* **EDL-3**, 338 (1982)

7.8. K. Lee, M. S. Shur, T. J. Drummond and H. Morkoç, *IEEE Trans. Electron Dev.* **ED-30**, 207 (1983)

7.9. T. Ando, A. B. Fowler and F. Stern, *Rev. Mod. Phys.* **54**, 437 (1982)

7.10. D. Delagebeaudeuf and N. T. Linh, *IEEE Trans. Electron Dev.* **ED-29**, 955 (1982)

7.11. P. Roblin, L. Rice, S. B. Bibyk and H. Morkoç, *IEEE Trans. Electron Dev.* **35**, 1207 (1988)

7.12. C. Chang and H. R. Fetterman, *IEEE Trans. Electron Dev.* **ED-34**, 1456 (1987)

7.13. D. Delagebeaudeuf and N. T. Linh, *Electron. Lett.* **18**, 510 (1982)

7.14. P. N. Tung, P. Delescluse, D. Delagebeaudeuf, M. Laviron, J. Chaplart and N. T. Linh, *Electron. Lett.* **18**, 517 (1982)

7.15. S. L. Su, R. Fischer, T. J. Drummond, W. G. Lyons, R. E. Thorne, W. Kopp and H. Morkoc, *Electron. Lett.* **18**, 794 (1982)

7.16. T. J. Drummond, S. L. Su, W. G. Lyons, R. Fischer, W. Kopp and H. Morkoç, *Electron. Lett.* **18**, 1057 (1982)

7.17. L. F. Lester, D. M. Smith, P. Ho, P. C. Chao, R. C. Tiberio, K. H. G. Duh and E. D. Wolf, *IEDM Tech.*,1988, p. 172

7.18. K. L. Tan, R. M. Dia, D. C. Streit, T. Lin, T. Q. Trinh, A. C. Han, P. M. Chow and H. C. Yen, *IEEE Trans. Electron Dev.* **ED-37**, 585 (1990)

7.19. H. Morkoc and H. Unlu, in *Semiconductors and Semimetals*, Vol. 24, R. Dingle, ed., Academic Press, New York, 1987, p. 135

7.20. R. Fischer, T. J. Drummond, J. Klem, W. Kopp, T. S. Henderson, D. Perrachione and H. Morkoç, *IEEE Trans. Electron. Dev.* **ED-31**, 1028 (1984)

7.21. T. P. Pearsall, R. Hendel, P. O'Connor, K. Alavi and A. Y. Cho, *IEEE Electron Dev. Lett.* **EDL-4**, 5 (1983)

7.22. K. H. G. Duh, P. C. Chao, S. M. J. Liu, P. Ho, M. K. Kao and J. M. Ballingal, *IEEE Microwave Guided Wave Lett.* **1**, 104 (1991)

7.23. U. K. Mishra, A. S. Brown, S. E. Rosenbaum, C. E. Hooper, M. W. Pierce, M. J. Delaney, S. Vaugn and K. White, *IEEE Electron Dev. Lett.* **EDL-9**, 647 (1988)

7.24. T. Enoki, K. Arai, Y. Ishii and T. Tamamura, *Electron. Lett.* **27**, 115(1991)

7.25. L. D. Nguyen, A. S. Brown, M. A. Thompson, L. M. Jelloian, L. E. Larson and H. Matloubian, *IEEE Electron Dev. Lett.* **EDL-13**, 143 (1992)

7.26. N. Pan, J. Elliot, H. Hendriks, L. Aucoin, P. Fay and I. Adesia, *Appl. Phys. Lett.* **66**, 212 (1995)

7.27. L. D. Nguyen, L. E. Larson and U. K. Mishra, *Proc. IEEE*, **80**, 494 (1992)

7.28. T. Enoki, M. Tomizawa, Y. Umeda and Y. Ishi, *Jpn. J. Appl. Phys.* **33**, 798 (1994)

7.29. P. Ho, M. Y. Kao, P. C. Chao, K. H. G. Duh, J . M. Balingall, S. T. Allen, A. T. Tessmer and P. M. Smith, *Electron. Lett.* **27**, 325 (1991)

7.30. D. C. Streit, K. L. Tan, R. M. Dia, A. C. Han, P. H. Liu, H. C. Yen and P. C. Chow, *Electron. Lett.*, **27**, 1149 (1991)

7.31. L. D. Nguyen, A. S. Brown, M. A. Thompson and L. M. Jellian, IEEE Trans. Electron Dev., **39**, 2007 (1992)

7.32. N. Pan, J. Elliot, H. Hendriks, A. Aucoin, P. Fay and I. Adesda, *Appl. Phys. Lett.* **66**, 212 (1995)

7.33. T. Suemitsu , T. Enoki, N. Sano, M. Tomizawa and Y. Ishi, *IEEE Trans. Electron Dev.* **45**, 2390 (1998)

7.34. A. A. Ketterson, W. T. Masselink, J. S. Gedymin, J. Klem, Chi- Kun Peng, W. F. Kopp, H. Morkoc and K. R. Gleason, *IEEE Trans. Electron Dev.* **ED-33**, 564 (1986)

7.35. R. Lai, P. K. Bhattacharya, D. Yang, I. Brock, S. A. Alterovitz and A. N. Downey, *IEEE Trans. Electron. Dev.* **39**, 2206 (1992)

7.36. J. H. Huang, T. Y. Chang and B. Lalevic, *Appl. Phys. Lett.* **60**, 733 (1992)

7.37. A. Chin and T. Y. Chang, *J. Vac. Sci. Technol. B*, **8**, 364 (1990)

7.38. J. Dickmann, K. Reipe, A. Geyer, B. E. Maile, A. Schurr, M. Berg and H. Daembkes, *Jpn. J. Appl. Phys.* **35**, 10 (1996)

7.39. K. B. Chough, T. Y. Chang, M. D. Feur, N. J. Sauer and B. Lalevic,*IEEE Electron Dev. Lett.* **EDL-13**, 451 (1992)

7.40. R. P. Schneider,Jr. and B. W. Wessels, *J. Appl. Phys.* **70**, 405 (1994)

7.41. J. B. Voos, W. Kruppa, B. R. Bennet, D. Park, S. W. Kirchoefer, R. Bass and H. B. Dietrich, *IEEE Trans. Electron Dev.* **45**, 1877 (1998)

7.42. M. Asif Khan, J. N. Kuznia, D. T. Olson, W. J. Schaff, J. W. Burm and M. S. Shur, *Appl. Phys. Lett*, **65**, 1121 (1994)

7.43. K. Van der Zanden, D. M. M-P. Schreurs, R. Menozzi and M. Borgarino, *IEEE Trans. Electron Dev.* **46**, 1570 (1999)

7.44. A. J. Valois, C. Y. Robinson, K. Lee and M. S. Shur, *J. Vac. Sci. Technol.* **B1**, 190 (1983)

7.45. S. Subramanian, *IEEE Trans. Electron Dev.* **ED-32**, 865 (1985)

7.46. Y. Gobert and G. Salmer, *IEEE Trans. Electron Dev.* **41**, 299 (1994)

7.47. T. Mizutani and K. Maezawa, *IEEE Electron Dev. Lett.* **13**, 8 (1992)

7.48. S. J. Zurek, R. B. Darling, K. J. Kuhn and M. C. Foisy, *IEEE Electron Dev.* **45,** 2 (1998)

7.49. M. C. Foisy, *A Physical Model for the Bias Dependence of the Modulation-doped Field-effect Transistor's High Frequency Performance*, Ph. D. Dissertation, Cornell Univ. Ithaca, N. Y. 1990

7.50. R. Singh and C. M. Snowden, *IEEE Trans. Electron Dev.* **45**, 1165 (1998)

7.51. H. Happy, S. Bollaert, H. Foure and A. Cappy, *IEEE Trans. Electron Dev.*, **45**, 2089 (1998)

7.52. S. Sen, M. K. Pandey and R. S. Gupta, *IEEE Trans. Electron Dev.* **46**, 1818 (1999)

7.53. S. Karmalkar and G. Ramesh, *IEEE Trans. Electron Dev.* **47**, 11 (2000)

7.54. J. J. Brown, J. A. Pusl, M. Hu, A. E. Schmitz, D. P. Docter, J. B. Shealy, M. G. Case, M. A. Thompson and L. D. Nguyen, *IEEE Microwave Guided Wave Lett.*, **6**, 91 (1996)

7.55. M. Aust, H. Wang, M. Biedenbender, R. Lai, D. C. Streit, G. C. Dow and B. R. Allen, *IEEE Microwave Guided Wave Lett.* **5**, 12 (1995)

7.56. S. W. Chen, P. M. Smith, S. J. Liu, W. F. Kopp and T. J. Rogers, *IEEE Microwave and Guided Wave Lett.* **5**, 201 (1995)

7.57. J. C. Huang, P. Saledas, J. Wendler, A. Platzker, W. Boulais, S. Shanfield, W. Hoke, P. Lymn, L. Aucion, A. Miquelarena, C. Bedard and D. Atwood, *IEEE Electron Dev. Lett.* **14**, 456 (1993)

7.58. M. H. Somerville, J. A. del Alamo and P. Saunier, *IEEE Trans. Electron Dev.* **45**, 1883 (1998)

7.59. K. Higuchi, H. Matsumoto, T. Mishima and T. Nakamura, *IEEE Trans. Electron Dev.* **46**, 1312 (1999)

7.60. T. Enoki, K. Arai, A. Kohzen and Y. Ishii, *Proc. IEEE 4th Int. Conf. InP Related Materials*, 1992, p. 371

7.61. G. Menghesso, A. Neviani, R. Oesterholt, M. Matloubian, T. Liu, J. L. Brown, C. Canali and E. Zanoni, *IEEE Trans. Electron Dev.* **46**, 2 (1999)

7.62. H. Maher, J. Décobert, A. Falcou, M. Le Pallec, G. Post, Y. I. Nissim and A. Scavennec, *IEEE Trans. Electron Dev.* **46**, 32 (1999)

7.63. A. J. Seeds and A. A. deSallers, *IEEE Trans. Microwave Theory and Technique*, **38**, 577 (1990)

7.64. D. Jager, *Tech Dig. Int. Topical Meeting on Microwave Photonics*, Dec. 1-2, 1996

7.65. C. Y. Chen, N. A. Olsson, C. W. Tu and P. A. Garbinski, *Appl. Phys. Lett.* **46**, 681 (1985)

7.66. D. Yang, P. K. Bhattacharya, R. Lai, T. Brock and A. Paolella, *IEEE Trans. Electron Dev.* **42**, 1056 (1995)

7.67. R. Lai, P. K. Bhattacharya and T. Brock, *Electron. Lett.* **27**, 1576 (1997)

7.68. Y. Takanashi, K. Takahata and Y. Muramoto, *IEEE Electron Dev. Lett.* **19**, 2279 (1998)

7.69. Y. Takahashi, K. Takahata and Y. Muramoto, *IEEE Trans. Electron Dev.* **46**, 2271 (1999)

7.70. G. Halkias and A. Vegiri, *IEEE Trans. Electron Dev.* **45**, 2430 (1998)

7.71. H.Fukuyama, K. Maezawa, M. Yamamoto, H. Okazaki and M. Miraguchi, *IEEE Trans. Electron Dev.* **46**, 281 (2000)

7.72. G. Menghesso, T.Grave, M. Manfredi, M. Pavesi, C. Canali and E. Zanoni, *IEEE Trans. Electron Dev.* **47**, 2 (1999)

CHAPTER 8

8.1. F. Capasso, ed., *Physics of Quantum Electron Devices* , Springer-Verlag, Berlin, 1990

8.2. L. L. Chang, L. Esaki and R. Tsu, *Appl. Phys. Lett.* **24**, 593 (1974)

8.3. R. Tsu and L. Esaki, *Appl. Phys. Lett.* **22**, 562 (1973)

8.4. S. Sen, B. R. Nag and S. Midday, in *Physical Concepts of Materials for Novel Optoelectronic device Applications II:Device Physics and Applications*, M. Razeghi, ed., Proc. SPIE 1362, 1991, p.750

8.5. M. Tsuchiya, H. Sakaki and J. Yoshino, *Jpn. J. Appl. Phys.* **24**, L466 (1985)

8.6. M. Tsuchiya and H. Sakaki, *Jpn. J. Appl. Phys.* **25**, L185 (1986)

8.7. M. Tsuchiya and H. Sakaki, *Appl. Phys. Lett.* **49**, 88 (1986)

8.8. M. Tsuchiya and H. Sakaki, *Appl. Phys. Lett.* **50**, 1503(1987)

8.9. S. Muto, T. Inata, H. Ohnishi, N. Yokoyama and S. Hiyamizu, *Jpn. J. Appl. Phys.* **25**, L577 (1986)

8.10. C. I. Huang, M. J. Paulus, C. A. Bozada, S. C. Dudley, K. R. Evans, C. E. Stutz, R. L. Jones and M. E. Cheney, *Appl. Phys. Lett.* **51**, 121 (1987)

8.11. S. Luryi, *Appl. Phys. Lett.* **47**, 490 (1985)

8.12. T. Weil and B. Vniter, *Appl. Phys. Lett.* **50**, 1281 (1987)

8.13. K. K. Choi, B. F. Levine, C. G. Bethea, J. Walker and R. J. Malik, *Phys. Rev. Lett.* **59**, 2459 (1987)

8.14. T. C. L. G. Sollner, W. D. Goodhue, P .E. Tannenwald, C. D. Parker and D. D. Peek, *Appl. Phys. Lett.* **43**, 588 (1983)

8.15. S. Sen, F. Capaso, A. L. Hutehinson and A. Y. Cho, *Electron. Lett.* **23**, 1229 (1987)

8.16. T. Inata, S. Muto, Y. Nakata, T. Fujii, H. Ohnishi and S. Hiyamizu, *Jpn. J. Appl. Phys.* **25**, L983 (1986)

8.17. P. D. Hadson, D. J. Robbins, R. H. Wallis, J. J. Davies and A. C. Marshall, *Electron. Lett.* **24**, 187 (1988)

8.18. J. E. Oh, I. Mehdi, J. Pamulapati, P. K. Bhattacharya and G. I. Haddad, *J. Appl. Phys.* **65**, 842 (1989)

8.19. S. Muto, T. Inata, Y. Sugiyama, Y. Nakata, T. Fujii, H. Ohnishi and S. Hiyamizu, *Jpn. J.Appl. Phys.* **26**, L220 (1987)

8.20. T. Inata, S. Muto, Y. Nakata, S. Sasa, T. Fujii and S. Hiyamizu, *Jpn. J. Appl. Phys. Lett.* **26** L1332 (1987)

8.21. E. T. Yu and T. C. McGill, *Appl. Phys. Lett.* **53**, 60 (1988)

8.22. L. F. Luo, R. Beresford and W. I. Wang, *Appl. Phys. Lett.* **53**, 2320 (1988)

8.23. J. R. Söderstrom, D. H. Chow and T. C. McGill, *IEEE Electron Device Lett.* **11**, 27 (1990)

8.24. T. C. L. G. Sollner, E. R. Brown, W. D. Goodhue and H. Q. Le, in *Physics of Quantum Electron Devices*, F. Capasso, ed., Springer-Verlag, Berlin, 1990, p. 147

8.25. F. Capasso, S. Sen, F. Beltram and A. Y. Cho, in *Physics of Quantum Electron Devices*, F. Capasso, ed., Springer-Verlag, Berlin, 1990, p. 181

8.26. E. R. Brown, J. R. Söderström, C. D. Parker, L. J. Mahoney,K. M. Molver and T. C. McGill, *Appl. Phys. Lett.*, **58**, 2291 (1991)

8.27. E. Ozbay and D. M. Bloom, *IEEE Electron Dev. Lett.* **12**, 480 (1991)

8.28. H. C. Liu and D. D. Coon, *Appl. Phys. Lett.* **50**, 1246 (1987)

8.29. R. Bouregba, O. Vanbesien, L. de Saint Pol, and D. Lippens, *Electron. Lett.*, **26**, 1804 (1990) *Electron. Lett.* **26**, 1804 (1990)

8.30. A. A. Lakhani and R. C. Potter, *Appl. Phys. Lett.* **52**, 1684 (1988)

8.31. J. F. Whitaker, G. A. Mouron, T. C. G. Sollner and Goodhue, *Appl. Phys. Lett.* **53**, 385 (1988)

8.32. P. England, J. E. Golub, L. T. Florez and J. P. Harbison, *Appl. Phys. Lett.* **58**, 887 (1991)

8.33. H. C. Liu, A. G. Steele, M. Buchanan and Z. R. Wasilewski, *IEEE Electron Dev. Lett.* **13**, 363 (1992)

8.34. T. K. Woodward, T. C. McGill, H. F. Chung and R. D. Burnham, *IEEE Electron Dev. Lett.* **9**, 122 (1988)

8.35. S. Sen, F. Capasso, A. Y. Cho and D. L. Sivco,*IEEE Electron Dev. Lett.* **9**, 533 (1988)

8.36. L. M. Lunardi, S. Sen, F. Capasso, P. R. Smith, D. L. Sivco and A. Y. Cho, *IEEE Electron Dev. Lett.* **10**, 219 (1989)

8.37. A. C. Seabaugh, Y. C. Kao and H. T. Yuan, *IEEE Electron Dev. Lett.* **13**, 479 (1992)

8.38. N. Yokoyama, K. Imamura, S. Muto. S. Hiyamizu and H. Nishi, *Jpn. J. Appl. Phys.* **24**, L853 (1985)

8.39. F. Capasso and R. A. Kiehl, *Appl. Phys.* **58**, 1366 (1985)

8.40. F. Capasso, S. Sen, A. C. Gossard, A. L. Hutchinson and J. H. English, *IEEE Electron Dev. Lett.* **EDL-6**, 636 (1985)

8.41. T. S. Moise, A. C. Seabaugh, E. A. Bean,III and J. N. Randall, *IEEE Electron Dev. Lett.* **14**, 441 (1993)

8.42. W. C. B. Weatman, E. R. Brown, M. J. Rooks, P. Maki, W. J. Grimm and M. Shur, *Electron Dev. Lett.* **15**, 236 (1994)

8.43. H. Fukuyama, K. Maezawa, M. Yamamoto, H. Okazaki and M. Muraguchi, *IEEE Trans. Electron Dev.* **46**, 281 (1997)

8.44. G. I. Haddad and P. Majumder, *Solid-St. Electron.* **41**, 1515 (1997)

8.45. D. K. Ferry, *FED Journal*, **10**, (Suppl. 1), 5 (1999)

8.46. J. P. A. Van der Wagl, A. C. Seabaugh and E. A. Beam, *IEEE Electron Dev. Lett.* **19**, 7 (1998)

8.47. S. L. Rommel, S. E. Dillon, M. W. Dashiell, H. Feng, J. Kolodzay, P. R. Berger, P. E. Thompson, K. D. Hobart, *Appl. Phys. Lett.* **73**, 2191 (1998)

8.48. H. C. Liu, A. C. Steele, M. Buchanan and Z. R. Wasilewski, *Electron Dev. Lett.* **13**, 363 (1992)

CHAPTER 9

9.1. B. R. Nag, *Appl. Phys. Lett.* **65**, 1938 (1994)

9.2. B. R. Nag, *Infrared Phys. Technol.* **36**, 831 (1995)

9.3. B. R. Nag, *Electron Transport in Compound Semiconductors*, Springer-Verlag, Berlin, 1980, p.260

9.4. K. Iga, *Fundamentals of Laser Optics* (Plenum Press, New York, 1994) p.65

9.5. M. Minorsky, *Non-linear Oscillations* (D. Van Nostrand Co., New York, 1962)

9.6. M. Asada, A. Kameyama and Y. Suematsu, *IEEE J. Quantum Electron* **QE-20**, 745 (1984)

9.7. R. Olshansky, C.B. Su, J. Manning and W. Powazinik, *IEEE J. Quantum Electron* **QE-20**, 838 (1984)

9.8. G. P. Agarwall and N. K. Dutta, *Long-wavelength Semiconductor Lasers*, Van Nostrand Reinhold, New York, 1986, p.57

9.9. A. R. Beattie and P. T. Landsberg, *Proc. Roy. Soc. (London)*, **Ser. A 249**, 16 (1959)

9.10. A. Sugimura, *IEEE J. Quantum Electron.* **QE-17**, 627 (1981)

9.11. O. Gilard, F. Lozes-Dupuy, G. Vassilieff, J. Barrau and P. Le Jeune, *J. Appl. Phys.* **84**, 2705 (1998)

9.12. T. L. Paoli, *Appl. Phys. Lett.* **34**, 652 (1979)

9.13. W. Streifer, R. D. Burnham and D. R. Scifres, *Appl. Phys. Lett.* **37**, 121 (1980)

9.14. D. Kasemset, C. S. Hong, N. B. Patel and P. D. Dapkus, *IEEE J. Quantum Electron* **QE-19**, 1025 (1983)

9.15. C. Weisbuch and B. Vinter, *Quantum Semiconductor Structures*, Academic New York, 1991, p.175

9.16. S. R. Chinn, P. S. Zory and A. R. Reisinger, *IEEE J. Quantum Electron* **24**, 2191 (1988)

9.17. M. Yamada, S. Ogita, M. Yamagishi, K. Tabata, N. Nakaya, M. Asada and Y. Suematsu, *Appl. Phys. Lett.* **45**, 324 (1984)

9.18. N. K. Dutta, *J. Appl. Phys.* **53**, 7211 (1982)

9.19. N. K. Dutta, *J. Appl. Phys.* **54**, 1236 (1983)

9.20. M. F. Lu, J. S. Deng, C. Juang, M. J. Jou and B. J. Lee, *IEEE J. Quantum Electron.* **31**, 1418 (1995)

9.21. P. S. Zory, Jr., ed., *Quantum Well Lasers*, Academic Press, New York, 1993.

9.22. R. D. Dupuis, P. D. Dapkus, N. Holonyak, Jr., E. A. Rezek and R. Chinn, *Appl. Phys. Lett.* **32**, 295 (1978)

9.23. W. T. Tsang, *Appl. Phys. Lett.* **39**, 786 (1981)

9.24. W. T. Tsang, *Appl. Phys. Lett.* **40**, 217 (1982)

9.25. Sei-Ichi Miyazawa, Y. Sekiguchi and M. Okuda, *Appl. Phys. Lett* **63**, 3583 (1993)

9.26. T. Hayakawa, K. Matsumoto, M. Morishima, M. Nagai, H. Horie, Y. Ishigame, A. Isoyama and Y. Niwata, *Appl. Phys. Lett,* **63**, 1718 (1993)

9.27. I. Zh. Alferov, A. M. Vasilev, S. V. Ivanov, P. S. Kopev, N. N. Lednestsov, M. E. Lutsenko, B. Ya. Melster and V. M. Ustinov, *Sov. Tech. Phys. Lett.* **14**, 782 (1988)

9.28. E. Kapon, S. Simhony, J. P. Harrison, L. T. Florez and P. Worland, *Appl. Phys. Lett.* **56**, 1825 (1990)

9.29. P. S. Zory, A. R. Reisinger, R. G. Waters, L. J. Mawst, C. A. Zmudzinski, M. A. Emanuel, M. E. Givens and J. J. Coleman, *Appl. Phys. Lett.* **49**, 16 (1986)

9.30. K. Uomi, M. Mishima and N. Chinone, *Appl. Phys. Lett.* **51**, 78 (1987)

9.31. P. L. Derry, T. R. Chen, Y. H. Zhuang, J. Paslaski, M. Mittelstein, K. Vahala and H. Yariv, *Appl. Phys. Lett.* **53**, 271 (1988)

9.32. Y. Suematsu and K. Iga, *Introduction to Optical Fibre Communication*, John Wiley & Sons, New York, 1978.

9.33. E. A. Rezek, N. Holonyak, Jr. and B. K. Fuller, *J. Appl. Phys.* **51**, 2402 (1980)

9.34. Y. Sasai, N. Hase, M. Ogura and T. Kajiwara, *J. Appl. Phys.* **59**, 28 (1986)

9.35. T. Yanase, Y. Kato, L. Mito, M. Yamaguchi, K. Nishi, K. Kobayashi and P. Lang, *Electron. Lett.*, **19**, 700 (1983)

9.36. N. K. Dutta and R. J. Nelson, *Appl. Phys. Lett.* **38**, 407(1981)

9.37. A. Sugimura, *Appl. Phys. Lett.* **39**, 21 (1981)

9.38. W. T. Tsang, F. S. Choa, R. A. Logan, T. Tanbum-Ek, M. C. Wee, Y. K. Chen, A. M. Sergent and K. W. Wecht, *Appl. Phys. Lett.* **59**, 3084 (1991)

9.39. H. Temkin, K. Alavi, W. R. Wagner, T. P. Pearsall and A. Y. Cho, *Appl. Phys. Lett.*, **42**, 845 (1983)

9.40. W. T. Tsang, *Appl. Phys. Lett.* **44**, 288 (1984)

9.41. W. T. Tsang, F. S. Choa, M. C. Wu, Y. K. Chen, A. M. Sergent and P. F. Sciortino, Jr. *Appl. Phys. Lett.* **58**, 2610 (1991)

9.42. E. Yablonovitch and E. O. Kane, *J. Lightwave Technol.* **4**, 504 (1986)

9.43. J. W. Mathews and A. E. Blakeslee, *J. Cryst. Growth* **27**, 118 (1974)

9.44. S. Smetona, B. B. Elenkrig, J. G. Simmons, T. Makino and J. D. Evans, *J. Appl. Phys.* **84**, 4076 (1998)

9.45. A. Gavini and M. Cardona, *Phys. Rev.* **B 1**, 672 (1970)

9.46. S. Shimada, *Optics Photon News* **1**, 6 (1990)

9.47. W. D. Laidig, P. J. Caldwell, Y. F. Lin and C. K. Peng, *Appl. Phys. Lett.* **44**, 653 (1984)

9.48. N. Yamada, G. Roos and J. S. Harris, Jr., *Appl. Phys. Lett.* **59**, 1040 (1991)

9.49. J. M. Kuo, Y. K. Chen, M. C. Wu and M. A. Chin, *Appl. Phys. Lett.* **59**, 2781 (1991)

9.50. S. D. Offsey, W. J. Schaff, L. F. Lester and L. F. Eastman, *Appl. Phys. Lett.* **58**, 1445 (1991)

9.51. J. P. Van der Ziel and Naresh Chand, *Appl. Phys. Lett.* **60**, 6 (1992)

9.52. Z. L. Liau, S. C. Palmateer, S. H. Groves, J. N. Walpole and L. J. Missagio, *Appl. Phys. Lett,* **60**, 6 (1992)

9.53. C. K. Sun, H. K. Choi, C. A, Wang and J. G. Fujimoto, *Appl. Phys. Lett,* **63**, 96 (1993)

9.54. Naresh Chand, E. E. Becker, J. P. Van der Ziel, S. N. G. Chu and N. K. Dutta, *Appl. Phys. Lett.* **58**, 1704 (1991)

9.55. R. L. Williams, M. Dion, F. Chatenoud, and K. Dzurko, *Appl. Phys. Lett.* **58**, 1816 (1991)

9.56. N. C. Fratechi, M. Y. Jow, P. D. Dapkus and F. J. Levi, *Appl. Phys. Lett.* **65**, 1748 (1994)

9.57. M. Okhubo, T. Ijichi, A. Iketani and T. Kikuto, *Appl. Phys. Lett.* **60**, 1413 (1992)

9.58. C. Lin, M. Wu, H. Shiao and K. Liu, *IEEE Trans. Electron Dev.* **46**, 1614 (1999)

9.59. D. Ahn, *Appl. Phys. Lett.* **66**, 628 (1995)

9.60. H. Tanaka, Y. Kawamura and H. Asahi, *Electron. Lett.* **22**, 707 (1986)

9.61. Jun-Ichi Hashimoto, T. Katsuyama, J. Shinkai, I. Yoshida and H. Hayashi, *Appl. Phys. Lett.* **58**, 879 (1991)

9.62. M. Mannoh, J. Hoshmia, S. Kamayama, H. Ohta, Y. Ban and K. Ohnaka, *Appl. Phys. Lett.* **63**, 1173 (1993)

9.63. M. Watanabe, J. Rennie, M. Okajima and G. Hatukoshi, *Appl. Phys. Lett.* **63**, 1486 (1993)

9.64. S. Nakamura, M. Senoh, N. Iwasa and S. Nagahama, *Jpn. J. Appl. Phys. Lett.* **34**, L797 (1995)

9.65. S. Nakamura, M. Senoh, N. Iwasa, S. Nagahama, T. Yamada and T. Mukai, *Jpn. J. Appl. Phys. Lett.* **34**, L1332 (1995)

9.66. S. Nakamura, N. Senoh, S. Nagahama, N. Iwasa, T. Yamada, T. Matsushita, H. Kiyoku and Y. Sugimoto, *Jpn. J. Appl. Phys.* **35**, L74 (1996)

9.67. J. D. Walker, D. M. Kuchta and J. S. Smith, *Appl. Phys. Lett.* **59**, 2079 (1991)

9.68. C. Lei, T. J. Rogers, D. G. Deppe and B. G. Streetman, *Appl. Phys. Lett.* **58**, 1122 (1991)

9.69. B. Tell, K. F. Brown-Goebler, R. E. Leibenguth, F. M. Baez and Y. H. Lee, *Appl. Phys. Lett.* **60**, 683 (1992)

9.70. S. S. Ou, M. Jansen, J. J. Yang and M. Sergant, *Appl. Phys. Lett.* **59**, 1037 (1991)

9.71. Ching-Ping Chao, Kwok-Keung Law and J. L. Merz, *Appl. Phys. Lett.* **59**, 1532 (1991)

9.72. S. S. Ou, M. Jansen, J. J. Yang, L. J. Mawst and T. J. Roth, *Appl. Phys. Lett.* **59**, 2085 (1991)

9.73. Y. Miyamoto, M. Cao, Y. Shingai, K. Furuya, Y. Suematsu, K. G. Ravikumar and S. Arai, *Jpn. J. Appl. Phys.* **26**, L 225 (1987)

9.74. E. Kapon, in *Quantum Well Lasers*, P. S. Zory, Jr., ed., Academic, New York, 1993, p.461

9.75. E. Kapon, S. Simhony, R. Bhat and D. M. Hwang, *Appl. Phys. Lett.* **55**, 2715 (1989)

9.76. S. Simhony, E. Kapon, E. Colas, D. M. Hwang, N. G. Stoffel and P. Worland, *Appl. Phys. Lett.* **59**, 2225 (1991)

9.77. S. Koshiba, H. Noge, H. Akiyama, T. Inoshita, Y. Nakamura, A. Shimizu, Y. Nagamune, M. Tsuchiya, H. Kano and H. Sakaki, *Appl. Phys. Lett.* **64**, 363 (1994)

9.78. T. Arakawa, M. Nishioka, Y. Nagamune and Y. Arakawa, *Appl. Phys. Lett.* **64**, 2200 (1994)

9.79. S. T. Chou, K. Y. Cheng, L. J. Chou and K. C. Hsieh, *Appl. Phys. Lett.* **67**, 2220 (1995)

9.80. D. E. Wohlert, S. T. Chou, A. C. Chen, K. Y. Cheng and K. C. Hsieh, *Appl. Phys. Lett.* **69**, 2386 (1996)

9.81. L. Li and Y. Chang, *J. Appl. Phys.*, **84**, 6162 (1998)

9.82. Zh. I. Alferov, *Phys. Scr.* **68**, 32 (1996)

9.83. D. Bimberg, M. Grundmann and N. N. Ledentsov, *Quantum Dot Heterostructures*, John Wiley,Chichester, 1999

9.84. D. Bimberg, N. N. Ledentsov, M. Grundmann, N. Kirstaeder, O. G. Schmit, M. H. Mao, V. M. Ustinov, A. Yu. Egorov, A. E. Zhukov, P. S. Kopév, Zh. I. Alferov, S. S. Ruvimov, U. Göesle and J. Heydenreich, *Jpn. J. Appl. Phys.* Pt. 1, **35**, 1311 (1996)

9.85. N. N. Ledentsov in *Proc. 23rd Intl. Conf. Phys. Semicond.*, eds., M. Scheffer and R. ZZimmerman, World Scientific, Singapore, 1996, p. 19

9.86. M. V. Maximov, Yu. M. Shernyakov, A. F. Tsatsulnikov, A. V. Lunev, A. V. Sakharov, V. M. Ustinov, A. Yu. Egrov, A. E. Zhukov, A. R. Kovsh, P. S. Kopev, L. V. Asyan, Zh. I. Alferov, N. N. Ledentsov, D. Bimberg, A. O. Kosogov and P. Werener, *J. Appl. Phys.* **83**, 5361 (1998)

9.87. R. Engelhardt, V. W. Pohl, D. Bimberg, D. Litvinov, A. Resenauer and D. Gerthsen, *J. Appl. Phys.* **86**, 5578 (1999)

CHAPTER 10

10.1. B. F. Levine, *J. Appl. Phys.* **74**, R1 (1993)

10.2. M. O. Manasreh (ed.), *Semiconductor Quantum Wells and Superllatices for Long-Wavelength Infrared Detectors* (Artech House, London, 1993)

10.3. L. C. West and S. J. Eglash, *Appl. Phys. Lett.* **46**, 1156 (1985)

10.4. B. F. Levine, A. Y. Cho, J. Walker, D. A. Kleinman and D. L. Sivco, *Appl. Phys. Lett.* **64**, 1481 (1988)

10.5. K. K. Choi, M. Taysing-Lara, P. G. Newman, W. Chang and G. J. Iafrate, *Appl. Phys. Lett.* **61**, 1781 (1992)

10.6. G. Hasnain, B. F. Levine, S. Gunapala and Naresh Chand, *Appl. Phys. Lett.* **57**, 608 (1990)

10.7. B. F. Levine, A. Zussman, S. D. Gunapala, M. T. Asom, J. M. Kuo and W. S. Hobson, *J. Appl. Phys.* **72**, 4429 (1992)

10.8. B. F. Levine, C. G. Bethea, G. Hasnain, V. O. Shen, E. Pelve, R. R. Abott and S. J. Hsieh, *Appl. Phys. Lett.* **56**, 851 (1990)

10.9. B. F. Levine, K. K. Choi, C. G. Bethea, J. Walker and R. J. Malik, *Appl. Phys. Lett.* **50**, 1092 (1987)

10.10. K. K. Choi, B. F. Levine, C. G. Bethea, J. Walker and R. J. Malik, *Appl. Phys. Lett.* **50**, 1814 (1987)

10.11. J. Y. Andersson and L. Lundquist, *J. Appl. Phys.* **71**, 3600 (1992)

10.12. G. Hasnain, B. F. Levine, C. G. Bethea, R. A. Logan and J. Walker, *Appl. Phys. Lett.* **54**, 2515 (1989)

10.13. J. Y. Anderson. L. Lundquist and Z. F. Pasku, *Appl. Phys. Lett.* **58**, 2264 (1991)

10.14. B. F. Levine, S. D. Gunapala, J. M. Kuo, S. S. Pei and S. Hui, *Appl. Phys. Lett.* **59**, 1864 (1991)

10.15. J. Katz, Y. Zhung and W. I. Wang, *Electron. Lett.* **28**,932(1992)

10.16. T. Cwik and C. Yeh, *J. Appl. Phys.* **86**, 2779 (1999)

10.17. G. Hasnain, B. F. Levine, D. L. Sivco and A. Y. Cho, *Appl. Phys. Lett.* **56**, 770 (1990)

10.18. S. D. Gunapala, B. F. Levine, D. Ritter, R. Hamm and M. B. Panish, *Appl. Phys. Lett.* **58**, 2024 (1991)

10.19. S. D. Gunapala, B. F. Levine, R. A. Logan, T. Tanbun-Ek and D. A. Humphrey, *Appl. Phys. Lett.* **57**, 1802 (1990)

10.20. S. D. Gunapala, K. M. S. V. Bandara, B. F. Levine, G. Sarusi, J. S. Park, T. L. Lin, W. T. Pike and J. K. Liu, *Appl. Phys. Lrett.* **64**, 3431 (1994)

10.21. J. S. Park, R. P. G. Karunasiri and K. L. Wang, *J. Vac. Sci. Technol B*, **8**, 217 (1990)

10.22. K. A. Harris, T. H. Myers, R. W. Yanka. L. M. Mohnkern and N. Otsuka, *J. Vac. Sc. Technol B*, **9**, 1752(1991)

10.23. M. S. Kiledjian, J. N. Schulman and K. L. Wang, *Phys. Rev. B* **44**, 5616 (1991)

10.24. G. Neu, Y. Chen, C. Deparis and J. Massies, *Appl. Phys. Lett.* **58**, 2111 (1991)

10.25. J. Shi and E. M. G oldys, *IEEE Trans. Electron Dev.* **46**, 83 (1999)

10.26. S. D. Gunapala, J. K. Liu, J. S. Psark, M. Sundaram, C. A. Shott, T. Hoelter, T. L. Lin, S. T. Massie, P. D. Maker and G. Sarusi, *IEEE Trans. Electron Dev.* **44**, 51 (1997)

10.27. C. G. Bethea, B. F. Levine, M. T. Asom, R. E. Leibenguth, J. W. Stayt, K. G. Glogovsky, R. A. Morgan, J. D. Blackwell and W. J. Parrish, *IEEE Trans. Electron Dev.* **40**, 1957 (1993)

10.28. L. J. Kozlewski, G. W. Williams, G. J. Sullivan, C. W. Farley, R. J. Anderson, J. Chen, D. T. Cheung, W. E. Tenant and R. E. DeWames, *IEEE Trans. Electron Dev.* **38**, 1124 (1991)

10.29. S. D. Gunapala, J. S. Park, G. Sarusi, T. L. Lin, J. K. Liu, P. D. Maker, R. E. Muller, C. A. Shott and T. Hoelter, *IEEE Trans. Electron Dev.* **44**, 45 (1997)

10.30. S. D. Gunapala, S. V. Bandara, J. K. Liu, W. Hong, M. Sundaram, P. D. Maker, R. E. Muller, C. A. Shott and R. Carralyo, *IEEE Trans. Electron Dev.* **45**, 1890 (1998)

10.31. M. Ershov and H. C. Liu, *J. Appl. Phys.* **86**, 6580 (1999)

10.32. M. Ershov, *J. Appl. Phys.* **86**, 7059 (1999)

10.33. T. H. Wood, C. A. Burrus, D. A. B. Miller. D. S. Chemla, T. C. Damen, A. C. Gossard and W. Wiegmann, *Appl. Phys. Lett.* **44**, 16 (1984)

10.34. K. W. Goosen, J. E. Cunninghnam and W. Y. Jan, *Appl. Phys. Lett.* **64**, 1071 (1994)

10.35. T. K. Woodward, Theodore Sizer, D. L. Sivco and A. Y. Cho, *Appl. Phys. Lett.* **57**, 548 (1990)

10.36. K. Fujiwara, K. Kawashima, K. Kobayashi and N. Sano, *Appl. Phys. Lett.* **57**, 2234 (1990)

10.37. J. S. Weiner, D. A. B. Miller, D. S. Chemla, T. C. Damen, C. A. Burrus, T. H. Wood, A. C. Gossard and W. Weigmann, *Appl. Phys Lett.* **47**, 1148 (1985)

10.38. K. Wakita, I. Kotaka, O. Mitomi, H. Asai, Y. Kawamura and M. Naganuma, *J. Lightwave Technol.* **8**, 1027 (1990)

10.39. F. Devaux, S. Muller, A. Ougazzaden, A. Mircéa, A. Ramdane, P. Krauz, J. Semo, F. Huet, M. Carré and A. Carenco, *Appl. Phys. Lett.* **64**, 954 (1994)

10.40. R. H. Yan, R. J. Simes, L. A. Coldren and A. C. Gossard, *Appl. Phys. Lett.* **56**, 1626 (1990)

10.41. K. Hu, Li Chen, A. Madhukar, Ping Chen, C. Kyriakakis, Z. Karim and A. R. Tanguay, *Jr.*, *Appl. Phys. Lett.* **59**, 1664 (1991)

10.42. J. Woodhead, P. A. Claxton, R. Grey, T. E. Sale, J. P. R. David, L. Liu, M. A. Pate, G. Hill and P. N. Robson, *Electron. Lett.* **26**, 2117 (1990)

10.43. B. Pezeshki, D. Thomas and J. S. Harris, Jr.,*IEEE Photon Technol. Lett.* **2**, 807 (1990)

10.44. B. Pezeshki, D. Thomas and J. S. Harris, Jr.,*Appl. Phys. Lett.* **57**, 1491 (1990)

10.45. R. H. Yan, R. J. Simes and L. A. Coldren, *IEEE Photon. Tech Lett.* **2**, 118 (1990)

10.46. M. Whitehead, A. Rivers, G. Parry, J. S. Roberts and C. Bas, *Electron. Lett.* **25**, 984 (1989)

10.47. K-K. Law, R. H. Yan, J. L. Merz and L. A. Coldren, *Appl. Phys. Lett.* **56**, 1886 (1990)

10.48. K-K. Law, R. H. Yan, L. A. Coldren and J. L. Merz, *Appl. Phys. Lett.* **57**, 1345 (1990)

10.49. J. Bleuse, G. Bastard and P. Voisin, *Phys. Rev. Lett.* **60**, 220 (1998)

10.50. J. Bleuse, P. Voisin, M. Voos, H. Munekata, L. L. Chang and L. Esaki, *Appl. Phys. Lett.* **52**, 462 (1988)

10.51. E. E. Mendez, F. Agulló-Rueda and J. M. Hong, *Phys. Rev. Lett.* **60**, 2426 (1988)

10.52. B. Pezeshki, S. M. Lord and J. S. Harris, *Jr.*, *Appl. Phys. Lett.* **59**, 888 (1991)

10.53. K. Hu, Li chen, A. Madhukar, Ping Chen, K. C. Rajkumar, K. Kaviani, Z. Karim, C. Kyriakakis and A. R. Tanguay, Jr.,*Appl. Phys. Lett.* **59**, 1108 (1991)

10.54. T. E. Sale, J. Woodhead, A. S. Pabla, R. Grey, P. A. Klaxton, P. N. Robson, M. H. Maloney and J. Hegarty, *Appl. Phys. Lett.* **59**, 1670 (1991)

10.55. D. A. B. Miller, D. S. Chemla, T. C. Damen, A. C. Gossard, W. Wiegmann, T. H. Wood and C. A. Burrus, *Appl. Phys. Lett.* **45**, 13 (1984)

10.56. D. Λ. B. Miller, D. S. Chemla, T. C. Damen, T. H. Wood, C. Λ. Burrus, Λ. C. Gossard and W. Wiegmann, *IEEE J. Quantum Electron* **QE-21**, 1462 (1985)

10.57. D. A. B. Miller, J. E. Henry, A. C. Gossard and J. H. English, *Appl. Phys. Lett.* **49**, 821 (1986)

10.58. A. L. Lentine, H. S. Hinton, D. A. B. Miller, J. E. Henry, J. E. Cunningham and L. M .F. Chirovsky, *Appl. Phys. Lett.* **53**, 1419 (1988)

10.59. H. M. Gibbs, S. L. McCall, T. N. C. Venkatesan, A. C. Gossard, A. Passner and W. Wiegmann, *Appl. Phys. Lett.* **35**, 451 (1979)

10.60. H. M. Gibbs, S. S. Tarng, J. L. Jewell, D. A. Wienberger, K. Tai, A. C. Gossard, S. L. McCall, A. Passner and W. Wiegmann, *Appl. Phys. Lett.* **41**, 221(1982)

10.61. S. S. Tarng, H. M. Gibbs, J. L. Jewell, N. Peyghambarian, A. C. Gossard, T. Venkatesan and W. Wiegmann, *Appl. Phys. Lett.* **44**, 360 (1984)

10.62. K. Nonaka, Y. Kawamura, H. Kawaguchi and K. Kubodera, *IEEE Photonics Tech. Lett.* **1**, 55 (1989)

10.63. S. M. Jenson, *IEEE J. Quantum Electron* **QE-18**, 1580 (1982)

10.64. M. Abramowitz and I. A. Stegun (Eds.), *Handbook of Mathematical Functions* (Dover Publications Inc. New York, 1965) p. 569

10.65. Y. Silberberg and P. W. Smith in *Nonlinear Photonics*, eds. H.M. Gibbs, G. Khitrova and N. Peyghambarian (Springer- Verlag, Berlin 1990) p.185

10.66. P. Li Kam Wa, J. E. Stich, N. J. Mason, J. S. Roberts and P. N. Robson, *Electron. Lett.* **21**, 26 (1985)

10.67. P. Li Kam Wa, P. N. Robson, J. P. R. Davids, G. Hill, P. Mistry, M. A. Pate and J. S. Roberts, *Electron. Lett.* **22**, 1129 (1986)

10.68. R. Jin, C. L. Chuang, H. M. Gibbs, S. W. Koch, J. N. Polky and G. A. Puban, *Appl. Phys. Lett.* **53**, 1791 (1988)

10.69. C. C. Barron, C. J. Mahon, B. J. Thibeault, G. WAng, W. Jiang, L. A. Coldren, J. E. Bowers, *IEEEE J. Quantum Electron* **31**, 1484 (1995)

10.70. P. Li Kam Wa, P. N. Robson, J. S. Roberts, M. A. Pate and J. P. R. David, *Appl. Phys. Lett.* **52**, 2013 (1988)

10.71. P. Li Kam Wa , A. Miller, C. B. Park, J. S. Roberts and P. N. Robson, *Appl. Phys. Lett.* **57**, 1846 (1990)

10.72. P. Li Kam Wa, A. Miller, J. S. Roberts and P. N. Robson, *Appl. Phys. Lett.* **58**, 2055 (1991)

Appendix

List of copyrighted figures used with the permission of one of the authors and the copyright owner

Fig.	Citation	Copyright
2.1.	C. T. Foxon and B. A. Joyce in *'Growth and Characterisation of Semiconductors'*, R. A. Stradling and P. C. Kleipstein eds. Adam Hilger, New York (1990), p. 36, Fig. 1.	IOP
2.2.	M. J. Ludowise, *J. Appl. Phys.* **58**, R31 (1985), Fig. 2.	AIP
2.3.	Y. Kawaguchi sand H. Asahi, *Appl. Phys. Lett.* **50**, 1243 (1987) Fig. 1.	AIP
2.4.	H. Kinoshita and H. Fujiyashu, *J. Appl. Phys.* **51**, 5845 (1988), Fig. 1.	AIP
2.5.	C. L. Goodman and M. V. Pessa, *J. Appl. Phys.* **60** , R65 (1986), Fig. 22.	AIP
2.6.	M. Notomi, M. Naganuma, T. Nishida, T. Tamamura, H. Iwamura, S. Nojima and M. Okamoto, *Appl. Phys. Lett.* **58**, 720 (1991), Fig. 1.	AIP
3.3.	C. G. Van de Walle and R. M. Martin, *Phys. Rev.* **B 35**, 8154 (1987), Fig. 1 & 2.	APS
3.5(a).	J. H. Marsh, J. S. Roberts and P. A. Claxton, *Appl. Phys. Lett.* **46**, 1161 (1991), Fig. 1.	AIP
3.5(b).	B. R. Nag and S. Mukhopadhyay, *Appl. Phys. Lett.* **58**, 1056 (1993), Fig. 1.	AIP
4.2.	J. R. Chelikowsky and M. L. Cohen,*Phys. Rev.* **B 14**, 1056 (1991), Fig. 1.	AIP
4.4.	B. R. Nag and S. Mukhopadhyay, *Phys. Lett. A* **166**, 395 (1992), Fig. 1	PL
4.8.	S. Gangopadhyay and B. R. Nag, *Phys. Stat. Sol.* **166**, 395 (1996), Fig. 1	PSS
5.4.	R. L. Greene, K. K. Bajaj and D. E. Phelps, *Phys. Rev.* **B 29**, 1807 (1984), Fig.1.	APS
5.5.	M. D. Sturge, *Phys. Rev.* **127**, 768 (1962), Fig. 3.	APS
5.6.	G. Livescu, D. A. Miller, D. S. Chemla, M. Ramaswamy, T. Y. Chang, N. Sauer, A. C. Gossard and J. H. English, *IEEE J. Quantum Electron.* **24**, 1677 (1988), Fig. 2.	IEEE
5.7.	R. C. Miller, A. C. Gossard, D. A. Kleinman and O. Munteanu, *Phys. Rev. B* **31**, 5569 (1985), Fig. 1.	APS
5.9.	E. J. Austin and M. Jaros, *Phys. Rev. B* **44**, 8054 (1985), Fig. 1.	APS
5.10.	Der-San Chuu and Yu-Tai-Shi, *Phys. Rev. B* 44, 8054 (1991), Fig. 2.	APS
5.11	D. A. B. Miller, D. S. Chemla, T. C. Damen, A. C. Gossard, W. Weigmann, T. H. Wood and C. A. Burrus, *Phys. Rev. B* **32**, 1043 (1985), Fig. 5.	APS
5.12	D. S. Chemla, D. A. B. Miller, P. W. Smith, A. C. Gossard and W. Weigmann, *IEEE J. Quantum Electron.* **QE-20**, 265 (1984),Fig. 5.	IEEE
5.14	H. Hillmer, A. Forchell, C. W. Tu and R. Sauer, *Semicond. Sci. Technol.* **7**, B235 (1985), Fig. 1.	IOP
5.15.	D. F. Weltch, G. W. Wicks and L. F. Eastman, *Appl. Phys. Lett.* **46**, 991 (1985), Fig. 2.	AIP
6.1.	B. R. Nag and S. Mukhopadhyay, *Jpn. J. Appl. Phys.* **31**, 3287 (1992), Fig. 1.	JJAP
6.2.	W. Walukiewicz, H. E. Ruda, J. Lagowski and H. C. Gatos, *Phys. Rev. B* **30**, 4571 (1984), Fig. 3 & 8.	APS
6.3.	S. Mukhopadhyay and B. R. Nag, *Phys. Rev. B* **48**, 17960 (1993), Fig. 3.	APS
6.4.	W. Ted Masselink, *Semicond. Sci. Technol.* **4**, 503 (1989), Fig. 8.	IOP

Fig.	Citation	Copyright
6.5.	D. S. Chemla, D. A. B. Miller, P. W. Smith, A. C. Gossard and W. Weigmann, *IEEE J. Quantum Electron.* **QE-20**, 265 (1984), Fig. 1.	IEEE
6.6.	D. Grützmacher, K.Wolter, H. Jürgensen, P. Balk and C. W. T. Bulle Liewma, *Appl. Phys. Lett.* **52**, 872 (1986), Fig. 2.	AIP
7.4.	T. J. Drummond, H. Morkoç, K. Lee and M. Shur, *IEEE Electron Dev. Lett.* **EDL-3**, 338 (1982), Fig. 4.	IEEE
7.5.	R. Lai, P. K. Bhattacharya, D. Yang, I. Brock, S. A. Alterovitz and A. N. Downey, *IEEE Trans. Electron Dev*, **39**, 2206 (1992), Fig. 1.	IEEE
7.6.	J. J. Brown, J. A. Pusl, M. Hu, A. E. Schmitz, D. P. Docter, J. B. Shealy, M. G. Case, M. A. Thompson and L. Nguyen, *IEEE Microwave Guided Wave Lett.*, **6**, 91 (1996), Fig. 1.	IEEE
9.4.	S. R. Chinn, P. S. Zory and A. R. Reisinger, *IEEE J. Quantum Electron.* **24**, 2191 (1988), Fig. 3.	IEEE
9.5.	N. K. Dutta, *J. Appl. Phys.* **53**, 7211 (1982), Fig. 1	AIP
10.2.	B. F. Levine, *J. Appl. Phys.* **74**, R1 (1993), Fig.20	AIP
10.3.	B. F. Levine, C. G. Bethea, K. K. Cho, J. Walker and R. J. Malik, *J. Appl. Phys.* **64**, 1591 (1988), Fig. 1.	AIP
10.4.	B. F. Levine, *J. Appl. Phys.* **74**, R1 (1993), Fig. 46.	AIP
10.5.	B. F. Levine, *J. Appl. Phys.* **74**, R1 (1993), Fig. 16.	AIP
10.6(a).	G. Hasanain, B. F. Levine, C. G. Bethea, R. A. Logan and J. Walker, *Appl. Phys. Lett.* **54**, 2515 (1989), Fig. 1	AIP
10.6(b).	J. Y. Andersson, L. Lundqvist and Z. F. Paska, *Appl. Phys. Lett.* **58**, 2264 (1991), Fig. 1.	AIP
10.8.	B. Pezeski, S. M. Lord and J. S. Harris, *Jr. Appl. Phys. Lett.* **59**, 888 (1991), Fig. 2.	AIP
10.10.	C. C. Barron, C. J. Mahon, B. J. Thibeault, G. Wang, W. Jiang, L. A. Coldren and J. E. Bowers, *IEEE J. Quantum Electron.* **31** , 1484 (1995), Fig. 4.	IEEE

Abbreviations of the copyright owner's names

AIP - American Institute of Physics, APS - American Physical Society
IEEE - Institution of Electrical and Electronic Engineers
IOP - IOP Publishing Co., PL - Physics Letters A
PSS - Physica Status Solidi, JJAP - Japanese Journal of Applied Physics

INDEX

SOLID-STATE SCIENCE AND TECHNOLOGY LIBRARY

KLUWER ACADEMIC PUBLISHERS – DORDRECHT / BOSTON / LONDON